# Trees: Their Natural History

**Second Edition**

Trees are familiar components of many landscapes and have been vital in determining the ecology of our planet as well as the development of human cultures and communities. Yet how much do we really understand about how they work? This updated and revised edition provides a comprehensive introduction to all aspects of tree biology and ecology, and presents the state-of-the-art discoveries in this area.

The wonders and mysteries of trees are explored throughout the book and questions such as why leaves turn spectacular colours in the autumn, how water reaches the top of the tallest trees or why the study of genetics has caused so many name changes in trees are all brilliantly answered.

Written with a non-technical approach, this book will be a valuable source of reference for students and those with a less formal interest in this fascinating group of plants.

**Peter Thomas** is Senior Lecturer in Botanical and Environmental Science at Keele University, UK with 30 years of experience in ecological aspects of trees and forest ecology in the UK, North and Central America, Europe, Africa, Russia and Australasia. He has written two other books for Cambridge University Press: *Ecology of Woodlands and Forests* (Thomas and Packham, 2007) and *Fire in the Forest* (Thomas and McAlpine, 2010).

# Trees: Their Natural History

**Second Edition**

PETER A. THOMAS
Keele University, UK

CAMBRIDGE
UNIVERSITY PRESS

# CAMBRIDGE
## UNIVERSITY PRESS

University Printing House, Cambridge CB2 8BS, United Kingdom

One Liberty Plaza, 20th Floor, New York, NY 10006, USA

477 Williamstown Road, Port Melbourne, VIC 3207, Australia

314-321, 3rd Floor, Plot 3, Splendor Forum, Jasola District Centre, New Delhi - 110025, India

79 Anson Road, #06-04/06, Singapore 079906

Cambridge University Press is part of the University of Cambridge.

It furthers the University's mission by disseminating knowledge in the pursuit of education, learning and research at the highest international levels of excellence.

www.cambridge.org
Information on this title: www.cambridge.org/9780521133586

First published 2000
Second edition 2014
6th printing 2018

*A catalogue record for this publication is available from the British Library*

*Library of Congress Cataloging in Publication data*
Thomas, Peter, 1957–
Trees: their natural history / Peter A. Thomas. – 2nd edn.
    p.  cm.
Includes index.
ISBN 978-0-521-13358-6 (Pbk.)
1. Trees.  I. Title.
QK475.T48 2014
582.16-dc23    2013035072

ISBN 978-0-521-13358-6 Paperback

But blessed is the man who trusts in the Lord, whose confidence is in him.
He will be like a tree planted by the water that sends out its roots by
  the stream.
It does not fear when heat comes; its leaves are always green.
It has no worries in a year of drought and never fails to bear fruit.

Jeremiah 17: 7–8


Your ghost will walk, you lover of trees,
(If our loves remain)
In an English lane

De Gustibus

Robert Browning (1812–1889)

# Contents

| | | Page |
|---|---|---|
| *Preface* | | ix |
| **1** | An overview | 1 |
| **2** | Leaves: the food producers | 13 |
| **3** | The trunk and branches: more than a connecting drainpipe | 51 |
| **4** | Roots: the hidden tree | 102 |
| **5** | Towards the next generation: flowers, fruits and seeds | 154 |
| **6** | The growing tree | 205 |
| **7** | The shape of trees | 245 |
| **8** | The next generation: new trees from old | 284 |
| **9** | Age, health, damage and death: living in a hostile world | 315 |
| **10** | Trees and us | 376 |
| *Further Reading* | | 387 |
| *Index* | | 389 |

# Preface

A lot has happened in the tree world since this book was first written. Some previously unanswered questions can now be addressed, and some things we thought we knew have proved to be wrong or not the whole story. These have been put right. Our understanding of trees has also made huge leaps in areas such as the role of genetics and genetic engineering, effects of climate change, hydraulic engineering (including the bitter debate over how water gets up a tree), factors limiting tree height and why trees have colourful leaves in the autumn. New material has been added on all these and in many other areas where new discoveries have been made. This ranges from the surreal (did you know that cosmic radiation and the ocean tides appear to affect tree growth?) to the practical (why have artificial wine bottle corks proliferated?). Many of the problems faced by trees are caused by us humans, so there is also a new chapter on our interactions with trees looking at just why they are good for us and therefore worth preserving. I hope you have as much fun reading this as I had writing it.

Following comments on the first edition, the number of references at the end of each chapter has been increased to help those who want to dig deeper. To keep the text flowing, however, I've only included reference to these in the text where it is not clear which source is being used. Also following reader feedback, more scientific names of trees have been included with the common names, which I hope you won't find too intrusive.

In addition to those thanked in the first preface, I am most grateful to the following for help in gathering data and clarifying ideas: Richard Hobbs, Leif Kullman, Norm Kenkel, Janet Sprent, Andy Hirons and Duncan Slater. Andy Lawrence did a wonderful job in drawing the new diagrams. Alan Lacey is also thanked for kindly providing access to his years of research on the sycamore.

The first stirrings of this second edition began during a Bullard Fellowship at Harvard University and the help this gave me is very gratefully acknowledged.

The collection of leaf data from Honduras that was used in Chapter 2 was kindly made possible by Operation Wallacea, and the following huge team are thanked for their help in braving the often inclement conditions to gather the data: Kelly Barber, Natalie Gérardy, Catherine Dieleman, Alex Rupp, Cassidy van Rensen, Stephanie Ryan, Heather Crochetiere, Nicole Meneses, Maja Goschorska, Matthew Levan, Amy Stypa, Sarah Jupp, Sophie Cunningham, Carlos Carias, Edwardo Antunez, Christie Rajtar, Matthew Sargeant, Yacyi Mi, Refika Mustafa and in particular Leanne Meadows, Amy Rushton, Nigel Taylor, Karly Harker and Gurjot Malhi for their stamina and fortitude. The canopy access guys in Honduras – Waldo Etherington, Cameron Alexander and Jim Fenn – are also heartily thanked for their good company, their help in collecting samples and safely getting me up and down.

All photographs are my own except where indicated otherwise.

Note: In the text sp. and spp. are used as shorthand for one or more species, respectively. Units used are hopefully self-explanatory but it might help to know that a micrometre (μm) is a thousandth of a millimetre, and ppm are parts per million (1000 ppm = 0.1%).

## Preface to the first edition

Why write a book on trees? The motivation for me came from the frustration of trying to teach a subject where much is known but is scattered over a huge range of journals and books from many countries. There are so many fascinating stories to be told about the ways in which trees cope with the world and the problems of being so large and long-lived: they are extremely well designed. *Someone* had to write this book!

My goal was to draw together strands of information to create a readable book that would answer common questions about trees, set right a number of myths and open up the remarkable world of how trees work, grow, reproduce and die. It is for you, the reader, to judge whether I have been successful. Please do let me know where you find errors or would wish to argue with the logic.

I am indebted to all those who helped with this book, especially Roger Davidson and Bill Williams who read and commented on the whole manuscript and numerous colleagues who commented on parts. Val Brown, Barry Tomlinson, Colin Black and K. J. Niklas kindly provided detailed information. Maria Murphy and Lynn Davy at Cambridge University Press are to be congratulated for their extreme patience with a faltering author. And, of course, my wife and sons are thanked for putting up with my pet project.

The following are gratefully acknowledged for their help with the diagrams: Ian Wright, Lee Manby (who drew Figures 4.2 and 7.1–7.14) and John Stanley (for drawing Figure 3.7).

May 1999

# Chapter 1: An overview

## What is a tree?

Everyone knows what a tree is: a large woody thing that provides shade. Oaks, pines and similarly large majestic trees probably come immediately to mind. Such big trees are characterised by the enormous changes in size from seed to mature tree: a mature giant sequoia (*Sequoiadendron giganteum*) is a billion, billion times heavier than the seed it came from (that's 1 with 12 zeros after it). A stricter, but more inclusive, botanical definition is that a tree is any plant with a self-supporting, perennial woody stem (i.e. living for more than 1 year). The first question that normally comes back at this point is to ask what then is a shrub? To horticulturalists, a 'tree' is defined as having a single stem more than 6 m (20 ft) tall which branches at some distance above ground, whereas a shrub has multiple stems from the ground and is less than 6 m. This is a convenient definition for those writing tree identification books who wish to limit the number of species they must include. In this book, however, shrubs are thought of as being just small trees since they work in exactly the same way as their bigger neighbours. Thus, 'trees' cover the towering giants over 100 m through to little sprawling alpine willows no more than a few centimetres tall.

Some plants can be clearly excluded from the tree definition. Lianas and other climbers are not self-supporting (although some examples are included in this book), and those plants with woody stems which die down to the ground each year, such as asparagus, do not have a perennial woody stem. Bananas are not trees because they have no wood (the trunk is made from leaf stalks squeezed together). Nor are bamboos since they are just hardened grasses even though they can be up to 25 m tall and 25 cm thick (see Box 1.1).

There are estimated to be 100 000 species of tree in the world, about 25% of all living plant species. An interesting feature of all these trees is how unrelated they are. It is usually easy to say whether a plant is an orchid or not because all orchids belong to the same family, have a common ancestor and share a similarity in structure (especially the flowers). This is true of most plant groups such as grasses and cacti (in their own families) and chrysanthemums (all in the same genus). But the tree habit has evolved independently in a wide range

| **Box 1.1** The range of trees found in different plant groups | |
|---|---|
| **Ferns (Pteridophytes):** | <u>Tree ferns:</u> mostly in the families of Cyatheaceae and Dicksoniaceae; rarely branched, no true bark and with a trunk containing woody strands; need frost-free shaded habitats |
| **Seed plants:** | Contains the Conifers and the Flowering plants. The ferns above produce spores but not seeds and so are not included here |
| **Conifers and their allies (Gymnosperms):** This term means 'naked seeds' (as in gymnasium where the Greeks exercised naked); the seeds are exposed to the air and can be seen in the cone or fruit without having to cut anything open | <u>Conifers:</u> 630 species in eight families <br><br> • Cupressaceae: cypress, junipers, and now including the former Taxodiaceae: redwoods <br> • Araucariaceae: including monkey puzzle (*Araucaria araucana*), kauri (*Agathis australis*) and the Wollemi pine (*Wollemia nobilis*) <br> • Podocarpaceae: more than 150 species in the southern hemisphere including the podocarps (*Podocarpus* spp.) and rimu (*Dacrydium cupressinum*) <br> • Pinaceae: pines, spruces, larches, hemlocks, firs, cedars <br> • Cephalotaxaceae: 11 species of *Cephalotaxus*, plum yews/cowtail pines <br> • Phyllocladaceae: 4 species of *Phyllocladus*, celery-pines <br> • Sciadopityaceae: only *Sciadopitys verticillata*, Japanese umbrella-pine <br> • Taxaceae: yews <br><br> <u>Ginkgo:</u> 1 species <br><br> • Ginkgoaceae: the ginkgo or maidenhair tree (*Ginkgo biloba*) <br><br> <u>Cycads:</u> palm-like with stiff leathery leaves <br><br> <u>Gnetales:</u> a strange group with a few interesting woody plants <br><br> • *Welwitschia mirabilis*: single species in SW Africa <br> • *Gnetum* spp.: mostly tropical climbers <br> • *Ephedra* spp.: 30+ low shrubs of dry deserts |

| **Box 1.1** (cont) | |
|---|---|
| **Flowering plants (Angiosperms):** This means hidden seeds: contained inside a fruit | <u>Dicotyledons</u> (two 'seed leaves' in the seed): The main group of trees such as oaks, birches, etc. Around 75 of the world's 180 families contain trees.<br><br><u>Monocotyledons</u> (one 'seed leaf'): A wide ranging set of trees concentrated in a few families<br><br>• Palmaceae (Arecaceae): palms; mostly tropical, a few temperate; nearly 3000 species<br>• Asparagaceae: a large family with a number of trees<br>    Dragon trees (*Dracaena* spp.); mostly N African<br>    Cordyline palms (*Cordyline* spp.); Australia and New Zealand<br>    European butcher's brooms (*Ruscus* spp.)<br>    Yuccas (including the Joshua tree, *Yucca brevifolia*)<br>• Pandanaceae: screw pines (*Pandanus* spp.); Old World Tropics; stilt roots supporting a stout forked trunk<br>• Xanthorrhoeaceae: grass trees (*Xanthorrhoea* spp.) from Australia with short trunk with forked branches and long narrow leaves, and aloes (*Aloe* spp.) from Southern Africa<br>• Sterlitziaceae: traveller's palm (*Ravenala madagascariensis*)<br><br>**Monocotyledons that are not trees**<br><br>• Musaceae: bananas (*Musa* spp.); the trunk is made from leaf stalks squeezed together<br>• Poaceae: bamboos (e.g. *Dendrocalamus* spp.) – are just hardened grasses with no wood |

of plants: at least 20 families in temperate areas and so probably hundreds worldwide. Given this wide range it is not surprising that the only common feature of trees is having a perennial woody skeleton. Box 1.1 illustrates how many major groups have evolved the tree habit. This is a superb example of 'convergent evolution' where a number of unrelated types of plant have evolved the same answer – height – to the same problem: how to get a good supply of light.

On the whole, this book is concerned with the two biggest groups of trees. These are the **conifers** and their allies, and the **hardwoods** like oak, birch and so on. (As you can see from Box 1.2 the terminology can be confusing so throughout this book we will stick to conifers and hardwoods as shorthand for gymnosperms and dicotyledon angiosperms.) The monocotyledon trees such as palms and dragon trees are mentioned in passing but on the whole they grow in a different way from conifers and hardwoods and the book can only be so long. Purists might indeed argue that since the trunks of these trees contain no real 'wood' (Chapter 3) they are not trees anyway. Tree ferns (Box 1.1) come into the same category.

## A short history of trees

Back in the Silurian, over 400 million years ago, the first vascular plants (those with internal plumbing) appeared on the earth. Initially this plumbing was just for conducting water up the plant with no structural strength. The tree habit took off once a way of making the plumbing (particularly the xylem; see *Parts of the tree* below for a definition of this) thicker and stronger had evolved; this was the cambium (again see below for a definition). The first trees (protogymnosperms) evolved in the early Devonian around 390 million years ago, capable of living for several decades and reaching up to 30 m tall and a metre wide. Within 100 million years, the coal-producing swamps of the Carboniferous (360–290 million years ago) were dominated by lush forests. We would have recognised the tree ferns from today's forests but the others – giant horsetails and clubmosses – have long since disappeared, leaving us just a few small relatives. The horsetails such as *Calamites* were up to 9 m tall and 30 cm in diameter but the clubmosses (notably *Lepidodendron*) must have been magnificent at up to 40 m high and a metre in diameter. In these forests the first primitive conifers appeared around 300 million years ago and by around 250 million years ago (the late Permian) trees such as cycads, ginkgos and monkey puzzles were recognisable: the sort of trees found fossilised in the petrified forest of Arizona from the late Triassic, 200 million years ago (Figure 1.1). The pines were not far behind, probably evolving around 180–135 million years ago (Jurassic) to share the earth with the dinosaurs. And by the end of the Cretaceous around 65 million years ago all the modern families of conifers had evolved.

Conifer domination was long and illustrious, from around 245 till 67 million years ago, but the early hardwoods were diversifying during the early Cretaceous around 120 million years ago. The hardwoods probably

**Box 1.2** Definitions that go with the two main groups of trees

Throughout this book the terms **Conifers** and **Hardwoods** will be used as shorthand for Gymnosperm and Angiosperm trees.

Gymnosperms                              Angiosperms

As explained in Box 1.1, these are the proper botanical terms but a little hard to digest. Both are seed plants but the angiosperms are also flowering plants; gymnosperms have no proper flowers.

**Conifers** and their Allies              All other trees

As you can see from Box 1.1 the gymnosperms include more than just the conifers but they are the major component.

Softwoods                                **Hardwoods**

The problem with these descriptive terms (which stem from the timber industry) is that although most gymnosperms *do* produce softer wood, there are many exceptions, and many hardwoods can be physically soft. Yew (*Taxus baccata*, a Softwood) produces very dense and hard wood whereas some Hardwoods, like balsa (*Ochroma pyramidale*), are very soft and easily broken or indented with a fingernail.

Evergreens                               Deciduous trees

It is often considered that conifers are evergreens and hardwoods are deciduous, losing all of their leaves at some point in the year. Exceptions can be found here as well. The dawn redwood (*Metasequoia glyptostroboides*), the swamp cypress (*Taxodium distichum*) and larches (*Larix* spp.), for example, are deciduous gymnosperms. In contrast, European holly (*Ilex aquifoilium*), rhododendrons and many tropical angiosperms are evergreen.

Needle trees                             Broad-leaved trees

Most conifers do indeed have needle-shaped leaves but again there are exceptions. The ginkgo tree (*Ginkgo biloba*) and monkey puzzle (*Araucaria araucana*) have definite broad flat leaves (admittedly these trees are easily identified oddities). Cycads, which are primitive gymnosperms, have long divided leaves that resemble palms. Some angiosperms have reverted to needle leaves or have largely lost their leaves and use their needle-like branches as leaves, e.g. gorses (*Ulex* spp.) and brooms (*Cytisus* spp.).

**Figure 1.1** Sections of petrified tree (in this case about 1 m in diameter) in the Petrified Forest National Park, Arizona, USA. These trees were growing in the late Triassic (200 million years ago) and became buried under river sediments which prevented rotting. Water flowing through the sediments deposited silica into the wood's tubes with other colourful minerals, such as iron, manganese and copper, and so preserved the original wood structure.

evolved from a now extinct conifer group that had insect-pollinated cones. The magnolias are some of the earliest types of hardwood that we still have around. During the Cretaceous period and into the early Tertiary (65–25 million years ago) the hardwoods underwent a massive expansion displacing the conifers, undoubtedly helped by the warm humid global climate of the early Tertiary. But further changes to the climate came to the rescue of the conifers with the development of polar ice caps at the end of the Eocene (35 million years ago), which allowed the northern pines to diversify, spread and take over the boreal forest. This was not without a price; others such as the dawn redwood (*Metasequoia glyptostroboides*), which had been very extensive around the world including on Axel Heiberg Island in the Arctic (79 °N) 60–50 million years ago, became severely limited in its distribution, now found natively only in eastern Asia.

At the end of the Permian period, around 250 million years ago, most of the earth's land masses were squashed together into the super-continent of Pangaea. By the time the hardwoods had evolved, Pangaea had broken into Laurasia (which gave rise to the northern hemisphere continents) and Gondwanaland (containing what is now Australia, Africa, S America, India and Antarctica) trapping the pines primarily in the northern hemisphere.

Laurasia and Gondwanaland themselves broke apart later, which goes some way to explaining why the hardwoods of the northern and southern hemisphere are so different from each other and yet remarkably similar around the globe within a hemisphere. It also explains why a number of genera are found throughout the northern hemisphere (their ancestors were found across Laurasia) but very few share species between the Old and New World; these evolved once Laurasia had separated. For example, the genus of tulip trees (*Liriodendron*) is found across the northern hemisphere but the Chinese tulip tree (*L. chinense*) is in the Old World and the tulip tree (*L. tulipifera*) in the New World. This similarity in genera but with different species leads to a striking similarity between south-east USA and south-east China. If you squint a bit so as not to notice exactly which species you are looking at, you could be in either.

By 95 million years ago (midway through the Cretaceous period) a number of trees we would recognise today were around: laurels, magnolias, planes, maples, oaks, willows and, within another 20 million years, the palms. When the dinosaurs were disappearing (by 65 million years ago) the hardwoods were dominating the world with the conifers exiled mostly into the high latitudes.

## Living fossils

Most of the types of tree we see every day have been around for a long time. Perhaps the most incredible are the growing number of rediscovered 'living fossils': trees know from the fossil record and which were thought to have become extinct and yet have been found hanging on in remote parts of the world. The most famous is the ginkgo (Japanese for 'silver apricot', named after the fruit) or maidenhair tree (*Ginkgo biloba*), a Chinese tree known from the fossil record back about 180 million years (the Jurassic era) and rediscovered in Japan by Europeans in 1690 (Figure 1.2).

The dawn redwood (*Metasequoia glyptostroboides*) was similarly refound in 1944 in China (Figure 1.2), and more recently in 1994 the Wollemi pine (*Wollemi nobilis*, a member of the monkey puzzle family, Araucariaceae) was found growing in two areas with less than 40 mature trees in Wollemi National Park near Sydney, Australia. The dawn redwood has fossils dating back 65 to 35 million years ago and the Wollemi pine from 200 to 2 million years ago. In these plants you can see real history (or prehistory) and touch plants that would have been familiar to the dinosaurs!

## Tree movement through history

As we've seen above, as climate changed through geological time, and as continents have moved around, so trees have moved around the world. Depending

Figure 1.2 Living fossils. (a) Leaves of the ginkgo or maidenhair tree (*Ginkgo biloba*) and (b) the distinctive shape of ginkgo trees along a road in Seoul Korea. (c) A branch of dawn redwood (*Metasequoia glyptostroboides*) with the small leaves and (d) trees in winter, having lost their leaves, showing the characteristic spire shape. Keele University, England.

upon their climatic needs (or tolerances really) they have moved as climate moves. And although this may look static now, the process is still happening (as is discussed in Chapter 9 looking at the likely consequences of climate change).

The most recent large change in tree distribution was after the last glaciation. There have been many waves of glaciation and warmer 'interglacial' periods, with the last ice age starting 2 million years ago and reaching a maximum 18 000 years ago with an ice sheet up to 5 km (3 miles) thick extending from the Arctic down to below the Great Lakes reaching New York, London and Berlin. At this point, with so much water bound up in ice, sea level dropped by 130 m (425 ft). In the southern hemisphere ice extended up from Antarctica to cover Chile, much of Argentina and Africa. This ice age started to lose its grip on the landscape 14 000 years ago and the ice was largely gone by 10 000 years ago following several periods of dramatic warming. The disappearance of the ice marked the end of the Pleistocene and the beginning of the Holocene (which we are still in).

The huge ice sheets pushed plants and animals further south into refuge areas. Much of Britain would have been devoid of plants as our trees found refuge in western France and northern Spain. As the climate warmed, animals, trees and other plants migrated polewards at a rate of 0.42 to 1 km per year (see Chapter 9 for a more detailed discussion on this). This migration produced some interesting patterns of trees. In N America the mountains tend to run north to south, so migration northwards was largely unimpeded and therefore eastern forests contain at least 10 major tree species making up the forests. In eastern Asia at similar latitudes, where the tree fauna is richer and the mountains run in a similar direction, forests can easily contain 20 major species. In Europe, however, the mountains (such as the Pyrenees and the Alps) tend to run east to west and were distinct barriers to migration, which explains why Europe has fewer tree species than eastern N America. Moreover in the British Isles, the land bridge joining us to Europe was submerged 8500 years ago leaving very little time for trees to reinvade; thus our forests have just five major trees species.

We've muddied the waters of natural distributions of trees by moving things around. The British Isles has around 35 tree species (depending upon what you count as a tree or shrub) that are regarded as native; that is they arrived in the islands by themselves after the last glaciation with no help from humans. Since then we've introduced another 500 species of tree that can be readily found in gardens and parks, and if you include rare species in botanic gardens then you can find upwards of 1700 species of tree in the British Isles. These new arrivals are described as non-native or exotic species. Of these an ever-increasing handful have become naturalised, that is they are non-natives that are sufficiently at home that they are reproducing and spreading by themselves. Sycamore (*Acer pseudoplatanus*) and Turkey oak (*Quercus cerris*) are in this category in the British Isles.

## Changing tree names and DNA

You will have undoubtedly noticed that after many years of just minor changes in plant names, we are now going through a period of rapid change. There is a reason! For hundreds of years plants that look similar have been put together in the same genus, and closely related genera have been put into the same family by looking at their detailed structure, particularly their flowers. This produced a fairly stable set of names for plants. The physical similarity was assumed to reflect their relatedness, so oaks in the same genus (*Quercus*) are closely related (like human siblings) while different genera lumped together in the same family (Fagaceae for the oaks along with beeches *Fagus* spp. and the chestnuts *Castanea* spp.) are a little less related (like cousins, aunts and uncles). In effect you can draw a family tree for these trees just as you can for your own family.

This relatedness also implies something about the evolution of these trees. Different oaks should share a fairly near common ancestor from which they all evolved, while members of a plant family may have split apart further back along the family tree. Since evolution is based on changes in genetic material inside the cells (the DNA), we should be able to say something about the relatedness of different trees by looking at the similarity of their DNA. Oaks should share more DNA than, say, an oak and a beech.

Since we can now look at the DNA of trees in detail, this has led to the discipline of molecular systematics which has produced so many of the recent name changes. From looking at the DNA that different trees share, new relationships have been worked out that sometimes change the old grouping of plants that were based just on physical features. Some plants that were thought to be closely related have been found to be less so. For example, the Kamtchatka rhododendron is not as closely related to other rhododendrons as was thought and has changed from *Rhododendron camtschaticum* to *Therorhodion camtschaticum*. On the other hand, *Ledum* species, such as the shrubby Labrador tea (*Ledum groenlandicum*), were found to be so closely related to the rhododendrons that that their name was changed to *Rhododendron groenlandicum*, etc. In a similar way, maples (*Acer* spp.), which were previously in their own family of Aceraceae, have now been put in the Sapindaceae family with other plants they are now known to be closely related to including the horse chestnuts (*Aesculus* spp.), the pride of India (*Koelreuteria paniculata*) and the lychee (*Litchi chinensis*)! As a last example, a whole group of trees that would appear to have little in common from just looking at them have now been moved together into the mallow family (Malvaceae) based on the similarity of their DNA. Thus the hibiscus is now joined by the cocoa tree (*Theobroma cacao*, previously in the Sterculiaceae), the baobab (*Adansonia* spp., Bombacaceae) and the limes/lindens (*Tilia* spp., Tiliaceae). If you have seen the first edition of this book, such changes also explain why Box 1.1 is somewhat different now from what it was. While this can all be very exasperating, these changes will slow as genetic relationships are sorted out once and for all and stability returns.

## Parts of the tree

Before we look at different aspects of trees in detail, we should start with an overview of the whole tree.

A tree lives in basically the same way as any other plant. The leaves produce the sugars which are the fuel used to run the tree and used to make the basic building blocks of cellulose and lignin which form the bulk of a tree (Chapter 2). The sugars are moved through the inner part of the bark (the phloem) to where they are needed around the rest of the tree. Sugars not required for immediate use are stored in the wood of the trunk, branches and roots (Chapter 3). The roots at the other end of the tree (Chapter 4) absorb water and minerals (such as nitrogen, phosphorous and potassium) from the soil. The water and dissolved minerals are pulled up through the wood of the tree (the xylem) to reach the leaves, the main users of water. The minerals are used with the sugars to build essential components of the tree, including the flowers and fruits needed to start the next generation (Chapter 5).

Trees get bigger in two ways. The buds scattered around the tree are the growing points for making existing branches longer or for making new branches (referred to as primary growth; Chapter 3). Once made, the woody skeleton gets fatter (secondary growth) by a thin layer of tissue (the cambium) beneath the bark adding new bark and new wood. In temperate areas where growth stops over winter these new layers are seen as the familiar annual rings in the wood (Chapter 3).

The pattern in which new buds are laid down on a developing branch, and which of these buds grow out and by how much, determines the characteristic shape of each tree (Chapter 7). Incidentally, many monocotyledons, such as palms, have one growing tip only and so inevitably live near the tropics where climate is less likely to kill the tip: if that growing point dies, so does the whole tree.

The essential difference between plants (including trees) and an animal is that most animals act as a whole unit (one heart, one set of eyes, one liver acting for the whole animal) whereas plants are modular, made up of similar parts added together, each acting largely independently, each replaceable. Thus a tree can lose and replace a branch or even the whole trunk, which in an animal would be equivalent to, for example, cutting me off at the feet and watching them regrow a new me. This modular organisation of trees makes for some interesting problems and solutions in keeping a woody skeleton going for hundreds and sometimes thousands of years. Generally, the living portion of the tree is a thin skin over a long-dead skeleton which nevertheless must be preserved from the attentions of fungal rot and animals in a number of ingenious ways (Chapter 9).

The story is not yet complete: we can see how a tree is organised within but it must still interact with its environment. It must start from a seed and do battle against a whole army of animals and fungi, and compete with its neighbouring plants. And since trees are long-lived compared to most other plants, they have some neat tricks for surviving (Chapters 6 and 8). As humans, we have a long history shared with trees and the value of trees to us is considered in Chapter 10.

## 🍃 *Further Reading*

Bonnicksen, T.M. (2000) *America's Ancient Forests*. Wiley, New York.

Coiffard, C., Gomez, B. & Thevenard, F. (2007) Early Cretaceous angiosperm invasion of Western Europe and major environmental changes. *Annals of Botany*, 100, 545–553.

Crawley, M.J. (1997) Biodiversity. *Plant Ecology*, 2nd edn. (edited by M.J. Crawley). Blackwell Science, Oxford, pp. 595–632.

Gerrienne, P., Gensel, P.G., Strullu-Derrien, C., *et al.* (2011) A simple type of wood in two early Devonian plants. *Science*, 333, 837.

Groover, A.T. (2005) What genes make a tree a tree? *Trends in Plant Science*, 10, 210–214.

Ingrouille, M.J. & Eddie, B. (2006) *Plants: Diversity and Evolution*. Cambridge University Press, Cambridge.

Sperry, J.S. (2003) Evolution of water transport and xylem structure. *International Journal of Plant Sciences*, 164 (3 suppl.), S115–S127.

Spicer, R. & Groover, A. (2010) Evolution of development of vascular cambia and secondary growth. *New Phytologist*, 186, 577–592.

Svenning, J.-C. & Skov, F. (2007) Ice age legacies in the geographical distribution of tree species richness in Europe. *Global Ecology and Biogeography*, 16, 234–245.

Willis, K.J. & McElwain, J.C. (2002) *The Evolution of Plants*. Oxford University Press, Oxford.

Yang, Z.H. & Rannala, B. (2012) Molecular phylogenetics: principles and practice. *Nature Reviews Genetics*, 13, 303–314.

# Chapter 2: Leaves: the food producers

Perhaps the most striking thing about tree leaves is their tremendous diversity in size. The Arctic/alpine willow (*Salix nivalis*), which grows around the northern hemisphere, can have leaves just 4 mm long on a sprawling 'tree' less than a centimetre high (Figure 2.1a). Smaller still, the scale needles of some cypresses are nearer a millimetre long. Amongst the largest of leaves are those of the foxglove tree (*Paulownia tomentosa*), which on coppiced trees can be over half a metre in length and width on a stalk another half metre long. Such large sail-like leaves are at risk of being torn by the wind (as in the traveller's palm – *Ravenala madagascariensis*; Figure 2.1b) so it is perhaps no surprise that big leaves are usually progressively lobed and divided up into leaflets to form a compound leaf. This can lead to even larger leaves: the Japanese angelica tree (*Aralia elata*) can have leaves well over a metre in length (Figure 2.1c). Many palms have feathery leaves over 3 m long and in the raffia palm (*Raphia farinifera*) up to 20 m (65 ft) long on a stalk another 4 m long.

The leaves are the main powerhouse of the tree. Combining carbon dioxide from the air with water taken from the soil they photosynthesise using the sun's energy to produce sugars and oxygen. These sugars (usually exported from the leaf as sucrose, the sugar we buy in packets) are the real food of a tree. They are used as the energy source to run the tree, and they form the raw material of starch and cellulose and, combined with minerals taken from the soil, allow the creation of all other necessary materials from proteins to fats and oils.

The role of leaves in producing food should not be underestimated. A large apple tree holds 50 000–100 000 leaves, a normal birch tree may average 200 000 leaves, while a mature oak can have 700 000 leaves. Even this pales somewhat in comparison to the 5 million leaves reported from mature American elms (*Ulmus americana*). Using these vast numbers of leaves a mature beech (*Fagus sylvatica*) can fix 2 kg of carbon dioxide per hour, producing as a by-product enough oxygen for 10 people every year. The world's trees have been estimated to produce 65 000–80 000 million metric tons of dry matter per year, two thirds of the total produced by all land plants.

**Figure 2.1** (a) Arctic willow (*Salix nivalis*) in the Canadian Rocky Mountains; (b) traveller's palm (*Ravenala madagascariensis*) on Tenerife, the Canary Islands, showing the large sail-like leaves that are readily torn by the wind; (c) a leaf of the Japanese angelica tree (*Aralia elata*) near Vladivostok, Russia (the part above the flower head is all one bipinnate leaf); and (d) Welwitschia (*Welwitschia mirabilis*) in the Namib Desert of Namibia. Photograph (d) by Geoff Smith.

(d)

Figure 2.1 (cont)

## The make-up of a leaf

Broadleaved trees, as the name suggests, have leaves with a broad leaf blade (or lamina), designed to catch as much light as possible. This limp, fragile material is permeated by a ramifying skeleton of veins that spread from the leaf base (Figure 2.2). These veins form the bulk of the leaf stalk (which botanists call the petiole). The veins contain strengthening tissue but also the conducting (vascular) tissue which brings water into the leaf (along the xylem; see Chapter 3 for more details) and takes away the manufactured sugars (through the phloem). Some leaves are accompanied by a pair of small scaly or leaf-like appendages at the base of the leaf stalk: the stipules (Figure 2.2b).

Some trees have simple leaves with no teeth or lobes (referred to as 'entire') but others are progressively more deeply lobed leading to the compound leaf where the blade has been divided up into separate leaflets (Figure 2.2 gives the names of the separate parts of a compound leaf). Trees have three common types of compound leaf: palmate where the leaflets are spread out like fingers from the palm of your hand (e.g. horse chestnut); pinnate where leaflets are attached either side of the stalk (the rachis) which is a continuation of the petiole (e.g. ash, walnut); and bipinnate where each leaflet of the pinnate leaf is further divided into leaflets (e.g. acacia and *Aralia* species – Figure 2.1c). But what differentiates a compound leaf from a branch with individual leaves? The easy answer is that in broad-leaved trees, there is a bud in the junction where a leaf

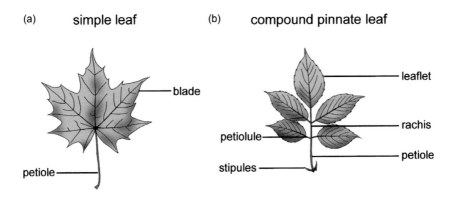

Figure 2.2 (a, b) Component parts of simple and compound leaves.

joins the twig (the leaf axil). In a compound leaf the only bud is at the very base of the petiole; there are no buds in the axils where the leaflets join the rachis.

As is usually the case, however, there are exceptions. In two related tropical genera, *Chisocheton* and *Guarea*, the compound leaves can have flower buds in the leaflet axils and a terminal bud that produces new leaflets. Conversely, and more commonly met in temperate regions, the deciduous dawn redwood (*Metasequoia glyptostroboides*; Figure 1.2) produces what appear to be pinnate leaves but are actually small simple leaves borne on a short-lived branch (the internal anatomy gives this away) that is normally shed at the end of the year. This may explain why the bud that should be in the axil between the 'leaf' and branch is usually to the side or below.

It is sometimes said that a compound leaf can be told from a branch of leaves because the compound leaf falls from the tree as a unit. Anyone looking at an ash tree in autumn will see that this is not necessarily so; leaflets often fall independently.

## Inside the leaf

Inside the leaf, the chlorophyll that captures the sun's energy is contained in tiny chloroplasts, normally just 4–6 μm[1] long. These are concentrated in the tightly packed cells of the upper part of the leaf where there is most light. Beneath each square millimetre of leaf there are up to 400 000 chloroplasts giving a total area exposed to light within a large deciduous tree of up to 350 square kilometres or more. A detailed study of a poplar tree (*Populus tremula*) in Sweden by Johanna Keskitalo and colleagues showed that each leaf contained

---

[1] μm = micrometre, a thousandth of a millimetre.

about 30 million cells and each cell held about 40 chloroplasts, and the whole tree contained around $10^{15}$ chloroplasts (that's 1 with 15 zeros after it, a million, billion, billion chloroplasts).

The leaf needs to be waterproof to maintain a moist working environment inside and prevent the leaf from wilting. To this end leaves have a skin (epidermis) of tightly packed cells overlaid by a waxy cuticle which has the property of resisting the passage of water without halting completely the movement of air. This is capitalised on in shoe polish which is traditionally based on the leaf wax of the Brazilian carnauba wax palm (*Copernicia cerifera*) (and which, incidentally, is also used to coat certain sugar-coated chocolate confectionery!). There is still a need for holes in this waterproof skin to allow effective exchange of gases. In trees, these holes or stomata (stoma, singular) are usually concentrated on the underside of the leaf away from the direct heat of sunlight (especially in glossy, evergreen leaves) but occur on the upper surface (although in low density) in about 20% of tree species (Box 2.1). Stomata can also be found on twigs and branches that have a green photosynthetic layer under the bark. The breathing holes of the more mature bark (the lenticels; see Chapter 3) are usually formed beneath old stomata.

Physical laws dictate that most gas can be exchanged for least water loss by having many small widely spaced holes. This is because more water is lost

| **Box 2.1** Examples of trees with stomata on different surfaces | | |
|---|---|---|
| **Both sides of the leaf** | Eastern cottonwood | *Populus deltoides* |
| | Black poplar | *P. nigra* |
| | European larch | *Larix decidua* |
| **Lower side only** | Hornbeam | *Carpinus betulus* |
| | English oak | *Quercus robur* |
| | Indian bean tree | *Catalpa bignonioides* |
| | Oleander | *Nerium oleander* |
| | Cherry laurel | *Prunus laurocerasus* |
| **Upper side only** | Common lime | *Tilia* x *europaea* |
| | Blue gum | *Eucalyptus globulus* |

Data from Bidwell, R.G.S. (1979) *Plant Physiology* (2nd edn). Macmillan, New York, and Fitter, A.H. & Hay, R.K.M. (2001) *Environmental Physiology of Plants* (3rd edn). Academic Press, London.

around the edges of the holes than in the middle so many small holes lose more water than would one big one of the same area. Because of these constraints, stomata in tree leaves are similar in size and number to other plants: typically around 0.01–0.03 mm long and one third that across with densities as low as 1400 per square centimetre ($cm^{-2}$) in larch to 25 000 $cm^{-2}$ in apple and over 100 000 $cm^{-2}$ in scarlet oak (*Quercus coccinea*) and almost this many in some cultivated varieties of red maple (*Acer rubrum*). Most leaves fall into the range 4000–35 000 $cm^{-2}$. Even with such high numbers the stomata still usually cover less than 1% of the leaf area. Despite the small area, the stomata will lose around 98% of the water taken up by the roots. Usually the stomata are evenly spread over the leaf but this is not always so. In European ash (*Fraxinus excelsior*) stomata are most plentiful in the middle of the underside while in horse chestnut (*Aesculus hippocastanum*) they are predominantly around the edge of the broadest parts of the leaflets.

## Sun and shade leaves

With the large number of leaves held by a tree it is almost inevitable that some will shade others. How the tree gets round this problem is dealt with in detail in Chapter 7. Part of the solution is that the leaves of most trees, even those with open canopies such as birch, can be divided into sun and shade leaves. Shade leaves are larger and thinner with a thinner cuticle, darker green (more chlorophyll per weight), and less lobed, with half to a quarter the density of stomata. They can work more efficiently at lower light intensities than sun leaves but cannot handle bright light for long periods. However, shade leaves have a quicker response time in opening stomata and so can utilise the brief moving sunflecks that dapple the ground under the canopy. Sun leaves work better in bright light but if put into deep shade their intrinsically higher respiration rates require more carbohydrate than they can produce so they lose weight and eventually die.

Bear in mind that each leaf develops according to the conditions under which it grows. The proportion of sun and shade leaves thus depends entirely on the growing conditions of the tree. Seedlings may be all shade leaves, acquiring sun leaves only as they emerge above surrounding competition. This can happen within the lifetime of an individual leaf. If a sun leaf is gradually shaded it reacts by 'plastic' changes in shape, becoming thinner and broader. But sudden changes can be more than a leaf can handle. Shade leaves of rainforest seedlings exposed by disturbance or, closer to home, inner leaves of hedges of hazel and hornbeam exposed by clipping can be subsequently scorched and damaged by bright exposed conditions.

Shade leaves are obviously effective but may contribute comparatively little to the overall growth of a tree. A study using 13-year-old wild cherry trees

(*Prunus avium*), 6–7 m tall, found that removing all but the top five whorls of branches (i.e. removing the bottom 22% of the canopy height) did not affect height growth and reduced growth of the trunk diameter by just 4% the next year. Removing half the canopy height, leaving just three whorls of branches also didn't reduce height growth and reduced the diameter by 5% in the first year and 9% in the next. The shade leaves at the bottom are useful but not as important as the upper sun leaves.

## Water loss by leaves

The loss of water by a tree (called transpiration) can be prodigious. A young 2 m tall apple tree or a more mature Mediterranean oak may use around 7000 l in a summer while a normal sized birch tree may use 17 000 l and a larger deciduous tree up to 40 000 l. This is equivalent to less than 100 l per day in conifers, to 20–400 l per day in eucalypts and temperate trees such as oak, reaching perhaps 500 l per day in a well-watered palm and as high as 1200 l per day in specimens of *Euperua purpurea* growing out of the top of the Amazonian rainforest canopy and bearing the full brunt of the hot sun. Water loss is, of course, dependent upon surrounding conditions. A well-watered plant in dry, warm, windy conditions leads to most loss and the uppermost leaves will normally transpire more water than those lower down (in conifers this can be 2.5 times higher) since they are more in the sun. In the same way, trees in dense temperate woodlands (or in 'urban canyons' between tall buildings) use about two thirds the water of a solitary tree due to lower evaporation. Similarly, a study by Zeppel *et al.* (2008) in Eastern Australia showed that *Eucalyptus crebra* used less than 25 l per day when it was dry but more than 250 l per day when more than 25 mm of rain fell. At the other end of the water use spectrum, a large woody cactus may lose only 25 ml (0.025 l) per day compared to 0.9 l per day in a drought-stressed cork oak (*Quercus suber*) and 9–18 l per day in acacias in arid environments.

Although transpiration is driven by the sun's energy, water loss is quite common in trees at night. Francis Darwin (Charles Darwin's son and a good botanist) estimated that 87% of terrestrial plants (not just trees) open their stomata at night. In sugar maples (*Acer saccharum*) in New York State night-time water flow in the trunk has been measured at about a tenth of that during the day although some of this may be refilling the water reserves of the trunk (see Chapter 3) rather than loss through the leaves. Similarly, the coastal redwood of California (*Sequoia sempervirens*) has poor stomatal control and rates of water loss at night are easily 10% of that during the day and on very dry nights can rise to 40%. This water loss may help to get nutrients to parts of the tree that are too short of water to allow the stomata to open during the day.

On the other hand, keeping the stomata open at night may be nothing to do with water loss but rather may help get photosynthesis off to an early start in the morning. Or it perhaps helps oxygen to diffuse into the leaves at night (it is needed for respiration) when there is no photosynthesis and so no oxygen production inside the leaf.

Although water loss is a side effect of having holes in the leaves for gas exchange, it does have its benefits. The flow of water up the tree is the main way of getting dissolved minerals taken up by the roots to where they are needed. Moreover, evaporation helps to keep leaves from overheating in hot sunshine. In fact, 98% of the light energy reaching the leaves may go in evaporating water. On a larger scale, the vast quantity of water evaporated into the air by forests produces cool moist air which influences rainfall patterns (especially as some trees release prolific quantities of chemicals into the air – think of the overpowering smell of eucalypts on hot days – which provide nuclei for water droplets and hence cloud formation). On the other side of the coin, most trees can little afford reckless water loss. As is explained in more detail in Chapter 4, water is carried up a tree by being pulled from moist areas (the roots) to drier areas (the leaves). This means that to get water, the leaves have to be constantly operating at a low moisture level: in tall trees there is a narrow safety margin between operating and wilting.

## Control of water loss

Since most water loss is through the stomata, this is an obvious place to control water movement. Most trees optimise the need for carbon dioxide uptake against needless water loss. Stomata usually close when it is too cold or dark for photosynthesis or when the leaves are in danger of losing too much water and wilting. For example, on a bright day photosynthesis often shows a mid-day dip when sunlight is most intense and the leaves are losing more water than they are supplied, so the stomata close to prevent wilting and damage. A nearby shrub in the shade may happily photosynthesise all day.

Some trees avoid this mid-day dip by keeping the most exposed leaves more or less permanently pointing downwards. (This also helps prevent leaves from overheating in the intense light of mid-day.) This can be seen in the tallest tropical forest trees, eucalypts and the desert shrub jojoba (*Simmondsia chinensis*, from which artificial sperm whale oil is produced). The leaves intercept only about half as much light during the hot part of the middle of the day than they do in the relative cool of the morning and evening. This explains why one finds remarkably little shade under a eucalyptus tree from the heat of the mid-day sun. Others, such as the false acacia/black locust

(*Robinia pseudoacacia*), fold up their leaflets at the top of the tree if conditions get too tough. In still air and full sun, a leaf can reach temperatures 20°C above ambient. Shape makes a big difference; leaf edges warm slower (they are cooled by air moving past by convection) and so smaller or very lobed leaves are common at the tops of trees (this also helps the upper leaves to cast less shade on leaves below). As such, pinnate leaves, with their long length of leaf edge, are often more common in places with high summer temperatures. Waxy and hairy surfaces can also reduce the amount of energy absorbed by up to 20%.

Control of stomata has to be quite complex. For example, evergreens need to prevent opening of their stomata in winter in warm sunny spells before the roots can warm sufficiently to take up water. In the early part of the 20th century two German foresters, Büsgen and Münch, found that if cut twigs of evergreens were placed in water in a warm room in winter, the stomata of holly (*Ilex aquifolium*) opened in a few hours, while those of yew (*Taxus baccata*) took a week, and those of ivy (*Hedera helix*) and box (*Buxus*) remained closed.

So delicate is the balance between gas movement and water loss that the density of stomata has been found to change in response to external conditions. Measurements from fossil leaves show that 140 000 years ago the dwarf willow (*Salix herbacea*) had 40% more stomata than the 12 500 stomata cm$^{-2}$ found today. Modern olive leaves (*Olea europaea*) have 52 000 stomata cm$^{-2}$ but leaves from Tutankhamun's tomb (1327 BC) had 35% more. In both cases the reduction in density over time appears to be in response to increasing carbon dioxide levels: fewer holes are needed to let in the same amount of gas. This decrease has also been seen to happen over just a few decades by comparing herbarium specimens from the 1920s with modern-day material. A similar response has been seen over a much shorter time in response to carbon dioxide and temperature. Compared to leaves produced in the comparative coolness of spring, oak leaves growing in a warm late summer have 15% fewer stomata.

Having said all this, trees vary widely in their ability to regulate water loss. Some trees, for example many eucalypts and European alder (*Alnus glutinosa*), cannot regulate water loss to any great extent by closing their stomata. If you want to drain a wet area, plant these! In other trees stomatal control is part of their strategy for dealing with drought. For example, the European Turkey oak (*Quercus cerris*) is very sensitive to internal water loss and during dry summer periods will quickly close its stomata (referred to as being drought tolerant). Under the same conditions, cork oaks (*Q. suber*) only partially close their stomata and keep internal water levels high by maintaining water uptake by the roots and so avoid drought to keep on growing.

In addition to stomatal loss, some water is also lost directly through the cuticle of the leaf. In well-watered leaves this 'cuticular transpiration' may be

(a)                                                      (b)

**Figure 2.3** (a) The hairy underside of holm oak leaves (*Quercus ilex*) and (b) in close-up. The star-shaped hairs act like miniature umbrellas helping to reduce water loss by reflecting light and by holding a layer of still air over the back of the leaf.

only 5–10% of total water loss, but in water-stressed leaves (where the stomata are closed) or harsh environments it may be significantly higher. Leaves in sunshine and dry air tend to have more wax in the cuticle and plugs of wax over the stomata. Hairs, such as the felted underside of holm oak (*Quercus ilex*) leaves (Figure 2.3), can help by reflecting light and by increasing the thickness of the boundary layer (the stationary layer of air held over the surface of the leaf) making it harder for water to escape. In the olive, a typical Mediterranean plant, a similar function is carried out by semi-transparent umbrella-like scales covering the lower surface of the leaf. The rhododendrons of the high mountains of Asia, facing cold, dry winters and hot, humid summers nicely illustrate the subtlety of adaptation. The underside of leaves of most species has a more or less dense covering of hairs, scales or small waxy pegs. During winter, this layer helps to prevent water loss, aided in some species by resins and gums secreted from glandular hairs which dry to further seal the underside of the leaf. In the humid summer, the leaf covering *aids* transpiration by keeping the surface of the leaf below the hairs free from drops of water and the stomata unblocked. The glandular hairs secreting resins may suck water from the leaf helping water loss and the flow of minerals up from the roots. Certainly, many plants, including a few trees, can exude drops of water from the leaves (called guttation). This guttation is driven by roots pushing water up the stem (see Chapter 4 for a discussion of this) and so is seen particularly in small trees (under 10 m tall) in humid environments where there is little evaporation. The droplets may appear from the stomata, from special glands at the ends of the leaf veins (hydathodes), or even from lenticels on branches.

Trees living in or near salty water, such as mangroves or tamarisk species, lose water through special salt glands where briny water is excreted as a way of getting rid of excess salt.

Antitranspirants, sprayed on plants by gardeners, are designed to reduce water loss either by hormonally causing stomatal closure or by adding an impermeable film over the plant. Used on cut Christmas trees they reduce water loss and prevent the needles falling prematurely. Treatment of fruit trees may increase fruit size (by making more water available to the fruit) but will generally reduce photosynthesis as well by preventing carbon dioxide getting into the leaf.

In a further attempt to reduce water loss, desert plants such as woody cacti have evolved to take up carbon dioxide at night when evaporation is least and to use this stored carbon dioxide during the day while keeping the stomata firmly shut (called Crassulacean Acid Metabolism or CAM). This is rare in trees and found only in some species of the tropical strangler genus *Clusia* in Venezuela, and then only when the individual is short of water. Since the seed germinates in the canopy of a host tree, it can be very short of water until the roots reach the ground. Different leaves (and perhaps even parts of the same leaf) can undergo normal and CAM photosynthesis at the same time.

Another variation is found in plants with $C_4$ photosynthesis (as opposed to the normal $C_3$ photosynthesis); the names come from how many carbon atoms are in the sugars made by photosynthesis. In $C_4$ photosynthesis, carbon dioxide is very strongly absorbed even at low concentrations so that photosynthesis can proceed at a high rate when the stomata are nearly closed, hence reducing water loss in hot dry environments. This modified way of growing is found in a number of shrubs in dry areas around the world but is rare in tall shrubs and trees, numbering just several species in the Chenopodiaceae and Polygonaceae in the Middle East and an astonishing diversity of euphorbias in Hawaii.

## Taking things in through the leaf

In *Rhododendron nuttallii* and other species, dye placed on the scales on the underside of the leaf was seen to quickly pass down the scales and into the leaf. Such leaves may therefore act to absorb moisture. This is seen even more vividly in welwitschia (*Welwitschia mirabilis*), a strange tree that grows in the coastal desert of Namibia (Figure 2.1d). It is really a type of underground conifer with just two leaves which grow strap-like from the top of the stem reaching 8 m long with an area of up to 55 square metres. As the leaves grow from the base they die off at the other end. Apart from the occasional downpour the coastal strip of desert where these plants grow is practically devoid of rain but it has fog on 300 days of the year. Water is absorbed through the stomata into the plant. The stomatal density at 22 200 per square centimetre is fairly average for trees but it has this many on both sides of the leaf. The waxy cuticle is surprisingly thin for a desert plant but it does have crystals of calcium oxalate in the cuticle, which probably play a role in reflecting the sun's energy.

Fog is also very important in supplementing rain in many arid parts of the world such as the laurel forests of the Canary Islands (Figure 2.4) and in semi-arid areas of the coastal mountains of Chile which have rainforest clumps surrounded by semi-arid scrub in areas with only 147 mm of rain. The rainforest intercepts fog giving the equivalent of at least another 200 mm of rainfall.

Perhaps the best example of fog uptake is seen in the coastal redwoods (*Sequoia sempervirens*) growing in a band of fog along the Californian coast. Water is absorbed directly through the bark of twigs but mainly through the cuticle of the leaves, especially older leaves, presumably because the cuticle is more worn. This water uptake is sufficiently good that the flow of water inside the tree (in the xylem; see Chapter 3) can be completely reversed in branches and possibly right through the trunk; water moving from the leaves down to the roots and perhaps even out into the soil instead of upwards as during the day. The amount of water involved is fairly small, amounting to around 5% of the water in the leaves. Some drips to the ground and can be taken up by the roots, but this still only amounts to around 20–30% of the water needs of a redwood. Although a small amount, it is of critical importance. Photosynthesis is higher in fog-exposed branches, and it may allow air-filled xylem tubes (which are useless for conducting water; Chapter 3) to be refilled and thus revitalised, and it could bring in nutrients dissolved from the leaf surface or contained in the fog (Simonin *et al.* 2009). The real value of the fog, however, may be that the humid environment created by the fog will not only rehydrate leaves but also slow water loss from the leaves (washing doesn't dry very quickly in damp weather since the air is already full of water vapour) and so help relieve the daily water stress at the top of such tall trees.

Some trees, such as the Utah juniper (*Juniperus osteosperma*), can take up water through the leaves directly from rain (as distinct from fog), which is especially useful in dry areas of the American Southwest where rain intense enough to recharge the soil is rarer than rain showers that are just enough to wet the soil and leaves.

Nutrients can also be taken up through the leaf. Nitrogen compounds in all forms (gaseous, liquid and particulate) have been seen to be taken up through the cuticle or stomata of trees. In Central European forests this absorption of nitrogen through leaves may contribute 30% of the total nitrogen needs of trees.

(a)

(b)

**Figure 2.4** (a) Cloud that habitually banks up against the NW end of Tenerife, Canary Islands (with Mount Teide in the background) and (b) the resulting wet laurel forest. The rainfall of 30 cm (1 ft) is supplemented by up to 20 times that amount from the water strained out of the fog as it condenses on the vegetation and drips to the ground.

## Leaf movements

Leaves are on the whole flexible enough to bend and give before the blast of rain and wind. But most woody plants can move their leaves by themselves, partly by growth of the shoot and partly by movements of the leaves themselves.

Perhaps one of the best known examples of leaf movement in trees, and one of the most rapid, is the sensitive plant (*Mimosa pudica* and several related species), a shrubby legume. Touching the plant results in immediate folding and collapse of the leaves and branches (Figure 2.5). The harder the touch the more of the plant collapses. In doing so the leaves are folded away between nasty-looking thorns, and one reason for this collapse may be to move the tender leaves away from browsing animals. Or it may be to dislodge herbivorous insects as they land, or to fold the leaves away from the pounding of rain drops.

The sensitive plant also shows 'sleep movements' (technically called nyctinastic movements; from the Greek words *nycto* = night and *nastic* = movement) where their leaves are folded down at night. This may have the advantage that since the leaves are not needed for photosynthesis there is value in tucking them away between thorns or away from accidental damage. Sleep movements can be seen in a wide variety of woody plants from the temperate laburnums, the tropical tamarind (*Tamarindus indica*) to the desert shrub creosote bush (*Larrea tridentata*). The rain tree (*Samanea saman*) of the tropics looks bare of leaves at night and yet has a dense canopy of leaves in the day due to these sleep movements. Most of these sleep movements seem to be under internal control since they carry on for at least a while if the plant is grown under an artificially constant environment. Movements can be quite complex. Darwin (1880) noted that in Indian laburnums (*Cassia* spp.) 'the leaflets which are horizontal during the day not only bend at night vertically downwards with the terminal pair directed considerably backwards, but they also rotate on their own axes, so that their lower surfaces are turned outwards'. The real value of sleep movements in unarmed woody plants is not really known but in herbaceous plants it has been found that folded leaves keep the buds marginally warmer, maybe by only 1°C, but a significant amount. In trees this could make the difference between frost damage and not. In herbaceous plants needing short days to flower (see Chapter 5), sleep movements cut down the absorption of light from the moon and stars preventing them from being misled into thinking the days are long (Bünning & Moser 1969).

Leaf movements during the day allow 'light tracking' where the leaves follow the sun around, resetting during the night to await the rising sun, to maximise the amount of light received by the leaves. Even the needles of conifers will move a little to maximise light catching. While some trees

**Figure 2.5** The sensitive plant (*Mimosa pudica*) has feathery compound leaves which can collapse in just 1/10 of a second when touched. If touched at (a) the signal travels in the direction of the arrows. The leaflets fold forward, then the whole leaf droops (as on the right). Movement is controlled by the swollen base of the leaflets and leaves (the pulvinus). From: Troll, W. (1959). *Allgemeine Botanik.* Ferdinand Enke, Stuttgart.

permanently dangle their leaves (see 'control of water loss' above), others fold up their leaves in response to too much light or to drought to reduce the amount of light hitting the leaves and hence water loss. For example, false acacia/black locust (*Robinia pseudoacacia*) folds its leaflets together like praying hands as the sun gets higher until by mid-day the leaflets are together and edge-on to the sun[2]. In the creosote bush the two leaflets of each leaf open

---

[2] For those who like long words, holding the leaves perpendicular to the sun to maximise light impact is called diaheliotropic positioning, and to dangle them to minimise light absorption is paraheliotropic positioning.

outwards during the day, but will stay closed or almost so if the plant is short of water. The leaves have chlorophyll both sides so the vertically closed leaves can still photosynthesise but with a reduced leaf area and reduced water loss. Rhododendrons and viburnums are well known for drooping and curling leaves in cold weather (thermonastic movement). Russell *et al.* (2009) have shown that the drooping protects leaves from high sunlight intensities in cold temperatures which would otherwise damage the leaf's photosynthetic machinery and the curling helps slow the rate of thawing in the morning, preventing damage during daily rapid rethawing. These examples demonstrate how leaf movements during the day help to optimise the compromise between photosynthesis and conserving water during changing daily conditions.

Some leaf movements are not easy to explain. A famous example is the Asian semaphore or telegraph tree (*Codariocalyx motorius*) 'whose leaves are in an irregular motion all day long. In this plant of the legume family, sometimes all the leaves move in circles and sometimes leaves on one side of the plant stem move up while those on the other side move down. At times it appears to run wild, with some leaves moving upward, others downward, and yet others moving in circles. It has a jerky movement and occasionally stops as if for a short rest.' (Sandved & Prance 1985).

So how are these movements made? Leaves can simply wilt (as in the drooping of rhododendron leaves in cold weather). Some trees possess the ability to alter the position by growing new cells and so become set in the optimum position, although this can be changed by further growth. But in most plants, repeated and reversible movements take place at bulges at the base of leaves and leaflets called pulvini (Figure 2.5). A pulvinus contains cells that can rapidly move water in and out. By shunting water from cells on one side of the pulvinus to the other, rapid movement of the petiole and leaf can be effected. This explains the movement of the individual pulvinus, but how is the message to collapse spread so quickly in the sensitive plant? The mechanism behind this animal-like movement is nerve-like electrical impulses carried not along nerves but through the vascular tissue (see Simons 1992 for details of how this and other plant movements work). The signal can travel at 1–10 cm per second, very impressive for a plant!

## Leaf size and shape

The conflicting needs to intercept light and take up carbon dioxide on the one hand, and conserving water on the other has resulted in a number of different leaf shapes in different environments.

A note of caution to begin. So many factors govern leaf shape that it can be difficult to make sense of what we see. Leaf size and shape vary according to the conditions under which they grow, including day length, temperature and moisture, nutrition, and herbivore damage to the leaf and elsewhere in the same or neighbouring trees. Leaves growing later in the summer can be progressively more deeply lobed as in sweet gum (*Liquidambar styraciflua*; Figure 2.6) or longer and narrower as in the lammas growth of oaks (Chapter 6). Leaf shape also changes through a tree canopy. Sweet gum leaves change from larger, shallowly lobed shade leaves at the bottom (similar in shape to leaf 1 in Figure 2.6) to smaller, more lobed leaves at the top (similar to leaf 8 in the figure). With all this going on, is it any wonder that trees can sometimes be very difficult to identify from a few leaves? Sometimes there is so much apparently random variation within a single tree or species that shape seems to have little significance. Despite this wealth of variation due to genetics and growing conditions, a number of generalisations can be made about leaf shape.

**Figure 2.6** The shape of sweet gum leaves (*Liquidambar styraciflua*) from the earliest leaves (1) in spring to the last formed (8) in the late summer. Note the deeper lobing through the season. From: Zimmermann, M.H. & Brown, C.L. (1971) *Trees: Structure and Function*. Springer, Berlin, Figure I-21, Page 49. With kind permission from Springer Science and Business Media.

Large flat leaves are obviously good at catching light. Moreover, they hold a thick still layer (boundary layer) of air over the leaf which thermally insulates the leaf leading to a temperature 3–10°C higher than surrounding air, helping them to photosynthesise in the cooler, moist inner regions of a forest. Large leaves tend therefore to be found in shaded areas (where the sun does not overheat the leaf) and where there is less risk of tearing from high winds, exactly the situation inside many tropical forests. Bigger leaves are held by bigger twigs (thicker and overall heavier), and thicker twigs hold proportionately less area of leaves as more material is invested in holding the whole thing up and less to supplying the leaves. This diminishing return may ultimately act to limit normal leaf size.

With small leaves (or leaflets of compound leaves, such as the very small leaflets of acacias), the boundary layer is thinner, air moves more easily, and the leaf is kept cool by convection rather than evaporation of water. These leaves are therefore more common in hotter, brighter areas where water is more precious, such as the less humid conditions of temperate areas, and in areas where nutrients are scarce and small leaves are cheaper to grow. Lobes and teeth on a leaf create turbulence to destroy the boundary layer making the leaf effectively smaller still; these are common in clearings in tropical forests where high light intensity and lower humidity require efficient cooling, and on larger-leaved temperate trees. This is further accentuated in trees such as aspen (*Populus tremula* and *P. tremuloides*) where the petioles are laterally compressed, causing the leaf to shake in the gentlest of breezes, further removing the boundary layer. This is akin to shaking your hands to dry them, which may account for poplars being some of the most profligate users of water and preferring moist alluvial soils. Smaller leaves can also be grown and opened quicker which would be an advantage where there is a high risk of being eaten. And presumably one small intact leaf is better than half a larger leaf with open wounds leaking water and letting in disease. It probably goes without saying that a tree with smaller leaves will usually have more of them. A greater number of small leaves also means more buds in their axils which gives the tree more options when it comes to responding to environmental conditions or damage than in a tree with fewer buds.

But this is not the whole story. For example, within rainforests leaves are generally smaller at higher altitudes. Figure 2.7a shows data from a high-altitude cloud forest in Honduras collected at two sites about 250 m elevation apart and with quite different rainfall and temperatures. Figure 2.7b shows the extreme range of leaf sizes possible in this forest. Leaves at the upper site are generally smaller and thicker than the lower site. The lower site leaves are 0.22 mm thick on average compared to 0.41 mm at the upper site, almost twice

(a)

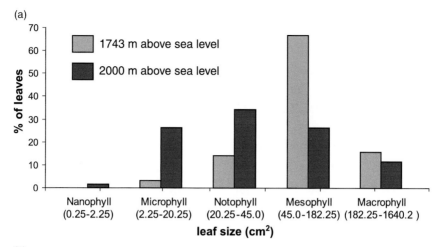

(b)

Figure 2.7 (a) The range of different leaf sizes from randomly selected trees at two areas
within the tropical cloud forest of Cusuco National Park, Honduras in 2009. The blue columns are from
an area 1743 m above sea level (with an annual rainfall of 1795 mm, and mean annual temperature of
19.8°C) and the red ones from an exposed ridge at 2000 m (annual rainfall 1913 mm, mean annual
temperature 13.2°C). The lower forest is dominated by mesophyll trees while the higher forest has a
much wider range of leaf sizes. Operation Wallacea's help in collecting the data is gratefully
acknowledged. (b) The photograph shows the wide range of leaf sizes. The forest contains a number of
trees that have large leaves (macrophylls) such as (*Oreopanax* sp.) on the left (which is 50 cm long),
*Elaeagia auriculata* in the centre and *Cecropia peltata* on the right, but at higher altitude many trees have
much smaller leaves (nanophylls), indicated by the arrow.

as thick. Curiously though, the shape (as defined by the ratio of width to length) does not change. The big question to all this is why? The research is still ongoing but it looks as if the main driver for these changes is the increase in precipitation up the mountain. This is likely to be because smaller, thicker leaves are less readily damaged by torrential rain. Also, the tough glossy leaves are designed to reduce the leaching of minerals by the abundant rain sloshing over the leaves.

Rain is also encouraged to run-off rainforest leaves by the elongation of the leaf tip into a 'drip tip'. This prevents water resting on the leaf and leaching minerals, and the growth of light-robbing epiphytes on the surface. As might be anticipated, tall rainforest trees that emerge from the canopy above others have thick leaves but without drip tips; they are dried rapidly by the sun. But you don't have to travel to the tropics to see drip tips; temperate trees in high rainfall areas, such as limes and birches, also have them.

Plants of dry sunny areas also tend to have thick leaves. Thick leaves are less efficient at photosynthesis because the chloroplast layer is thicker and they shade each other and compete for carbon dioxide, but they produce more food without significant extra transpiration costs. The olive tree (*Olea europaea*) seems to have found a partial solution to the internal shading: the leaves have hard T-shaped stony cells penetrating the leaf-like drawing-pins stuck into the surface. These prongs appear to act as miniature optical fibres piping light into the leaf.

In much drier areas, such as Mediterranean scrub where rainfall is seasonal, or northern regions where unfrozen water can be in short supply (especially in spring) leaves become smaller still. Small leaves may also be a response to poor or wet soils, which limit root growth and therefore ability to take up water and nutrients. This is seen in ericaceous plants such as heathers (*Calluna vulgaris* and *Erica* spp.).

Most leaves are intricately folded inside the bud but soon unfurl to start photosynthesis as soon as possible. In tropical lowland rainforest, a number of trees (including many *Ficus* species) keep their leaves tightly rolled or folded until they are at least 50% of their final length. This seems to be a mechanism to reduce herbivory on young leaves since much of the delicate structure is protected inside the folds or rolls. This comes at a cost of lost photosynthesis but these leaves are found mainly in shade-tolerant species which typically have the longest-lived leaves. So over the life span of a leaf this loss in sugar production is a small price to pay if the leaf is more likely to survive unscathed. Rolled leaves are also found in palms growing in the open but here it would seem that because they have so few leaves, each one is important and the loss in photosynthesis is outweighed by the greater leaf area that survives the early high herbivory stages.

## Needles and scales

Reduction in leaf size with inhospitable conditions can be seen beautifully in conifers, where needles and scales are the commonest type of leaf. Needles usually occur singly but in pines they are in bundles of two, three or five (which fit together to make a cylinder) borne on dwarf branches (as always seems to be the case in plants, there are exceptions; there can be up to eight needles in a bundle or as in the pinyon pine, *Pinus monophylla*, and occasionally others, single, cylindrical needles). Internally, the needle is no more than a compact version of a simple leaf (Figure 2.8), beautifully compact and adapted to withstand harsh dry conditions (the benefit of evergreen and deciduous needles is discussed below). The chlorophyll-bearing tissue surrounds the central vein, and is in turn surrounded by a thick-walled epidermis and thick waxy cuticle. The relatively few stomata are set in rows (sometimes on both surfaces, more often only on the lower surface) and are sunken into pits (to hold still air over the stomata), often covered in wax to further reduce water loss. The stomata are seen as lines of white waxy dots along the leaf. Needles therefore lose little water when it is in short supply. They are also long and thin to shed snow, and contain little sap for freezing.

In most cypresses and in a few others such as the red pines in the Podocarpaceae (*Dacrydium* spp.), the Tasmanian cedars (*Athrotaxis* spp.) and Japanese red cedar (*Cryptomeria japonica*) in the redwood family (Taxodiaceae), the leaves are even more reduced to scales just a few millimetres long which clasp the stem. In some, including many junipers, the free tip of the scale is elongated to form a needle.

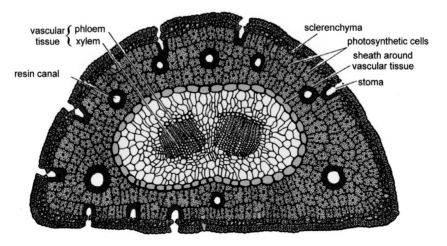

**Figure 2.8** Cross-section of a pine needle. From: *Comparative Morphology of Vascular Plants* (2nd edn) by Foster *et al.* ©1989 by W.H. Freeman & Company. Reproduced with permission.

## Phyllodes, phylloclades and no leaves at all

Reduction in size of leaves is carried to even greater lengths by some trees. In many acacias, the leaf blade is small or absent and photosynthesis is carried out by broad flattened petioles (phyllodes); Figure 2.9. The part of the leaf that loses most water is shed, allowing the plant to continue to function in dry sunny conditions. Some trees and shrubs use leaves when conditions allow but have green photosynthetic stems as a backup during dry periods. This is seen in the creosote bush (*Larrea tridentata*) and other shrubs of N American deserts, and the brooms (*Cytisus* spp.) of Europe.

Abandonment of leaves as green photosynthetic organs is permanent in some trees and shrubs. In she-oaks (*Casuarina* spp.), which grow naturally in dry areas of the southern hemisphere, all the leaves are reduced to toothed sheaths surrounding articulations of the stem, producing trees which look like giant, branched horsetails: photosynthesis is carried out by the green cylindrical branches. Similarly, the leaves of the Japanese umbrella pine (*Sciadopitys verticillata*) and the celery-topped pines (*Phyllocladus* spp.) are reduced to small brown scales leaving photosynthetic branches (Figure 2.9c). Branches in the celery-topped pines have taken on the role of leaves in a convincing manner by becoming flattened into green leaf-like shoots (called either phylloclades, cladophylls or cladodes, but all meaning essentially the same thing), which look remarkably like pieces of leafy celery. Phylloclades (with and without leaves) are found in a number of other woody plants around the world including the butchers'-broom (*Ruscus aculeatus*) in southern England and the Mediterranean. In this, the flowers grow out of the middle of the 'leaves' and appear very strange but really the flowers are growing on a stem just like in any other plant (Figure 2.10). It is sometimes possible to find a normal tree, such as a holly (*Ilex aquifolium*), with the occasional flattened branch looking like a ribbon with leaves. This 'fasciated' shoot is an error in the growth of the branch – an accidental phylloclade – which seems to do little harm.

Many temperate trees and shrubs have chlorophyll in the bark and can photosynthesise during the dormant season. This helps offset the respiration cost of the living tissue, and so save on stored food. Gram for gram, photosynthesis in bark has been measured in the European spindle (*Euonymus europaeus*) bark to reach 17% of the sugar produced by leaves. To put this into context, desert shrubs which spent most of their year leafless can produce up to half of their annual food from their green branches.

## Other modifications: stipules, tendrils and climbing leaves

In trees that have them, the stipules (Figure 2.2) are often small, scale-like and fall from the tree early in the life of the leaf. In others, such as the

(a)

(b)

(c)

Figure 2.9 (a) Leaves of the Australian coastal wattle (*Acacia sophorae*) reduced to just a broad flattened petiole (phyllode) and in (b) the swamp wattle (*Acacia retinoides*) with leaflets still growing at the end of the phyllode. In (c) the celery-topped pine (*Phyllocladus trichomanoides*, called tanekaha in New Zealand) the leaves are reduced to small brown scales and the branches are flattened into photosynthetic branches called phylloclades. Photographs (a) Barronjoey Head Aquatic Reserve, Sydney, Australia, (b) Keele University, England, (c) Tongariro National Park, New Zealand.

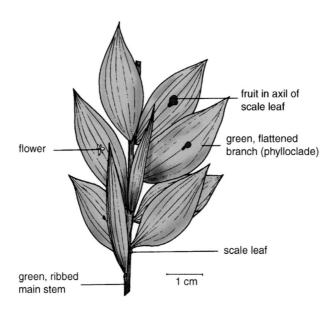

**flower** —

**fruit in axil of scale leaf**

**green, flattened branch (phylloclade)**

**scale leaf**

**green, ribbed main stem**

**1 cm**

Figure 2.10 The butcher's broom (*Ruscus aculeatus*) which has flattened stems (phylloclades) as the main photosynthetic organs. The leaves are reduced to small scales. This arrangement leads to the flowers appearing to spring from the middle of the 'leaves'. From: Bell, P.R. & Woodcock, C.L.F. (1983) *The Diversity of Green Plants.* Edward Arnold, London.

Japanese quince (*Chanomeles* spp.), they are large and leaf-like and add considerably to the photosynthesis of the leaf during its whole life. Yet, stipules can fulfil other functions. In the tulip tree (*Liriodendron tulipifera*) and magnolias, the stipules at the base of an expanding leaf stay closed around the young developing leaves at the end of the shoot. Once the next leaf expands the stipules of its protective big brother soon fall. In many trees the stipules form bud scales, e.g. beech, oak, limes, magnolias and many members of the fig family. But in false acacia/black locust (*Robinia pseudoacacia*), desert ironwood (*Olneya tesota*) of western United States, and acacias, to mention a few, stipules become woody and persist on the twig as two thorns for defence (Figure 9.4c).

Leaves can also be used to help woody climbers on their way. In Virginia creeper (*Parthenocissus quinquefolia*) and grapevines (*Vitis* spp.), the climbing tendrils are modified stems (they grow out from buds). In clematis, as described by Darwin, it is the leaf stalk that twists. The leaves are normal but the sensitive petioles are clasping and wrap around an object. Rattan palms (*Calamus* spp.) of tropical Asia use their leaves in a similar way. The upper leaflets are barbed 'grappling hooks' that catch onto the bark and stems of trees helping the vine-like palms scramble to the top of the forest.

# Why have compound leaves?

Imagine two trees, one of which has branches with 'ordinary' leaves and the other has compound leaves the same overall size and shape as the first tree's branches. What is the advantage of the one over the other? The answer is that compound leaves are cheap disposable branches which give the tree an advantage under two conditions.

The first is that they help reduce water loss in arid seasons. Small twigs (with a large surface area to volume ratio and thin bark) are the biggest source of water loss once leaves have fallen. So compound leaves are common in areas with severe dry summers such as savannahs, thorn forests and warm deserts on such trees as acacias and the baobab tree (*Adansonia digitata*) of African savannahs, the mesquites (*Prosopis* spp.) and palo-verdes (*Cercidium* spp.) of N American desert grasslands, and horse chestnuts (*Aesculus* spp.) of various dry parts of the world. One might also include the ash trees (*Fraxinus excelsior*) that grow on dry limestone screes in central England. Such water loss is unlikely to be a significant factor in most temperate trees and indeed these areas have fewer compound-leaved trees.

The other advantage of having cheap disposable branches is where rapid vertical growth is important. Since the rachis of a compound leaf is held up primarily by water pressure (turgidity of the plant cells) and fibrous material it is not as costly to build as a woody branch. A tree that is growing rapidly can display leaves on cheap branches which will only be needed for a short time before they too are shaded, and so can invest the saved energy in height growth. The tree thus avoids making an unnecessarily durable and expensive investment. Compound leaves are seen in many species that invade open areas and need to grow tall quickly to stay ahead of the competition, especially in tropical trees that invade gaps. Some trees in the family of Meliaceae grow for months or years without branches, using large compound leaves as branches. In temperate regions, sumacs (*Rhus* spp.), tree of heaven (*Ailanthus altissima*) Kentucky coffee-tree (*Gymnocladus dioica*) and the Japanese angelica tree (*Aralia elata*; Figure 2.1c) also fill this niche and all bear few main branches and numerous compound leaves as throw-away 'branches'. Note that others such as Manitoba maple (*Acer negundo*) and ashes (*Fraxinus* spp.) are suggested to have compound leaves because they are predominantly floodplain trees and are therefore disturbed area specialists that need to cope with periodic damage.

These proposals are obviously not the whole answer because there are examples of evergreen compound-leaved trees in the moist tropics, and invaders of open areas with sparse branches, capable of rapid growth that have large simple leaves and few branches, especially when young. These include the catalpas (*Catalpa* spp.) and the foxglove tree (*Paulownia tomentosa*).

Non-compound-leaved trees have found their own answers. Rapid growing species such as poplars which have simple leaves do actually shed twigs very readily (Chapter 6); these twigs are thin, not very woody and so fairly cheap to produce.

## Juvenile leaves

Changes in leaf shape can also occur as a tree matures. Certain trees go through a number of completely different leaf shapes and arrangements. Leaves of young plants sometimes differ from those of the mature plant in both form and arrangement. For example, many eucalypts, pines and junipers produce juvenile leaves before the mature foliage. In eucalypts the juvenile leaves are round and stem-clasping in comparison to the slender stalked mature leaves (Figure 2.11). A young tree often has a cone of juvenile leaves inside the canopy showing that the branch ends all changed to mature foliage at the same time, usually at around 5 years of age. In many hardwood trees, juvenile foliage is not a completely different type from mature foliage but a different size and shape. The tulip tree (*Liriodendron tulipifera*) illustrates this nicely. It goes through a juvenile stage up to about 10 years of age during which the leaves can be 25–30 cm long, reducing to 10–15 cm on a mature tree. The shape also changes from four-lobed through

Figure 2.11 Juvenile (left) and adult foliage (right) of the cider gum (*Eucalyptus gunnii*). The juvenile foliage lasts for 2–3 years. London, England.

to leaves with an extra pair or two of lobes (this may be related to sun and shade leaves discussed earlier). The same sequence is restarted with coppicing, sometimes with even greater disparity in size.

## Evergreen versus deciduous (and leaf age)

The difference between deciduous and evergreen trees is quite clear but holds the seeds for tremendous confusion! Deciduous trees hold their leaves for less than 1 year and are bare during an unfavourable season (winter in temperate areas, summer in Mediterranean climates). This includes most temperate and some tropical species, and a number of conifers including dawn redwood (*Metasequoia glyptostroboides*), swamp cypress (*Taxodium distichum*) and larches (*Larix* spp.). Evergreen trees by contrast have some leaves all year round. Evergreens include most conifers and also many hardwood species of the tropics, mountains shrubs (e.g. rhododendrons) and temperate trees (e.g. holly, *Ilex aquifolium*). Most conifers and hardwood temperate trees (such as holly) keep leaves for 3–5 years but yews, firs and spruces may keep needles for up to 10 years. Some are even longer-lived; both the bristlecone pine (*Pinus longaeva*, the oldest trees in the world; see Chapter 9) and the monkey puzzle (*Araucaria araucana*) regularly keep leaves for 15 years and exceptionally for over 30 (the life span of a horse). Evergreen leaves usually live for longer (by 2–3 years) at higher altitudes. There can also be quite a variation within closely related trees: cork oak leaves (*Quercus suber*) live for only just over a year while those of the holm oak (*Q. ilex*) live for 1–3 years and the shrubby kermes oak (*Q. coccifera*) for 5–6 years. Also, leaves at the top of a tree tend to live for a shorter time. At the other end of the size spectrum, evergreen shrubs such as heathers keep leaves for just 1–4 years.

An evergreen canopy is therefore made up of leaves of differing age, particularly at increasing latitude and altitude, and on poorer soils, where leaves are generally kept longer. In most trees the age of a leaf can be worked out by counting back the number of years of annual growth. It is harder to age leaves in tropical trees with continuous growth but marking experiments have shown that 6–15 months is probably the normal age limit with some living up to 2–3 years.

Long-lived evergreen leaves do not necessarily stop growing in the first year they are made. Pine needles can continue to grow in length and thickness from the base for many years. Monkey puzzle leaves are attached to the branch along a long base and, living for several decades, they have to grow wider at the base to remain attached to the expanding branch.

The distinction between deciduous and evergreen habit is not always straightforward. Evergreens such as eucalypts and citrus species can be partially summer deciduous; they may shed some leaves when under the stress of a hot dry summer. As one might expect, it is the older shaded leaves that go in competition with younger more vigorous leaves. In Britain the native privet (*Ligustrum vulgare*) is similar: in mild winters more than half the leaves may survive, but in severe winters they are all shed. (In many evergreen species it is spring drought that causes most problems; see Chapter 9.) Temperate and tropical trees can further flaunt our nice definitions by being 'leaf exchanging'. Here new leaves appear around the time that the majority of old leaves fall, either just before or just after (this can vary even within the same species), so some specimens are almost completely leafless for a brief period. This happens in a number of tropical and Mediterranean species including evergreen magnolia (*Magnolia grandiflora*), camphor tree (*Cinnamomum camphora*), avocado (*Persea americana*) and various oaks includ-ing the cork oak (*Quercus suber*). When these leaf-exchanging species are grown in temperate conditions they tend more towards being truly evergreen, but some leaf exchangers can be found under temperate conditions. The Spanish or Lucombe oak (*Quercus* x *hispanica*; Figure 2.12) and Mirbeck's oak

(a)   (b)

Figure 2.12 (a) Leaves of the Spanish or Lucombe oak (*Quercus* x *hispanica*) which is a hybrid between the deciduous Turkey oak (*Q. cerris*) and the evergreen cork oak (*Q. suber*) and tries to be both deciduous and evergreen, dropping its leaves in spring just before it grows a new set, described as leaf-exchanging. Leaf shape is also somewhat intermediate between the round prickly leaves of the cork oak and the long lobed leaves of the Turkey oak. (b) In the UK many early trees were grafted as this one shows, with a root stock of English oak (*Q. robur*). The corky bark from its cork oak parent is characteristic of this hybrid. Keele University, England.

(*Q. canariensis*) from Spain and North Africa keep their leaves until early spring (by which time they can be pretty ragged), losing them all just before the new set grows. Leaf-exchanging trees could qualify as evergreen because they have leaves for near enough the whole year but also qualify as deciduous because the leaves live around 12 months and change all at once every year. Many young trees in the tropics show continuous evergreen growth but as they get bigger they become leaf-exchanging or deciduous.

Why are some trees evergreen and some deciduous? Starting at the beginning, if there is no distinctly unfavourable period for growth during the year as in the tropics, a plant will usually be evergreen: there is no reason to lose leaves that are still working. In climates with a period unfavourable for growth – a winter or a hot dry summer – deciduous trees have the upper hand. It is cheaper to grow relatively thin, unprotected leaves which are disposed of during the inhospitable season than to produce more robust leaves capable of surviving the off season. These seem to be optimum for warm temperate areas (but from the tropics to the Arctic there are always at least a few hard-wood deciduous trees and shrubs to be found).

In areas with even worse growing conditions, however, evergreen leaves will reappear. Firstly, this includes areas where the growing season is very short such as towards the poles and alpine areas. Evergreen leaves start photosynthesising as soon as conditions allow and no time is wasted while a new set of leaves is grown. For the same reason, in European woodlands holly (*Ilex aquifolium*) and ivy (*Hedera helix*) benefit from being evergreen; they are usually shaded by taller trees and need to do a lot of growing in early spring and late autumn when the trees above are leafless. Secondly, evergreen leaves are found where nutrients are in short supply on poor soils (such as on sandy soils and much of the tropics) or very wet soils where rooting is restricted. Here it is too expensive to grow new leaves each year and it is cheaper to pay the cost of producing more robust leaves which can survive the inhospitable period. Finally evergreen leaves are also found in dry areas such as in Mediterranean climates, for example as in the holm and cork oaks. Thick leathery leaves are needed during the fairly dry winter growing season just to survive (thick leaves can grow more sugar for the same water loss compared to thin leaves during hot, dry periods). This makes them tough enough to withstand the hot dry summer as well. So much energy and nutrients are invested in the leaves that it makes economic sense to hang onto them for as long as possible.

Last of all, if the growing conditions become even more severe, such as near the northern tree line and at upper alpine areas, trees may again revert to being deciduous. Despite the problems of having to cope with a short growing season and few nutrients, the winter is so severe that being deciduous is still a cheaper strategy than trying to keep leaves alive over the winter. Thus, the northern-most trees in the Arctic are birches and the final trees up mountains are often birches, willows and larches.

## Death and senescence

Deciduous trees in temperate areas shed the bulk of their leaves before winter primarily in response to decreasing day length. This is why autumn colours appear in different species in more or less the same order each year. But the precise timing and duration of leaf shedding can also be influenced by temperature, soil moisture and nutrient supply. That's why it is bad policy to give woody plants nitrogen-rich fertilisers late in the summer since it delays leaf senescence and winter-hardiness. Inevitably, autumn senescence is a compromise. If the leaves start falling too early they miss the end of the growing season but have plenty of time to recover nutrients from the leaves. If they leave it too late, they can produce more sugars but risk being killed by bad weather before the nutrients can be reclaimed.

Tropical trees are more complicated. Trees may lose their leaves together as a species or as an isolated tree (although there is a tendency for trees within a species to become more synchronised as they become older). Or they may shed leaves from one or more branches at a time giving the tree a mottled look. The loss may be at irregular times or regular but not necessarily linked to the 12 month calendar. Trees may be leafless for a few weeks to 6 months. Leaf shedding may be in response to an approaching dry season or even a rainy season (cloudier and darker?).

There is the probably apocryphal story told by Edlin (1976) about the enterprising park keeper who said he would only plant evergreens since they didn't lose their leaves and he would therefore be saved the chore of sweeping. He would have been somewhat disappointed! They may lose only a proportion of their leaves each year but they do shed. In some evergreens leaf shedding is spread through the year (with perhaps a mid-summer peak as in *Cupressus* species) and so goes largely unnoticed. In contrast, others drop their oldest leaves in one go, such as holly in spring and early summer. In evergreen and 'leaf-exchanging' species leaf shedding is probably not in response to day length but to competition from new leaf growth that stimulates senescence of oldest leaves. This is an important point: leaves act as independent units, similar to a block of apartments. If a tenant is not paying their rent, they are thrown out; if a leaf is a net drain on the tree – it is using more energy than it produces – it is shed.

Leaf death in the autumn is preceded by a very organised breaking down of the contents of the leaf using enzymes to allow the useful bits to be taken back into the tree for future use. In a study of trembling aspen (*Populus tremula*) in Sweden by Keskitalo *et al.* (2005), the internal breaking down of the leaves took 18 days. Within a week of this starting, chlorophyll levels had fallen by 50% and by abscission (after 3 weeks) less than 5% of the original amount was left. Mobilisation of nutrients out of the leaf continued until several days before the leaf was shed by which time 80% of the nitrogen and phosphorus, 50% of the sulphur and 20% of the iron had been moved out of the leaves and leaf weight had halved. Most trees are capable of eventually removing more than 99% of nitrogen and phosphorus from the leaf before it is shed and the amount remaining can be as low as 0.3% for nitrogen and 0.01% for phosphorus.

Once nutrients have been recovered, leaf shedding is not just a case of leaf death: if a branch is snapped or the leaves are killed by sudden stress, the leaves wither in place but are remarkably hard to pull off. Leaf fall itself involves a carefully executed severance similar to the way other bits of trees such as fruits, flower parts and even branches are shed. Across the base of the leaf and leaflet stalk there runs a line of weakness – the abscission zone – made up usually of small cells lacking lignin (Figure 2.13). At leaf fall the 'glue' holding these cells together weakens (controlled hormonally; ethylene speeds up and auxin slows down the process[3]) until finally the leaf is held on by just the cuticle and the plumbing tubes (vascular tissue). At this point the leaf is usually torn off by wind or forced off by frost. With the latter, expanding ice crystals force the leaf off which drops when the ice melts: this explains the sudden deluge of leaves on a warm autumn morning following a frost. As the leaf falls or even before, the abscission zone seals off the broken end of the branch by producing a corky layer.

Certain trees, including beeches, oaks, hornbeams, American hop hornbeam (*Ostrya virginiana*) and sugar maple (*Acer saccharum*), are renowned for holding on to some of the dead withered leaves. Leaves are normally kept on the bottom 2 m of the tree but occasionally up to 10 m. The blade and most of the petiole die but the very base remains alive until the following spring when abscission is completed and the leaf falls. Wind may rip away some leaves in the winter but the petiole base doggedly hangs on. It seems that these trees are rather slow at beginning abscission in the autumn and it gets halted halfway through by cold temperatures, only to resume where it left off in the spring. But why? It may act to keep the buds of young growth protected from frosts. But Otto and Nilsson (1981) found that the spring-shed leaves of beech and oak were richer in soluble minerals than autumn-shed leaves which had been lying

---

[3] Useful tip No. 1: ethylene speeds the shedding of needles from cut Christmas trees, so keep ripening bananas, a potent source of ethylene, away from your tree.

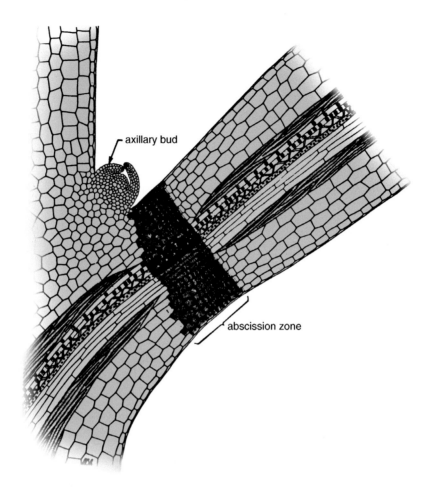

axillary bud

abscission zone

Figure 2.13 Abscission zone made up of corky cells at the base of a leaf. From: Weier, T.E., Stocking, C.R., Barbour, M.G. & Rost, T.L. (1982) *Botany: An Introduction to Plant Biology* (6th edn). Wiley, New York. Reprinted by permission of Dr. M.G. Barbour.

on the ground over winter, presumably because leaves held on the tree were drier and therefore less readily attacked by fungi and microbes. When the leaves fall in spring, the roots are active and able to take in these nutrients as they leach from the leaves. Keeping the leaves only on the lower branches improves the chances that they fall within reach of the tree's own roots. Retention of dead leaf bases is also a normal feature of palms and tree ferns where they help protect the valuable growing point from damage and environmental stress.

## Leaf colours

Healthy leaves look green because photosynthesis is driven by blue and red light and it is the green that is left over to be reflected or passed through the leaf. But not all leaves are green. New leaves, especially in shade-tolerant tropical trees but also a number of temperate trees, can be very pale green to white, various pinks and brilliant reds (red in maples and northern birches, brown in cherries, beech, southern beeches, poplars, the katsura, tree of heaven and *Catalpa* species). Here other pigments (particularly anthocyanins which also gives the rich colours of *Begonia rex* and *Coleus*) are formed before the chlorophyll. Delayed greening due to lack of chlorophyll appears to be a protection mechanism. Tender young leaves are prone to being eaten (they typically receive 60–80% of their lifetime herbivory when young) so the trees do not put valuable chlorophyll into leaves until they are fully expanded and tough, and the low nutrition value will in turn help to reduce the amount of herbivory. This helps explain the lack of chlorophyll, but pale leaves are quite rare and most young, non-green leaves are highly coloured. So why do trees manufacture expensive pigments such as anthocyanins and put them into the young leaves? The traditional view is that these pigments prevent light damage to sensitive tissues, particularly from ultraviolet light. This may certainly be true for a number of trees including birches that produce red leaves in the Arctic with long hours of sunlight. Also, yellow pigments in evergreen species in Mediterranean areas are known to help protect the leaves against high light levels at mid-day and when it is too dry for photosynthesis to occur. But protection from light is not the whole story. Nathaniel Dominy and colleagues (Dominy *et al.* 2002) found that tropical leaves at the top of the canopy, and so in high light conditions, were no more frequently red than those in the deeper shade below. They concluded that the main reason for the red pigments was to hide the leaves from insects. Most insects do not register long wavelength light at the red end of the spectrum and so to them red leaves look very dark or possibly dead and so unattractive (more on this below under autumn colours).

In naturally occurring variants of some trees, such as the copper beech (*Fagus sylvatica* 'Purpurea') and sycamore (*Acer pseudoplatanus* 'Purpureum'), anthocyanins occur sufficiently intensely to mask the chlorophyll for the life of the leaf. Anthocyanin production requires light so the reddest leaves are on the outside of the canopy and the inner shaded ones can be almost a normal green. This masking of the chlorophyll does not appear to be much of a handicap to growth.

Pigments are used in other ways. Leaves below flowers (officially called bracts) are modified in dogwoods and the dove (handkerchief) tree (*Davidia involucrata*) to a brilliant white to help attract insects to the diminutive flowers. Variegated leaves are much prized by gardeners. The cause is either genetic mutations (including chimaeras; see Chapter 8), viral (as in tulips) or mineral deficiencies.

Perhaps the most familiar colouring of leaves is in temperate trees in the autumn. Some leaves drop while green (e.g. alder and ash). In the case of alder (*Alnus* spp.) this appears linked to the abundant nitrogen 'fixed' by their roots (see Chapter 4) and so they have little need to reabsorb materials from leaves before they are shed. But many trees produce a mixture of bright yellow, orange and red. Archetti (2009a) found yellow colouration in 16% and red in 12% of 2368 species growing in temperate areas around the world, including deciduous conifers such as larch, dawn redwood and ginkgo (*Ginkgo biloba*). In eastern Canada and New England the autumn colours are spectacular and 70% of the 89 woody species examined at Harvard Forest contained red anthocyanins (see Lee *et al.* 2003). Yet the same species grown in Europe rarely attain the same intensity. Why? Some of the blame can be put down to genetic differences between individuals. A row of sibling seedlings can vary tremendously in the intensity and duration of autumn colour. In Europe, we may be just unlucky with our choice of seedlings. But autumn colours are so reliably less good in Europe no matter what we plant that there must be something else.

As noted above, as a leaf starts to die in the autumn, substances that can be reused, such as proteins and chlorophyll, are broken down and taken back into the tree. At the same time as leaves are systematically stripped of useful assets they are filled with unwanted things such as silicon, chlorine and heavy metals. (It has been suggested that leaves are used as a dumping ground for waste products like filling rockets with rubbish and firing them off into space: 'excretophores'.) The yellow pigments (carotenoids, i.e. carotenes and xanthophylls), normally found in a leaf but previously masked by the chlorophyll, are also broken down but at a much slower rate than chlorophyll and so now show through (a similar process happens in the ripening of bananas and citrus fruit). Studies on aspen (*Populus tremula*) in Sweden, mentioned above, found that when the leaves turned yellow, chlorophyll levels had decreased by 75% and carotenoids had decreased by only 50%.

Red pigments (anthocyanins) mixed with the yellow in different proportions produce the oranges, reds, purples and sometimes blues. These anthocyanins are commonly produced in great quantities in dying tissue from sugars that remain in the leaf. But this requires warmth and bright light during the day for the remaining chlorophyll to work, and cold nights to slow the transport of sugar out of the leaf. These are common features of the sunny autumns of eastern N America but less so in the often cool and overcast autumns of Britain and western Europe, explaining why even with the right trees we rarely get good colours. The brown colours so typical of the grey shores of Britain are due to other pigments (proanthrocyanins), which also give the dark colour to heartwood and the pink tinge to some sapwoods.

For many years it was accepted that the production of anthocyanins was just a chance, if aesthetically pleasing, mistake caused by the biochemical pathways of the leaf gently falling apart in the senescing leaf. More recently, however, evidence is accumulating that these pigments may have an important role. This evidence concentrates around two main ideas.

The first is that, as in spring colours, the anthocyanins protect the inside of the leaf from high light levels and free radicals while the photosynthetic apparatus is dismantled, giving more time for sugars and minerals (particularly nitrogen) to be reabsorbed as described above. This certainly seems to be true in a range of trees, including sugar maples (*Acer saccharum*), since Schaberg and colleagues (2008) found that red leaves maintained better vascular links with the twig than yellow leaves, allowing nutrient recovery to go on for longer. Anthocyanins are also often more abundant in plants deficient in nitrogen so a longer recovery period would be a distinct advantage. Other evidence is ambiguous. In the study of trembling aspen leaves mentioned above, anthocyanin was not very abundant in leaves until around 25% of the chlorophyll had disappeared despite the preceding 12 days being all sunny but very mild. However, after this, anthocyanin accumulation was highest on cold but very sunny and bright days, and more than half of it disappeared in the milder and rainy week that followed, while frost and sun after this resulted in a new accumulation (Keskitalo *et al.* 2005). This suggests that high light levels were important but only towards the end of the senescence process.

Not all evidence supports the idea of anthocyanins as light protection, and a number of other ideas have been floated (see Archetti 2009b for a list). From these, the most plausible second idea is that the red colour produced by anthocyanins is a signal to aphids and other insects, either telling them that the leaves are chemically defended and thus unpalatable, or it is camouflage since, as noted above for young leaves, they appear dark and dead to most insects (see White 2009). It has been counter-argued that since aphids can't see red light, how can they react to red leaves? Perhaps the signal they pick up is more subtle, such as differences in brightness or even chemicals associated with red leaves. To support this hypothesis, autumn colours have been found to be most intense in tree species that host the highest diversity of specialist aphids (Hamilton and Brown 2001). But this raises the question of why it matters if dying leaves attract insects if they are going to be shed soon anyway. The answer is that the aphids that land will be laying eggs on the twigs, which will produce a new crop of aphids to infest the spring leaves. By repelling autumn aphids, trees are offering some advance protection for next spring's leaves.

In Japan, Koike (1990) has pointed out that on trees which invade open areas and produce leaves continually through the year (such as birches; see

Chapter 6 for more details) autumn colours begin in the inner part of the canopy and move outwards. Conversely in trees that produce all the year's leaves in one flush in spring (maples for example), autumn colours start on the outside of the canopy. But it can be more complicated since in maples colours can first form on branches in full sunlight while trees such as aspen have been seen to synchronise colouration through all of the canopy.

## 🍃 *Further Reading*

Archetti, M. (2009a) Phylogenetic analysis reveals a scattered distribution of autumn colours. *Annals of Botany*, 103, 703–713.

Archetti, M. (2009b) Classification of hypotheses on the evolution of autumn colours. *Oikos*, 118, 328–333.

Beerling, D.J. & Chaloner, W.G. (1993) Stomatal density responses of Egyptian *Olea europaea* L. leaves to $CO_2$ change since 1327 BC. *Annals of Botany*, 71, 431–435.

Brodribb, T. & Hill, R.S. (1993) A physiological comparison of leaves and phyllodes in *Acacia melanoxylon*. *Australian Journal of Botany*, 41, 293–305.

Bünning, E. & Moser, I. (1969) Interference of moonlight with the photoperiodic measurement of time by plants, and their adaptive reaction. *Proceedings of the National Academy of Sciences, USA*, 62, 1018–1022.

Büsgen, M. & Münch, E. (1929) *The Structure and Life of Forest Trees* (3rd edn). Chapman & Hall, London.

Chabot, B.F. & Hicks D.J. (1982) The ecology of leaf life spans. *Annual Review of Ecology and Systematics*, 13, 229–259.

Cooney, L.J., van Klink, J.W., Hughes, N.M., *et al.* (2012) Red leaf margins indicate increased polygodial content and function as visual signals to reduce herbivory in *Pseudowintera colorata*. *New Phytologist*, 194, 488–497.

Darwin, C. (1880). *The Power of Movement in Plants* (2nd edn). J. Murray, London.

Dawson, T.E., Burgess, S.S.O., Tu, K.P., *et al.* (2007) Nighttime transpiration in woody plants from contrasting ecosystems. *Tree Physiology*, 27, 561–575.

Dominy, N.J., Lucas, P.W., Ramsden, L.W., *et al.* (2002) Why are young leaves red? *Oikos*, 98, 163–176.

Dörken, V.M. & Stützel, T. (2011) Morphology and anatomy of anomalous cladodes in *Sciadopitys verticillata* Siebold & Zucc. *(Sciadopityaceae)*. *Trees*, 25, 199–213.

Edlin, H.L. (1976) *The Natural History of Trees*. Weidenfeld & Nicolson, London.

Ezcurra, E., Arizaga, S., Valverde, P.L., Mourelle, C. & Flores-Martínez, A. (1992) Foliole movement and canopy architecture of *Larrea tridentata* (DC.) Cov. in Mexican deserts. *Oecologia*, 92, 83–89.

Feild, T.S., Lee, D.W. & Holbrook, N.M. (2001) Why leaves turn red in autumn. The role of anthocyanins in senescing leaves of red-osier dogwood. *Plant Physiology*, 127, 566–574.

Ford, B.J. (1986) Even plants excrete. *Nature*, 323, 763.

Givish, T.J. (1978). On the adaptive significance of compound leaves with particular reference to tropical trees. In: *Tropical Trees as Living Systems* (edited by P.B. Tomlinson & M.H. Zimmermann). Cambridge University Press, New York, pp 351–379.

Grubb, P.J. & Jackson, R.V. (2007) The adaptive value of young leaves being tightly folded or rolled on monocotyledons in tropical lowland rain forest: an hypothesis in two parts. *Plant Ecology*, 192, 317–327.

Hamilton, W.D. & Brown, S.P. (2001) Autumn tree colours as a handicap signal. *Proceedings of the Royal Society of London, Series B, Biological Sciences*, 268, 1489–1493.

Hoch, W.A., Singsaas, E.L. & McCown, B.H. (2003) Resorption protection. Anthocyanins facilitate nutrient recovery in autumn by shielding leaves from potentially damaging light levels. *Plant Physiology*, 133, 1296–1305.

Karabourniotis, G., Papastergiou, N., Kabanopoulou, E. & Fasseas, C. (1994) Foliar sclereids of *Olea europaea* may function as optical fibres. *Canadian Journal of Botany*, 72, 330–336.

Karlsson, P.S. (1992) Leaf longevity in evergreen shrubs: variation within and among European species. *Oecologia*, 91, 346–349.

Keskitalo, J., Bergquist, G., Gardeström, P. & Jansson, S. (2005) A cellular timetable of autumn senescence. *Plant Physiology*, 139, 1635–1648.

Kikuzawa, K. (1995) The basis for variation in leaf longevity of plants. *Vegetatio*, 121, 89–100.

Kleiman, D. & Aarssen, L.W. (2007) The leaf size/number trade-off in trees. *Journal of Ecology*, 95, 376–382.

Koike, T. (1990) Autumn coloring, photosynthetic performance and leaf development of deciduous broad-leaved trees in relation to forest succession. *Tree Physiology*, 7, 21–32.

Kursar, T.A. & Coley, P.D. (1992) Delayed greening in tropical leaves: an antiherbivore defense? *Biotropica*, 24, 256–262.

Lee, D.W., O'Keefe, J., Holbrook, N.M. & Feild, T.S. (2003) Pigment dynamics and autumn leaf senescence in a New England deciduous forest, eastern USA. *Ecological Research*, 18, 677–694.

Lovelock, C.E., Jebb, M. & Osmond, C.B. (1994) Photoinhibition and recovery in tropical plant species: response to disturbance. *Oecologia*, 97, 297–307.

Lüttge, U. (2008) *Clusia*: Holy Grail and enigma. *Journal of Experimental Botany*, 59 (7), 1503–1514.

MacDonald, M.T., Lada, R.R., Martynenko, A.I., *et al.* (2010) Ethylene triggers needle abscission in root-detached balsam fir. *Trees*, 24, 879–886.

Marks, C.O. & Lechowicz, M.J. (2007) The ecological and functional correlates of nocturnal transpiration. *Tree Physiology*, 27, 577–584.

Otto, C. & Nilsson, L.M. (1981) Why do beech and oak trees retain leaves until spring? *Oikos*, 37, 387–390.

Pearcy, P.W. & Troughton, J. (1975) $C_4$ photosynthesis in tree form *Euphorbia* species from Hawaiian rainforest sites. *Plant Physiology*, 55, 1054–1056.

Rolshausen, G. & Schaefer, H.M. (2007) Do aphids paint the tree red (or yellow) – can herbivore resistance or photoprotection explain colourful leaves in autumn? *Plant Ecology*, 191, 77–84.

Russell, R.B., Lei, T.T. & Nilsen, E.T. (2009) Freezing induced leaf movements and their potential implications to early spring carbon gain: *Rhododendron maximum* as exemplar. *Functional Ecology*, 23, 463–471.

Sandved, K.B. & Prance, G.T. (1985) *Leaves.* Crown Publishers, New York.

Schaberg, P.G., Murakami, P.F., Turner, M.R., Heitz, H.K. & Hawley, G.J. (2008) Association of red coloration with senescence of sugar maple leaves in autumn. *Trees*, 22, 573–578.

Simonin, K.A., Santiago, L.S. & Dawson, T.E. (2009) Fog interception by *Sequoia sempervirens* (D. Don) crowns decouples physiology from soil water deficit. *Plant, Cell & Environment*, 32, 882–892.

Simons, P. (1992) *The Action Plant.* Blackwell, Oxford.

Springmann, S., Rogers, R. & Spiecker, H. (2011) Impact of artificial pruning on growth and secondary shoot development of wild cherry (*Prunus avium* L.). *Forest Ecology and Management*, 261, 764–769.

Taulavuori, K., Pihlajaniemi, H., Huttunen, S. & Taulavuori, E. (2011) Reddish spring colouring of deciduous leaves: a sign of ecotype? *Trees*, 25, 231–236.

Thomas, P.A., El-Barghathi, M. & Polwart, A. (2011) Flora of the British Isles *Euonymus europaeus* L. *Journal of Ecology*, 99, 345–365.

van der Berg, A.K., Vogelmann, T.C. & Perkins, T.D. (2009) Anthocyanin influence on light absorption within juvenile and senescing sugar maples leaves – do anthocyanins function as photoprotective visible light screens? *Functional Plant Biology*, 36, 793–800.

Vogel, S. (2009) Leaves in the lowest and highest winds: temperature, force and shape. *New Phytologist*, 183, 13–26.

White, T.C.R. (2009) Catching a red herring: autumn colours and aphids. *Oikos*, 118, 1610–1612.

Williams, A.P., Still, C.J., Fischer, D.T. & Leavitt, S.W. (2008) The influence of summertime fog and overcast clouds on the growth of a coastal Californian pine: a tree-ring study. *Oecologia*, 15, 601–611.

Winter, K. (1981) $C_4$ plants of high biomass in arid regions of Asia – occurrence of $C_4$ photosynthesis in Chenopodiaceae and Polygonaceae for the Middle East and USSR. *Oecologia*, 48, 100–106.

Wullschleger, S.D., Meinzer, F.C. & Vertessy, R.A. (1998) A review of whole-plant water use studies in trees. *Tree Physiology*, 18, 499–512.

Zeppel, M., Macinnis-Ng, C.M.O., Ford, C.R. & Eamus, D. (2008) The response of sap flow to pulses of rain in a temperate Australian woodland. *Plant and Soil*, 305, 121–130.

# Chapter 3: The trunk and branches: more than a connecting drainpipe

## The woody skeleton

What makes a tree different from other plants is the trunk (or bole) and branches making up the woody skeleton. The main job of this tough, long-lasting skeleton is to display the leaves up high above other lesser plants in the battle for light. As well as support, though, the trunk and branches have two other important jobs: getting water from the roots to the leaves and moving food around the tree to keep all parts, including the roots, alive. But is the trunk just a large connecting drainpipe that keeps the two ends of the trees apart? In many senses 'yes' but its structure allows it to do many other things that no mere drainpipe could do.

Starting from the outside is the outer bark, a waterproof layer, over the inner bark or phloem (Figure 3.1). The phloem is made up of living tissue that transports the sugary sap from the leaves to the rest of the tree. Inside the bark is the cambium which, as will be shown, is responsible for the tree getting fatter. Inside this again is the wood proper or xylem. Although seemingly 'solid wood' it is the part of the tree responsible for carrying water from the roots to the rest of the tree. The water moves upwards through dead empty cells. But wood is not entirely dead. Running from the centre of the tree are rays of living tissue (made up of thin-walled 'parenchyma' cells) which reach out into the bark (and in some trees there are lines of these living cells running up through the wood as well). As will be seen later these living cells are involved with movement and storage of food and the creation of heartwood, the dense central core of wood that (reputedly) supports the tree as it becomes larger. At the very centre of the tree, some trees, but not all, have a core of pith (the strengthening tissue when the shoot was very young and soft).

## How the trunk grows

New branches grow longer, as in all plants, from the growing point at the end (the apical meristem). Inside, the central pith and overlying tissue (called the cortex) make up the bulk of the new growth. Arranged around the pith are

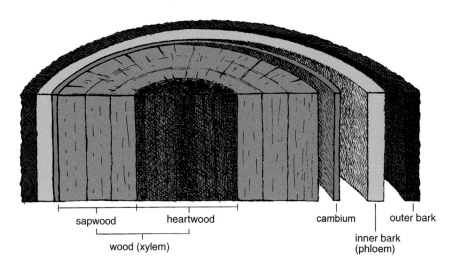

sapwood     heartwood           cambium | outer bark

wood (xylem)                        inner bark
(phloem)

**Figure 3.1** Tree cross-section.

strands of 'plumbing' (the vascular tissue) which have the phloem on the outside and xylem inside (Figure 3.2). Herbaceous plants normally develop no further than this 'primary' growth but woody plants specialise in not just getting longer but also getting fatter. This is called 'secondary' growth: it happens after the twig has grown longer. A branch gets *longer* by adding new primary growth on the end and gets *fatter* by secondary growth. (I finished my primary growth some years ago but I now seem to be undergoing secondary growth as over the years my waist expands.) In the young twig, a tissue capable of producing new cells, the cambium (Figure 3.1), forms first between the phloem and xylem and then grows around the twig to join up and form a continuous band of cambium around the whole of the inside of the twig, forming a continuous sheath over the entire tree under the bark. In theory the cambium is just one cell thick (around 5 μm), resembling the gossamer-like sheets of tissue found between the shells of an onion, but in practice it can be a dozen cells or so thick. This fragile, thin, translucent sheet is all that is needed to produce new 'secondary' phloem and xylem (and so strictly speaking should be called the vascular cambium: it produces vascular tissue) and is responsible for small seedlings growing in girth to giants of the forest. New xylem (wood) is added onto the inside of the cambium and new phloem on the outside. You will see that as the wood accumulates and the tree grows in girth, it stretches the cambium and the bark. The cambium copes by growing sideways. For the bark (including the phloem) things are a little more complicated (see Bark, below).

Several points are worth making. Note that the oldest wood is in the middle of the tree and gets younger towards the outside; that's why the pith at the centre of a young twig can be found at the centre of a trunk many metres

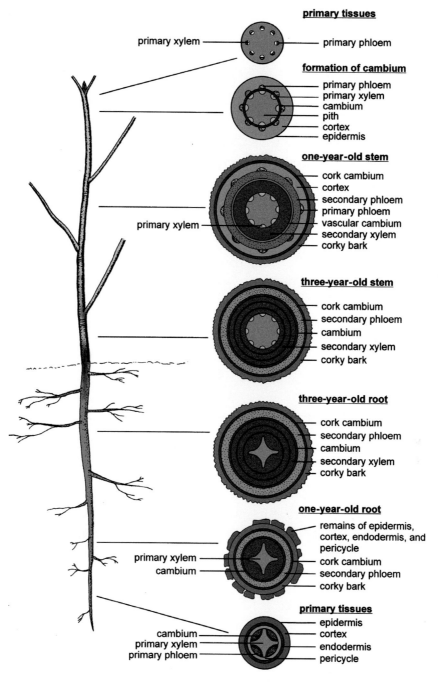

**primary tissues**

primary xylem — — primary phloem

**formation of cambium**
- primary phloem
- primary xylem
- cambium
- pith
- cortex
- epidermis

**one-year-old stem**
- cork cambium
- cortex
- secondary phloem
- primary phloem
- vascular cambium
- secondary xylem
- corky bark

primary xylem —

**three-year-old stem**
- cork cambium
- secondary phloem
- cambium
- secondary xylem
- corky bark

**three-year-old root**
- cork cambium
- secondary phloem
- cambium
- secondary xylem
- corky bark

**one-year-old root**
- remains of epidermis, cortex, endodermis, and pericycle
- cork cambium
- secondary phloem
- corky bark

primary xylem —
cambium —

**primary tissues**
- epidermis
- cortex
- endodermis
- pericycle

cambium —
primary xylem —
primary phloem —

Figure 3.2 Development of a young tree showing how the young shoots and roots develop the woody structure (secondary thickening).

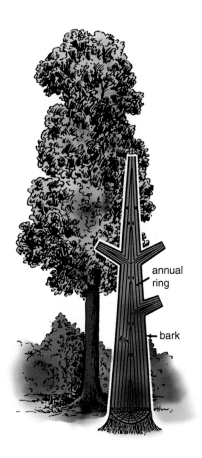

annual
ring

bark

**Figure 3.3** A vertical section through a tree showing that it grows by adding on new shells of wood over the old tree. From: Kozlowski, T.T. (1971) *Growth and Development of Trees: Vol. I, Seed Germination, Ontogeny, and Shoot Growth.* Academic Press, New York.

across. In effect as a tree grows it lays down a new shell of wood over the whole tree and the old tree is left fossilised inside. It is possible to dissect a tree and see how it used to be through its life (of course, broken branches will have been lost so the complete tree may not still be there; see Figure 3.3). There's an important corollary to this: trees do not grow by stretching upwards, they put on new layers over the old. If I stab a knife into a tree and come back in 10 years' time, the tree will have got fatter (and may be burying the knife) but since the sharp end is embedded into an old part of the tree, which is the same size as it ever was, the knife will still be at the same height above ground. If you find this hard to believe, try stretching a plank of wood. If you can't do it, neither can the tree! Stories like the proverbial 'my swing tied to a branch is further away from the ground than when I was a child' can be the result of the ground getting lower by erosion. But also, branches, being sideways, will lift things further away from the ground as they get fatter.

(a)

Figure 3.4 (a) A cut trunk of hawthorn (*Crataegus monogyna*) showing flutes in the bark which have led to patches of included bark, some of which are now completely enclosed inside the wood (upper right of the picture). Kent, England. (b) In the case of this yew (*Taxus baccata*), the stone surrounded by bark became embedded as two trunks grew fatter, squeezed together and eventually joined (the two centres are indicated by arrows). The stone didn't do the chainsaw much good! Photograph (b) by Gerald Burgess.

Not all trunks grow as perfect cylinders (or cones), for example, hawthorns (*Crataegus* spp.) and yews (*Taxus* spp.) tend to have very fluted trunks, resembling a series of trees fused together. In these trees some parts of the cambium are much more active, growing wood more quickly, leading to that part of the tree bulging out. This can become so extreme that two bulges meet up, press together and fuse leaving a patch of bark included in the wood (Figure 3.4). The reason for fluting is largely unknown but does appear to be often genetically determined. In lianas, where the biggest problem is not supporting the weight of the foliage but withstanding the twisting pressure, fluting is very common. Lianas often have flat ribbon-like or star-shaped stems caused by the relative inactivity of parts of the cambium, allowing them to twist and turn without damage.

(b)

Figure 3.4 (cont)

## Do all trees get wider?

The general answer is almost all do, one way or another. Tree ferns, which you might argue are not really trees, are a major exception and don't. Monocot plants (see Box 1.1) usually have needle-like bundles of vascular tissue scattered across the stem (hence palm wood is often called porcupine wood). Some, like palms and screw pines (*Pandanus* spp.) do not get fatter by secondary thickening as described above but they may get fatter at the base by growing extra parenchyma cells (but without any new vascular tissue) or by hiding their skinny nature under layers of old leaf bases or masses of adventitious roots (see next chapter). Other monocots, such as the Joshua tree (*Yucca brevifolia*), dragon trees (*Dracaena* spp.), century plants (*Agave* spp.), cabbage palms (*Cordyline* spp.) and the grass trees of Australia (*Xanthorrhoea* spp.), are capable of secondary growth of a sort. Here, new needles of vascular tissue arise vertically through the stem, making the stem fatter like pushing pencils down into a cylinder of foam rubber. The palm-like cycads (really primitive conifers) do have proper secondary growth from a succession of cambia and develop loose-textured growth rings around a large pith, but it is a very slow process.

Incidentally, a tree fern or a palm 20 cm thick cannot start this thick as a seedling, so how do they get to be this size without secondary thickening? As the seedling grows successive bits of stem between leaves (the internodes) are slightly wider than those below: the growing point grows simultaneously in width and length. As these internodes are relatively short the mature stem diameter is quickly reached near the base and stability is maintained by adventitious roots.

## What is wood made of? Wood structure

As new plant cells grow, cellulose is added around the cell to form a wall, which is what distinguishes all plant cells from those of animals. Many cells inside a tree are non-woody (such as in the leaves, flowers and rays in the wood) and this is where they finish, having just a primary wall made up mostly of cellulose. If a cell is to be made 'woody', as xylem cells are, then the cells grow another (secondary) wall inside the first that is thick and made up of several layers. This secondary wall contains the majority of the lignin and gives the required strength to the cell. Thus, cellulose is the building material of all plants; lignin is the main material that makes plants woody. Overall, the main constituents of wood are: 40–55% cellulose (the main constituent of paper), 25–40% hemicelluloses (a very diverse group of cellulose-like compounds) and 18–35% lignin. Conifers tend to have more lignin and less cellulose/hemicellulose than hardwoods; similarly hardwoods that grow slowly and have a high wood density also contain more lignin. Lignin is a remarkable compound which is very hard to decompose by fungi. It is what makes brown paper brown (and stronger than white paper) and forms the majority of the black humus found in soils. It also, of course, helps to make wood very resistant to decomposition.

As explained in Chapter 1 there are many differences between conifers (by which is really meant the gymnosperms) and the hardwoods (angiosperms). One profound difference is in the structure of the wood.

### Conifers

As can be seen from Figure 3.5, the wood of conifers is composed chiefly of dead empty tubes called tracheids that fit tightly together. These tracheids are the tubes through which water flows but since they make up 90–94% of the wood volume they also hold the tree up. The other 6–10% of the wood is made up of rays (these are sheets of cells, one cell wide in gymnosperms, that run along the radii of the wood from the centre of the tree).

If you look at a polished piece of pine, it is hard to believe that there are tubes in what looks like just 'solid wood' (any obvious holes seen, either running parallel to the tracheids or through the rays, are probably resin canals: see below). This demonstrates that the tubes are very narrow. In N American woods the diameter of these tubes ranges from 0.025 mm in Pacific yew (*Taxus*

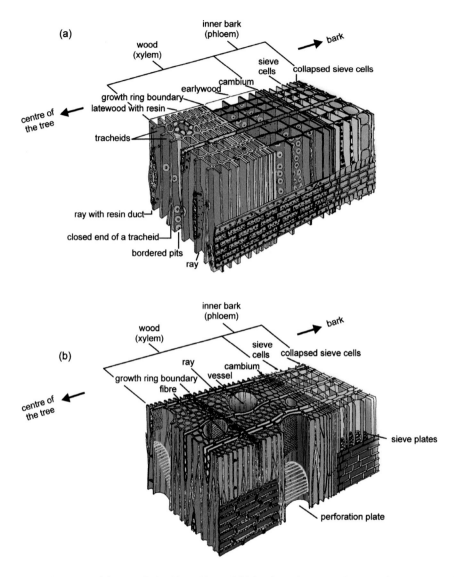

**Figure 3.5** Structure of the wood of a (a) conifer and (b) hardwood. From: Schweingruber, F.H. (1996) *Tree Rings and Environment: Dendroecology*. Haupt, Berne, Switzerland.

*brevifolia*) up to 0.080 mm in coastal redwood (*Sequoia sempervirens*): that's 40 to 12.5 tubes to the millimetre, respectively). Nor are the individual tubes very long, varying from 1.5–5 mm long on average, reaching up to 11 mm in the monkey puzzle (*Araucaria araucana*). Indeed a cubic inch of Douglas fir wood (*Pseudotsuga menziesii*) can contain 3 million tracheids! Thus, in conifers water has to pass through several tens if not hundreds of thousands of tracheids on its journey from the roots to leaf. This is an even more amazing feat when you consider that the individual tracheids have closed ends (as can be seen in

Figure 3.5). The trick is that they are joined together by a variety of 'pits'. The commonest type are bordered pits where there is a raised doughnut around a central hole. Inside, a thickened lump (the torus) is suspended by a web of cellulose strands (called the margo) just like a trampoline suspended by a web of elastic cords. These pits act as valves. Water can flow freely through the strands of the margo but if damage to a tracheid lets in air on one side, the pressure difference between the two sides pulls the torus over and blocks the hole. The value of this to the tree is discussed below but it can cause us humans a problem. Timber from trees like spruce, where these sealed (officially called 'aspirated') pits are common in dried timber, is particularly difficult to treat with preservative solutions unless the wood is treated green. There are a variety of other pit types but they are much less common; for example, Western red cedar (*Thuja plicata*) is one of the few N American conifers with simple pit membranes lacking torus and margo. Instead the pits consist of many closely packed strands of cellulose acting like a sieve but still acting like a valve in stopping air flowing from one tracheid to another.

## Hardwoods

The wood of angiosperms is more complex, being made up of vessels which conduct water (6.5–55%, average 30%, of the wood volume), small-diameter fibres which provide strength (27–76%, average 50%) and rays (6.5–30%, average 20%). Relieved of having to provide mechanical strength, the vessels can be very large in diameter (varying from 0.05 to 0.8 mm in diameter) and so are often easily visible to the naked eye on the cut surface (where they are referred to as pores) as in Figure 3.5. The vessels start life as vertical series of cells (called vessel elements) each no more than a millimetre long. As they die the cross-walls largely disappear to leave longer tubes that might be just a few centimetres long or may run the whole length of the tree. The old cross-walls are left as 'perforation plates' which vary from a single round hole to complex 'multiple perforations' in patterns of slits or holes. Where neighbouring vessels meet each other and rays, they are joined by pits similar to those in conifers.

The fibres are long narrow cells with thick walls, a small cavity and tapering ends, closely bound together to form a matrix holding the vessels. The rays are also more complicated in hardwoods. As in conifers they may be just one cell wide (uniseriate) as in willows and poplars, but are commonly many cells thick (multiseriate) and up to several millimetres wide as in oak. In addition, many hardwoods have strands or sheets of ray material (parenchyma cells) running vertically through the wood between or around the vessels (called axial parenchyma). The rays may carry canals filled with gums, resins, latex or other material used in the defence of the tree (Chapter 9). Rays also tend to hold the oils that give wood their characteristic smell and taste.

Crystals are also found quite commonly in hardwoods (and sporadically in the pine family), both in the wood and bark. These white crystals, usually calcium oxalate, are formed in the parenchyma cells when excess calcium taken up with water from the soil combines with oxalic acid, a common constituent of cell sap. Tropical woods can also contain silica as small round beads that increase in size from sapwood to heartwood: you can imagine how quickly they blunt saws!

### Exceptional wood structure

Nothing ever seems clear cut in biology. For the record, both *Ephedra* and *Gnetum* species (primitive 'conifers'; see Box 1.1) have vessels. To even things up, a few hardwoods have tracheids instead of vessels such as the South American and New Zealand genus of *Drimys* (we grow Winter's bark, *Drimys winteri*, in Britain) and several East Asian genera.

## Growth rings

A feature of both hardwoods and conifers is the presence of rings, each one normally marking a temporary halt in growth at an unfavourable time of the year: winter in temperate areas. In conifers the rings appear as repetitive bands of wide light and narrow dark wood (Figure 3.6). Wood cells grown early in the spring (earlywood) are large and thin walled, just right for carrying the large quantities of water needed during the active growth in spring and early summer. As the year progresses, the cells become progressively smaller with thicker walls; the tree needs less water once it has finished putting out new growth and the emphasis changes to producing strong wood. So, a growth ring

Figure 3.6 Scots pine (*Pinus sylvestris*) showing rings of different width but each one being made up of alternating bands of light-coloured earlywood and dark-coloured latewood. Shugborough, England.

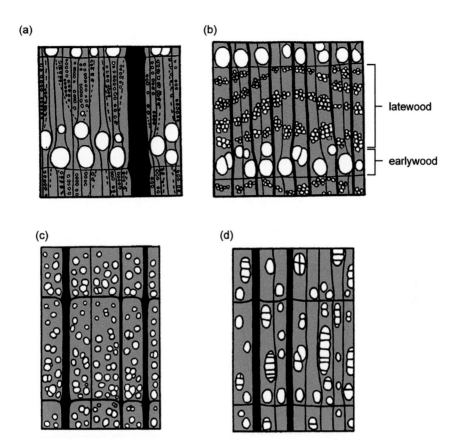

**Figure 3.7** Ring-porous woods (a) oak and (b) elm; and diffuse-porous woods (c) beech and (d) alder. The centre of the trees is towards the bottom of the page; horizontal lines mark the edge of a year's growth; vertical lines are rays.

starts off with less dense wood which looks lighter from a distance, merging into denser and darker latewood followed by an abrupt join (representing winter) before the earlywood of next year.

As is usual, things are a little more complicated in hardwoods! They can be divided into ring- and diffuse-porous trees. In ring-porous trees (like ash, elm, oak, black locust, hickory, catalpa and teak), the earlywood is dominated by huge vessels which contrast distinctly with the smaller ones of the latewood (Figure 3.7). Looking at the cleaned cut end of one of these trees, the large holes or pores in the earlywood can be seen to follow each ring around the tree: hence ring-porous. In diffuse-porous trees (like birch, maple, beech, poplar, lime, mahogany and eucalypts) the pores are more even in size and evenly distributed throughout the ring making the rings less obvious. But the boundary between rings can still usually be seen as a thin dark line made up of small thick-walled cells that border on the earlywood. To add variety, some eucalypts have 'reverse latewood' where there is a sharp transition between early- and latewood followed by a gradual transition between the latewood of one year and the earlywood of the next spring. Just why is unknown.

Which trees are ring- and which diffuse-porous is not always clear cut. It can vary within one tree; in a ring-porous tree, the innermost rings (i.e. the first rings grown), or ones grown while the tree is under great stress, can be diffuse-porous. Related trees, and even individuals in the same species, may be ring-porous in the north and diffuse-porous further south. Indeed, ring-porous trees are more common in mid-latitudes and diminish down to only 1% of species in the tropics. The ecological reason behind this apparent confusion is discussed in the next section.

In temperate areas, with a regular unfavourable period each year, we are well used to a tree forming one ring per year, enabling us to age a tree by counting the rings. Although this usually holds true, it is not always so. More than one ring may form in a year where growth has been interrupted by frost, fire, flooding, drought, defoliation, and then resumed. Citrus trees have been seen to produce 10 rings in 1 year! In other cases, rings may be missing, sometimes from just part of the tree; yew is famous for its missing rings, sometimes hundreds. Localised damage to the cambium is one potential cause but the most usual reason is linked to the way a tree grows a ring.

In early spring, the cambium becomes active at the base of swelling buds, spreading down the branches and trunk to the roots. In ring-porous trees this wave of new wood growth spreads rapidly downwards within days, giving the impression of being almost synchronous along the whole tree. In diffuse-porous trees the downward spread is slower with new wood appearing at the base of the trunk several weeks (sometimes almost 2 months in older trees) after its initiation in the twigs. Conifers are intermediate, usually taking about a week. The whole process is governed by hormones produced by the buds. Under difficult growing conditions (cold, drought) the ring may never make it to the bottom of the trunk, or may be restricted to one side of the trunk. Indeed, the incidence of missing rings at the base of a trunk increases in trees growing in stressful environments. It is tempting to predict that missing rings should be more common in diffuse-porous trees and conifers. To make life even more interesting, you should note that whereas earlywood starts from the top down, latewood begins at the base of the tree and moves upward, and ceases growing from the top down. Latewood is therefore thicker near the base of tree and may disappear altogether in the upper reaches of a tree.

In the more equable climate of the tropics, trees may grow continuously and so will not produce obvious rings and even if they do, they need not be annual. But if there is a distinct dry season, then they can be annual. In leaf-exchanging trees (Chapter 2), which may be leafless for just a few days, the cambium may not become fully dormant and so rings may be indistinct or missing. Half the species in the Amazon Basin produce no rings; in India the figure rises to three quarters. In some cases, subtle annual change in growth linked to slight environmental variations through the year may be detectable as changes in carbon isotopes[1] or calcium levels in the wood (see Ohashi *et al.* 2009).

### *Variation in ring width*

Ring width varies tremendously due to variations in the environment, rings getting narrower in years with less favourable weather. This is considered further in Chapter 6. Suffice it to say here that in conifers, as the ring gets narrower, it is the earlywood that reduces while the latewood varies less (look back at Figure 3.6). So as the rings get narrower, the wood gets denser overall. Conversely, in ring-porous trees like oak the latewood gets proportionately less in narrow rings, so, because of the great vessels in the earlywood, wood with narrow rings is less dense. (The situation in diffuse-porous trees is less clear.) The conclusion is that to grow dense, strong oak, it should be grown fast while a conifer-like pine should be grown more slowly. Indeed, most conifers in the UK (grown fast because of the warming influence of the gulf stream) are structurally unsuited for building, and we import such timber from the continent (see Panshin & de Zeeuw 1980 for further details).

## How water gets up a tree

The tubes that carry the water up through wood are dead tubes. So how can water rise to the top of the tallest trees, over 100 m tall? There are three possible answers: it can be pushed up, pulled up or some combination of the two. We will consider these in order.

### *Pushing water up trees: root pressure*

The water going up a tree is usually fairly pure with about 0.001% (a thousandth of one per cent) of its weight consisting of dissolved minerals such as nitrogen, potassium, etc. However, in trees such as maples and birches this changes and a positive pressure builds up in the wood (xylem) several weeks before the leaves open in spring. Root cells (and sometimes cells in the trunk)

---

[1] This includes carbon-14 bomb peak dating. The nuclear experiments of the late 1950s and 1960s raised the levels of carbon-14 (a radioactive form of carbon with an atomic mass of 14) which have been declining since. The amount found in wood can therefore be used to give a fairly accurate date after 1955 (see Soliz-Gamboa *et al.* 2011).

secrete extra minerals and sugars into the xylem and water follows by osmosis, creating the positive pressure that forces water up the tree. Having said this, spring sap movement in maples, birches and walnuts (*Juglans* spp.) is a little more complicated. Root pressure appears to be supplemented by a positive pressure in the stem since the best flow occurs on warm sunny days above 4°C with sharp frosts at night (below −4°C). It is suggested that gas in the fibres around the water-filled vessels is compressed on cold nights as ice builds up inside the fibre walls, giving the gas less room. In the morning, the gas warms and expands, pushing the melting water back into the vessels under pressure (see Cirelli *et al.* 2004 for a review). Whatever the cause, can root pressure be used to explain the ascent of water in most trees? No, for several reasons. Firstly, many trees (e.g. pines) rarely if ever develop root pressure and in many deciduous temperate trees the root pressure disappears once the leaves emerge. Secondly, the pressure developed is rarely more than one to three atmospheres (0.1–0.3 MPa); enough to move water 10–20 m up a tree but clearly not enough to get water to the top of tall trees.

The question then arises, what use does root pressure serve? There are several answers. In trees and climbers of humid tropical regions where evaporation is very low, forcing water up through the plant ensures a supply of minerals to the top. This is enough to push water 5 m up the tree or vine when there is no transpiration. Excess water is lost as beads of moisture that exude from the leaf tip and edges (guttation), and many species have special pores (hydathodes) at the ends of the veins for this purpose. I have a Swiss-cheese plant (*Monstera deliciosa*) which produces large drops around the edges of the leaves in the morning; very unpleasant to brush past. A second reason is to refill tubes that have become air filled (more later). Thirdly, it is a way in some temperate trees of getting an extra boost of stored sugars from the roots up the tree in early spring to supplement that moved through the phloem. The sugars are added to the xylem sap which pulls the water into the xylem. Thus, birches, maples and even the butternut of eastern N America (*Juglans cinerea*) can be tapped in late winter and early spring to collect the sugary sap (2–3% w/v sugar in maples, weaker in the butternut) to make wine and syrup (60% sugar). A birch may yield 20–100 litres and a sugar maple (*Acer saccharum*) 50–75 litres. The Canadian maple syrup output of 41.4 million litres in 2009 required around two billion litres of sap.

## Pulling water up trees: the Tension-Cohesion theory

It has been known since the work of Henry Dixon and John Joly in 1894 that water is transported under tension (or suction), implying that it is pulled not pushed. Capillary action is probably the most obvious mechanism but is sufficient to get water only a metre up a tree. The mechanism to get water higher is quite simple in its elegance. The 'pump' in the system is the leaves. As the leaves lose water through their stomata (transpiration; Chapter 4) and the internal cells become drier they pull water from the next wettest cell and so on until the pull reaches a vein and exerts suction (or tension) on the water of the xylem. Because water is extraordinarily cohesive (it is a bipolar molecule with a positive and negative end which enables water molecules to cling together like magnets) the tension is passed on down the tree and has the effect of pulling a continuous column of water up through the xylem tubes. In fact, water is so cohesive that intact water columns inside trees can theoretically be lifted to 120–130 m, which agrees with the tallest tree being currently 115.7 m (379 ft 7 in); see Chapter 6. Water can thus be moved up tall trees in copious quantities (Chapter 2) by a combination of tension and cohesion: the Tension-Cohesion theory. This is an amazingly elegant, energy efficient way of transporting water since it uses the energy of the sun not that of the tree. If trees had to use just their own energy to pump water to the top, they would not exist.

The estimated tension (suction) required to move water to the top of the tallest trees is about –30 atmospheres (–3 MPa). This is no mean tension. So much so that if the width of a tree is carefully monitored, the trunk is seen to get thinner during a hot dry day as the suction pulls in the sides of the tree and fatter during the night as it recovers. The fact that trees do not completely collapse under this enormous strain is a reflection of the strength of the individual tubes. This reduction in tree diameter is small but measurable (Figure 3.8). Measured under the bark (so it is just the wood)

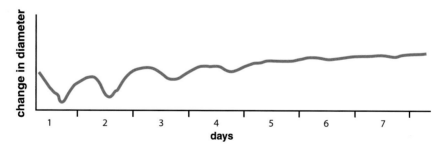

Figure 3.8 Variation in trunk diameter over several days in a hypothetical tree. The left-hand end of the line shows that the diameter of a trunk reduces in the afternoon when evaporation of water from the leaves causes great suction inside the trunk, pulling the sides of the trunk inwards. At night, the trunk relaxes outwards because without the sun, less water is evaporated and the suction is reduced. The right-hand end does not show such pronounced shrinkage during the day because rain and high humidity is reducing water loss from the leaves, and all that is left is the gentle upward trend due to diameter growth. Based on data from Kozlowski, T.T. (1968) *Water Deficits and Plant Growth*. Academic Press, New York.

the shrinkage in diameter varies from 0.01 to 0.25 mm on a tree 30–40 cm in diameter, which represents a change of just 0.01–0.07% (see Offenthaler *et al*. 2001 for more details).

## A combination of mechanisms?

It has been suggested by a variety of plant physiologists that the amount of suction needed to supply the leaves with water is far more than any column of water could stand and it just would not work. Instead they suggest that getting water to the tops of trees may be more complicated and involve a number of other mechanisms. These include help from the living cells in the trunk (rays, axial parenchyma and phloem) which exert a pressure on the xylem system. Thus the high tensions developing in the xylem would be lowered by the living cells pressing in on them in much the same way that a jet-pilot's suit inflates to squeeze his/her lower body to prevent all the blood pooling there during tight turns. In conifers this may be helped by the positive pressure that develops in the resin system. Other ideas include use of a film of negatively charged ions lining the tubes which provides an electrical gradient to pull water upwards, osmotic water-lifting in the same way as described above for root pressure, and slime-like films of sugar polymers maintaining a continuous film of water inside the tubes helping to attract water upwards. These and other suggested mechanisms are included in the 'multi-force' theory of Zimmermann and others.

In 2004, Ulrich Zimmermann[2] and colleagues wrote a review in the journal *New Phytologist* saying that the Tension-Cohesion theory left a lot to be desired and particularly suggested a number of flaws in published work, including that 'many plant physiologists still view the Cohesion Theory as the absolute and universal truth because clever wording from the proponents of this theory has concealed... [that some of the well-used methods do not really work]'. Zimmermann tries to show 'that the arguments of the proponents of the Cohesion Theory are completely misleading'. Unsurprisingly, this upset many people and in the next issue a letter signed by 45 mostly very prominent researchers (see Angeles *et al*. 2004) included: 'We, the undersigned, believe that this review is misleading in its discussion of the many recent papers which demonstrate that the fundamentals of the Cohesion-Tension theory remain valid'. The controversy continues but despite the elegance of the multi-force ideas, the Tension-Cohesion theory

---

[2] Not to be confused with the late Martin Zimmermann of Harvard University who also wrote much on water movement, including the seminal *Xylem Structure and the Ascent of Sap* published in 1983.

is still the best explanation of how water gets to the top of tall trees. This was reinforced in 2008 when Tobias Wheeler and Abraham Stroock of Cornell University invented a 'synthetic tree' made of hydrogel (an artificial jelly filled with water) which, although just a millimetre thick, was used to demonstrate that very high suction was possible without the water columns snapping.

## Air in the system

Sucking water up a tree only works if there is no air in the system. This can be demonstrated using a thin plastic tube. If the tube is dangled in a bucket of water and suction is applied at the top, water will rise only about 9 m (30 ft) and this does take some sucking! No matter how much extra suction is applied the water will not rise to the top. This is because the column of water becomes so heavy that the extra suction simply acts to pull apart the air at the top of the tube creating a partial vacuum. If the tube is full of water at the beginning, however, then water can be pulled up well over 100 m. This is why water pumps have to be primed and why tracheids and vessels begin life full of water: ready primed!

Trees then, have to ensure that the columns of water remain free of air. A broken column can no longer lift water. Air can come from one of several sources: too much tension, or damage, or freezing. Under hot dry conditions leaves can lose water faster than the tree can take it up and there is increasing tension in the xylem (stomata in the leaves will close as a first line of defence; see Chapter 2). The tension can become so great that individual columns of water break with an audible snap; this 'cavitation' can take less than one hundredth of a second (producing the snap) and leaves behind an embolism: an air-filled tube. Once a tube is air-filled, the negative pressure inside surrounding water-filled tubes can pull air through the porous pits (valves) between tubes, a process called air seeding. If tension gets too high, for example if there is no stomatal control to stop further water loss, this extreme tension can lead to 'runaway embolism' causing hydraulic failure (because all the tubes are air-filled and so can't conduct water) and death of the tree. Wider columns are more prone to cavitation (there is less friction from the wall to help hold the column together) so, as you might expect, tube diameter is smaller higher up a tree where the tensions are highest. Root pressure can help refill tubes when it is available, as can water absorbed through the leaves (see Chapter 2). Perhaps most useful is 'Mütch water'. When sugars are removed from the phloem in the roots and lower trunk, the water is recycled back into the xylem. While this is only 1–3% of all water moved up the xylem, it is released primarily into the outer rings where it may make up 45% of the total water content: enough to help refill cavitated tubes.

Winter is a distinct problem. Despite a widespread belief that trees drain in winter ('the sap going down in the autumn and rising in the spring'), with the exception of some vines, they do actually stay full of water. As the water in the xylem cools and freezes it can hold less of the gas that is dissolved in it (ice holds 1000 times less gas than liquid water), which appears as air bubbles. In the spring these bubbles must be redissolved if the tube is to function. There are two distinct approaches to overcoming problems of air bubbles, the safe and the efficient.

In conifers and diffuse-porous trees the water-carrying tubes tend to be narrow and safe (see Wood structure above). If they are less than $30\,\mu m$ in diameter, they are less likely to cavitate during hot summer spells. Also, any bubbles forming in the winter are likely to remain small and in the spring will usually redissolve within minutes or hours with the aid of root pressure (although those trees without significant root pressure, such as poplars and beech, can take much longer, perhaps several days). Thus, by the time the leaves start sucking hard, the xylem tubes are again water-filled. Here the apparently useless perforation plates and pits play a role; they catch bubbles when the ice thaws and prevent small bubbles coalescing into larger, longer-lived bubbles (the percentage of diffuse-porous species with complex perforation plates, which act as better valves, increases with altitude and latitude where it is colder in winter). In this way each individual tube may remain functional for many years. This is the safety in numbers strategy: lots of narrow tubes each taking a comparatively small amount of water at a slow speed. In conifers water can be conducted up through 30–40 years' worth of rings. Water flow in the xylem is usually fastest close to the cambium (the tubes here are newest) and usually decreases inwards with something like 60% of the volume flowing in the outer 1.6 cm and 80% in the outer 2.4 cm of the trunk. Using multiple small tubes is ideally suited to cold climates where freezing is common. It obviously works because Italian alder (*Alnus cordata*) has been measured as losing more than 80% of hydraulic conductivity during the winter but by early spring the loss was down to less than 20% (i.e. the tubes had refilled) and remained < 30% during the summer.

The alternative to the conservative, safe approach of the conifers and diffuse-porous trees is seen in the more efficient but riskier throw-away approach of the ring-porous trees. The amount of water that vessels conduct is determined partly by their length (the more interconnecting pits, the greater the friction and so the slower the water flow) but more important is the vessel diameter. Ring-porous woods have wide and long vessels in the earlywood which allows the movement of prodigious quantities of water. Flow increases

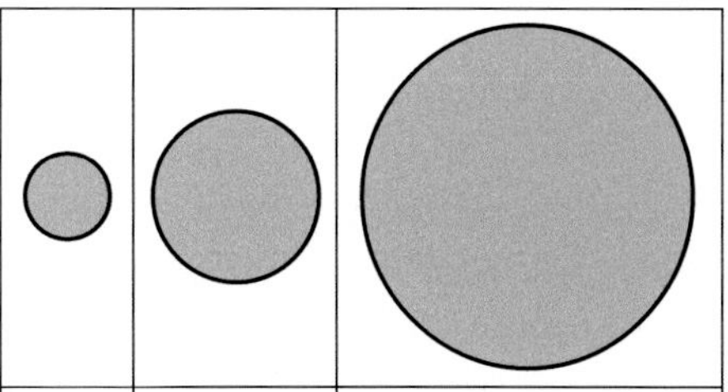

| Diameter (cm) | 1 | 2 | 4 |
|---|---|---|---|
| Cross-sectional area (cm²) | 1 | 4 | 16 |
| Relative flow rate | 1 | 16 | 256 |
| Per cent flow rate | 0.4 | 5.9 | 93.7 |

Figure 3.9 If we take three drinking straws of the diameters shown, the cross-sectional area is the square of the diameter but because there is proportionately less friction with the tube wall in bigger straws, the flow rate of a liquid is proportional to the fourth power (the square of a square) of the diameter. So if I had a straw 1 cm in diameter and you had one 4 cm in diameter and we both dipped them into a milk shake, and both sucked with the same force, I would get just 0.4% of the milk shake. Thus in trees, wider tubes may be risky but they can move a disproportionately large amount of water.

by the fourth power of the diameter[3]: so making a tube four times wider increases the flow volume by 256 times (Figure 3.9), up to an upper limit of probably around 0.5 mm diameter.[4] Speeds of water flow up ring-porous trees can thus be up to 1–4 m h$^{-1}$ (metres per hour) compared to usually less than 30 cm h$^{-1}$ in conifers and diffuse-porous trees like eucalypts. Flow rates are much lower at night, typically 0–15 cm h$^{-1}$ regardless of tree type.

But these super highways of water are at great risk from air bubbles. Many of the vessels will cavitate under hot dry conditions because in wide tubes there is proportionately less friction from the walls compared to the volume of water and so less help in holding the water column together. Moreover, the vessels that do survive until the autumn are likely to develop large air bubbles during the winter which can take days or weeks to disappear in spring, effectively making them useless. In one study with red oak (*Quercus rubra*) 20% of vessels were embolised by August and 90% after the first hard frost. Some refilling of tubes by root pressure may be possible in some species but the main solution

---

[3] Called the Hagen–Poiseuille law.

[4] Experiments were once tried where coal gas was forced through branches and ignited at the other end. The length of branch through which this could be done gave an idea of the relative permeability to water: minimum lengths were: ash (ring-porous) branches 3 m (10 ft) long; maple (diffuse-porous) 0.6 m (2 ft); softwoods even 4 cm (1½ in) stopped gas flow. It is said that if you take a 1 in x 1 in x 2 in block of red oak (*Quercus rubra*) and dip one end in detergent, you can blow bubbles with it!

to this problem can be found in how the trees start-up in spring. Ring-porous trees tend to leaf-out later than diffuse-porous trees but if you measure trunk growth in early spring you find that ring-porous trees are already fatter with new spring wood. Ring-porous trees depend on the growth of the new early-wood in early spring (which starts ready filled with water) before the leaves emerge to provide the necessary water during the growing season. Rarely, some diffuse-porous trees such as European beech (*Fagus sylvatica*) and *Prunus* species do the same to supplement other mechanisms of having water-filled tubes, but for the ring-porous trees it is the only mechanism. Thus most of the water moved up a ring-porous tree like oak is through the newest growth ring which may be just a few millimetres wide.

Impressive though this is, it does seem very risky. Whereas a diffuse-porous tree may be sucking water through tens of rings, a ring-porous tree, using just one ring, is putting all its proverbial eggs in one basket. Or so it appears. But this may be less risky than it looks; so good are the tubes at conducting water that trees such as the English oak (*Quercus robur*) may theoretically need only about 2% of all the vessels to supply water even for rapid transpiration. Also, ring-porous trees have an insurance policy in the form of small vessels in the latewood. If the big vessels of the outer ring become air-filled and stop con-ducting, the suction from the leaves becomes so great that water will move up the less-efficient smaller vessels scattered in the latewood over several rings. It's a bit like an accident on a motorway that forces cars onto smaller roads, which may be much slower and less efficient but will get them to their destination eventually. In a tree this may provide enough water to keep it alive until it can grow a new ring next year. A single ring, however, has important consequences for elms and Dutch elm disease (see Chapter 9).

The strategies of slow-but-safe (conifers and diffuse-porous hardwoods) and fast-but-risky (ring-porous hardwoods) are, of course, advantageous under different environmental conditions. The slow-but-safe trees are superior at high latitudes and/or dry areas since they are resistant to cavitation and to the problems of freezing. They can also start growing early in the growing season to make the most of good conditions. The fast-but-risky trees can grow very fast and efficiently but being more prone to cavitation and freezing are superior where the climate is mild and the growing season is long, as in temperate latitudes.

## Hydraulic architecture of the whole tree

On average, vessel lumens (the holes down the middle that conduct water) make up 8% of the cross-sectional area of hardwood trees (the rest being fibres) whereas tracheid lumens occupy 41% of a conifer. This huge difference shows how efficient the vessels are at conducting water and why conifers generally need a greater number of rings over which to conduct water up the tree. But it does hide just how good the bordered pits in conifers are at conducting water. It has been calculated that the resistance to water flow caused by the valves (bordered pits of conifers and the perforation plates in hardwoods) in its journey through the tree is pretty much the same in both sorts of tree for tubes of similar diameter: around 64% reduction in conductivity in conifers and 56% in hardwoods. Yet tracheids are around 10 times shorter than vessels and so water moving to the top of a conifer has to flow through at least 10 times as many valves as it would in a hardwood tree. So efficient are these pits that if conifers had perforation plates, they would experience over 95% reduction in conductivity. Bordered pits offer almost 60 times less resistance to water flow than perforation plates, which is down to the superb design of the torus-margo system. Perforation plates, having no moving parts, are not very good seals against air seeding and they offer a high resistance to the flow of water. By contrast, bordered pits are much better valves. Conifers that are good at resisting air seeding and cavitation tend to have a very flexible margo (the cellulose springs around the edge) and a large overlap between the torus (the solid lump in the middle) and the pit aperture, to ensure a good tight seal is formed (see Delzon *et al.* 2010 for more details).

So, the reason why conifers need so many tubes through which to conduct water is not due to the valves slowing things down, but to the high friction of the water against the sides of the narrow tubes; narrow to help resist cavitation and keep air bubbles small. These efficient valves help conifers compete with hardwoods particularly in colder areas that favour diffuse-porous hardwoods. But, as discussed above, hardwoods can win out in climates that allow them to have much wider vessels.

Water supply through the wood can sometimes be linked to growth rates. Trees with wide tubes, or lots of them, will generally have lower density wood since, when dried as timber, there is more air in a plank. Since these tubes will deliver a lot of water to the growing parts of a tree, they should also grow faster. So trees with low density wood often grow faster. This has certainly been found to be true in the tropics (see Putz *et al.* 1983) but not in New Zealand hardwoods and conifers (see Russo *et al.* 2010).

It is interesting that the vessels and tracheids in roots are often up to 2–3 times wider than in trunks, especially in the deepest roots. This seems to be due to a number of reasons. Firstly, it may reduce hydraulic resistance to flow from deep in the soil. Secondly, roots do not need to support themselves in the same way as a trunk and so can have wider tubes and lower overall wood density. Roots are normally largely immune to freezing so there is no need for small tubes to keep bubbles small. Finally it may be because water is

**Box 3.1** Water supply in Douglas fir

Supplying water to the top of a tree can be a slow process. A large study using the 87 m tall Wind River Canopy Crane in a forest of Washington State (see Čermák *et al.* 2007) found that water may take two and a half days to pass through a 13.5 m tall Douglas fir (*Pseudotsuga menziesii*) and up to 21 days in conifers more than 50 m tall. With water moving this slowly, there can be a long delay between leaves experiencing water loss and this stress being felt by the roots. This has been measured as 4.5 hours in a 51 m tall Douglas fir and 6 hours in young Sitka spruce (*Picea sitchensis*). In one particular Douglas fir in Washington, measurements showed that when the upper canopy began losing water in the morning, more water was leaving the top of the trunk than was entering at the bottom until around 10.00 am when this was reversed and 'refilling' of the trunk occurred during the afternoon and night. This depletion of water at the top of the tree early in the day puts the foliage under a potentially large drought stress and could lead to periods of lost growth. But fortunately the tree has an internal store of water which is used to supply the leaves during these vulnerable times. Most of this water is stored in cavitated tubes and between woody cells, and some comes from the phloem. Using the large Douglas firs again, Čermák and colleagues found that the outer bark contained 5% water by volume, the inner bark (phloem) 32%, sapwood 44% and the heartwood 10% water. The sapwood was less than 10 cm thick and made up about one third of the wood volume and contained 1200 litres of water, around 85% of the tree's 'free' water (that is, water that is not chemically bound to the wood and so able to move around). In these large Douglas firs, an average of 45 litres of this stored water is used in a day, representing about 20–25% of the daily 150 to 300 litres used by a tree. This represents about 3% of the free water in the tree. This amount of water may not seem very much but it resulted in 18% and 10% more photosynthesis each day in 60 m and 15 m tall trees, respectively.

comparatively more available in soil than at the top of a tree so there is less water stress and less need for narrow tubes to hold the water column together.

In passing, if water takes a number of days or weeks to reach the top of a tall tree (see Box 3.1), this also means that the plant hormones carried in the water will take an equally long time to move around the tree. This is why each part of the tree is able to respond independently to environmental pressures. Leaves will respond to water stress long before chemical signals sent by roots in response to low soil water levels reach them. Most long-distance signalling in trees is to do with seasonal processes such as leaf shedding or such things as bending of branches by reaction wood.

## Hydraulic network

Getting water reliably to the top is one problem, making sure that every part of the tree gets its share of water is another. The leaves at the top of a tree are important because they are best placed in the competition for light. Since they are higher up, the water pathway is longer, requiring greater suction to get water (hindered further by the tubes becoming narrower to reduce the risk of air embolisms), and they may not always be the only ones in bright sun and so evaporating most water. An analogy is two children with straws in a thick milkshake; one straw is 20 cm long and the other 2 m; if they both start sucking the one with the shortest straw will get more than his share since there is overall less friction to overcome. So how can leaves at the top of a tree get their share of water when competing with those nearer the roots?

The answer is that a tree has constrictions at the junctions of the trunk and branches. These are created by a decrease in tube diameter, and, in hardwoods, by more frequent vessel endings. Lev-Yadun and Aloni, Israeli specialists in tree structure, also believe that the circular patterns that develop at branch junctions in softwoods and hardwoods, which contain non-functional circular vessels increase the hydraulic segmentation of a branch from the stem (Figure 3.10). Conductivity through these junctions is often less than half that of the branch itself. This is like putting a crimp in the short straw to even up the share of milkshake. The mechanism is not perfect however; when water is short, it is usually the top leaves that suffer first.

During drought, these constrictions may also prevent air embolisms that develop in branches from getting back easily into the main stem. In all trees, leaves and even branches are more readily renewable and it makes sense to sacrifice them in times of water shortage and keep the trunk working. Indeed, palms, which have only one growing point and one set of xylem tubes through their life, have considerable hydraulic resistance at their leaf bases to ensure that any water stress is felt by the leaves and not the trunk and its growing point.

Water can also be moved around the tree to even out supply and demand. Leonardo da Vinci considered a tree to be like a series of drainpipes where each set from the root gives water to a set part of the canopy. Although this idea has its uses (and has been championed by the Japanese in the Pipe Model of water uptake), it is evident that water-conducting tubes intermingle like spaghetti hanging from a fork and are abundantly linked together. This allows water moving up a tree to fan out around a growth ring (and to some extent between

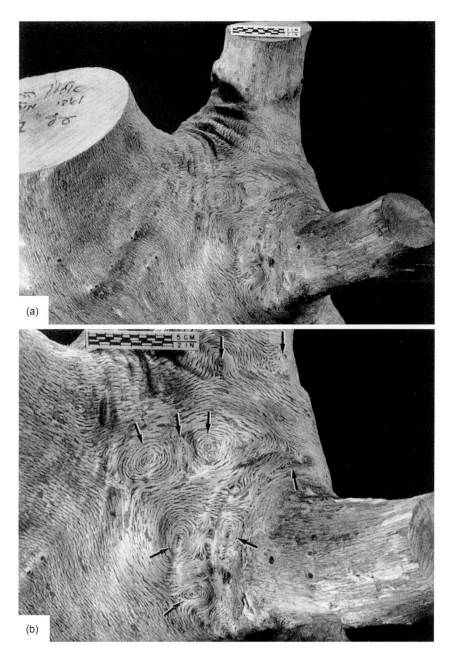

**Figure 3.10** Branches from an oak (*Quercus ithaburensis*) with the bark removed showing the non-functional circular vessels which increase the hydraulic segmentation of a branch from the stem (arrows indicate the circular regions in the close-up photograph). These prevent a branch sucking up more than its share of water. From: Lev-Yadun, S. & Aloni, R. (1990) Vascular differentiation in branch junctions of trees: circular patterns and functional significance. *Trees*, 4, 49–54, Figure 1A, page 50, with kind permission of Springer Science and Business Media.

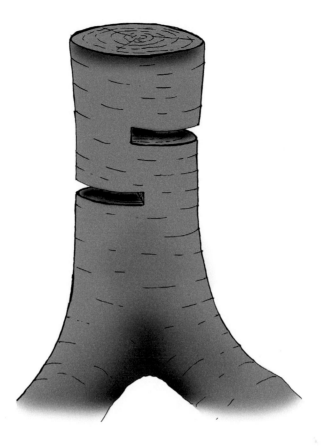

**Figure 3.11** The famous double saw-cut experiment where two cuts are made one above the other from opposite sides of the tree. Despite the fact that all xylem tubes should have been cut, water can still be sucked up the tree provided the cuts are far enough apart.

rings) with an angle of tangential spread of usually around 1 degree. Water from any one root spreads out as it goes up the tree, reaching not just one branch but a large part of the crown. This is a good safety measure since the loss of one particular root does not affect an individual branch but rather just reduces the overall water supply to the crown.

Such is the complexity of possible pathways of water up the trunk that water can still be sucked up under the most trying circumstances. This can be illustrated by the famous double saw-cut experiments. Here two saw cuts are made into the trunk parallel to the ground, one above the other on opposite sides of the tree and overlapping so that looking from the top, the tree is completely cut through (Figure 3.11). Yet water still ascends providing the cuts are more than a critical distance apart. One reason is that since the vessels grow crooked some may weave their uninterrupted way around the two cuts. More important, however, is how air is sucked into the cut tubes. When a tube is cut by the saw the suction from above pulls air into the tube. But this air only goes as far up and down from the cut as the first complex perforation plate or end of vessel (which are rapidly blocked by tyloses or gums; see Chapter 9), or tracheid with its valve-like pits. Beyond these barriers the tubes remain full of water. So, providing there are water-filled tubes between the two saw cuts, water will be sucked up by being passed sideways between tubes. In short-vesseled trees such as maples, the cuts can be quite close (around 20 cm) but in oak, where the vessels are long with simple perforation plates, the cuts need to be more than 0.5 m apart.

## Water and tree height

Despite the clever hydraulic design of trees, there are inescapable problems met in very tall trees. The need for air-free tubes creates special problems for tall trees and may be one of the main factors that ultimately limit tree height. As trees get taller the negative pressure in the tubes increases (gets more negative) partly because the water column is longer and thus heavier and is pulled back downwards by gravity, but also because of the increased friction of the water with the walls making the movement of water slower against the suction to move it.

Due to these higher negative pressures with height, there is much more risk of the water column snapping and thus not working. Paradoxically, as already mentioned, this is partly overcome by the water-conducting tubes becoming narrow towards the top of the tree. Even though this creates yet more friction, this is exactly what helps to hold the water column together. In conifers it has been found that the size of the hole through bordered pits also gets smaller with height; this again introduces extra friction and slows water flow but makes the valve-like pits much better at holding back air bubbles from getting from one tracheid to the next. There are, however, compensations to help water move through these tree tops. For example, the cross-sectional area of sapwood (which conducts water) gets higher in proportion to leaf area (so each leaf is supplied from more water-conducting tubes), and stomata can stay open under greater stress, so evaporating more water and maintaining suction. At the very top of the tallest trees, such as the coastal redwood (*Sequoia sempervi-rens*), the normal trend is reversed and tracheids become wider or longer (i.e. fewer pits in a length of tube) as presumably this is the only way to get water to flow under such extreme conditions. This is also helped by water being absorbed from the frequent fog (described in Chapter 2).

The overall consequence is that generally at the tops of tall trees, water movement is slow and less water is available to a leaf over time. As a result, the top leaves of a tall tree are usually smaller and thicker as if they are growing in a drought. And this is exactly what is happening. The leaves at the bottom and top of a tall tree are like having two plants in the garden side by side with one well-watered and one given hardly any water. Certainly in coastal redwoods, growth rates at the top of the tallest trees appear to be dictated by water stress not by light. The theoretical limit for Douglas fir (*Pseudotsuga menziesii*) beyond which water transport is impossible is around 138 m (Domec *et al.* 2008), consistent with the suggested height records of 140 m. This is referred to as the 'hydraulic limitation hypothesis'; the reason why it's called a hypothesis is that there is a good deal of debate as to what limits the height of ordinary trees, not just the giants. There are a number of other potential reasons which are discussed in Chapter 6.

## Grain in wood

The vertical orientation of most of the cells in wood is responsible for the 'grain' of the wood: cut across a tree and you are cutting across the grain. But the grain is rarely truly vertical, rather it spirals up the trees from the roots to the branches (Figure 3.12) generally by just a few degrees off the vertical although it can be up to 40° or almost horizontal near the base of a tree. In a young conifer the grain starts off by usually spiralling to the left (like the middle of the letter S placed on the bark), straightens up to almost vertical at around the age of 10 years before spiralling to the right (like the middle stroke of Z) at around 10–30 years of age. Hardwoods are more complicated including many with interlocked grain, where the direction alternates every few years. Spiral grain helps prevent a tree splitting between rings (like pulling sheets of paper apart) under wind or snow load since whichever way the tree twists one part is being tightened up and will remain strong. It also appears to aid a more even distribution of water and food around the trunk and hence over more of the canopy and roots (see above). Wood grown in the first 10–30 years tends to be of lower density than that grown after and these two phases of wood growth are sometimes referred to as juvenile and adult wood.

The desirable 'figure' or surface pattern on cut wood can be highly influenced by the grain. In birds-eye maple the grain grows in small pimples pointing in towards the pith which when cut through by the saw produce the pattern. Similarly, 'fiddle-back' in maples (commonly used for violin backs) is formed where the grain grows in vertical waves. Other features of the grain, induced by excessive strain, include internal cracks along the rays (usually

**Figure 3.12** Lodgepole pine (*Pinus contorta*) killed by fire in the Canadian Rockies. Where the bark has peeled off, it reveals the 'grain' of the wood (xylem) spiralling up the trees.

called heart shakes) or around the rings (cup shakes, from the cupped pattern made when a plank from an afflicted tree falls apart).

## Wood movement

Cellulose, the main constituent of wood, is a hygroscopic material (i.e. it absorbs water) and so swells and shrinks as it absorbs and loses moisture. Wood in a living tree is mostly always wet and so we only tend to notice this movement when the tree dies. Telegraph poles with excessive spiral grain can twist as they wet and dry, with imaginable effect on wires at the ends of the cross-arms. Cells shrink and expand in girth rather than length so wood is fairly stable along the grain. In the same way, rays act as restraining rods reducing radial shrinkage, so most shrinkage is tangential (around the tree). A dried slice of tree nearly always has a 'pie-slice' crack running to the centre. Such changes in dimension are significant even over the normal temperatures and humidities expected over a year. For example, the pattern used for shaping the mould for the 16–17 t anchor heads of the cruise ship *QEII* was made of pine and the shrinkage between winter and summer (0.5–1.0%) made a difference of 100–150 kg (2–3 hundredweight)!

(e)

Figure 3.13 Frost cracks: (a) at the time the stem first cracked; (b) after the crack closed at the end of winter; (c) the crack covered by one summer's wood growth; and (d) covered by several years regrowth by which time enough wood has been grown to heal over the crack, leaving a frost rib down the trunk, as shown in the oak tree (*Quercus petraea*) in (e). (a)–(d) Modified from Kubler, H. (1988) Frost cracks in stems of trees. *Arboricultural Journal*, **12**, 163–175. (e) Kindrogen, Scotland.

Drying of wood can be significant to the living tree. This is where frost cracks come in, especially common in hardwoods such as oaks, planes (*Platanus* spp.) and ashes but also firs (*Abies* spp.). Even in the relatively mild climate of Britain you can sometimes hear gun-shot like reports in the woods in winter as a tree suddenly cracks open. Once you get up off the ground, you can see what appears to be an open crack running down the trunk (Figure 3.13). On cold nights the cell walls in the outer part of a trunk freeze-dry in the same way that unwrapped food desiccates in a freezer (in the woody cells the water is driven inwards to join the ice already inside the tube). As it dries it contracts over the still-wet and unshrinking centre. This is aided by contraction of the wood due to the cold, especially after very cold but sunny days when the tree has warmed up in the sun and the outside cools and contracts much quicker than the centre when the sun goes in. The pressure developed is eventually relieved by the wood cracking open. This may well be aided by weaknesses in the wood where a branch has died or it has been wounded. Cracks on big trees in continental climates can be several centimetres wide in mid-winter. In the spring, the crack closes and the new wood formed that summer papers over the crack (Figure 3.13b, c), but this may not be strong enough to resist cracking next year (the old wood in the crack never, of course, rejoins). Several mild years are usually needed to build a strong enough bridge over the crack to prevent future cracking (Figure 3.13d). Repeated opening and healing results in the formation of protruding lips of callus along the edges called frost ribs (Figure 3.13e).

## Rays

We have seen that rays are strips of living tissue that run from the centre of the tree out through the cambium and into the phloem and so join the phloem and xylem into a giant circuit. The cells of rays live for several years, even decades, and one of their prime roles in the tree is to store food (they are also involved in the creation of heartwood; see below). When more food is made by photosynthesis than can be used straight away, and when minerals such as nitrogen are reabsorbed from leaves in the autumn, the surplus is stored in the living ray cells of the wood and bark. The most favoured storage places are the wood of the roots and base of the trunk but even small twigs are used. Indeed, the sapwood of the sugar pine (*Pinus lambertiana*) from California and the Mexican white pine (pino de azucar, *Pinus ayacahuite*) both yield a sugary sap if damaged. Sugars transported by the phloem are converted into starch or fats when they arrive at the storage site. Around 1900 trees were classified as either starch trees (most ring-porous hardwoods plus some conifers like firs and spruces) or fat trees (most diffuse-porous trees plus pines) depending upon what they primarily stored, but it is not always such a clear-cut distinction since both starch and fat can be found within the same species or even the same tree. Generally, low temperatures favour storage as fats.

As you might expect, an annual cycle of storage is most pronounced in deciduous trees. They need to grow a new set of leaves in spring before they can start making new food, and, as seen above, ring-porous trees have to grow a new ring of wood before that. The energy and materials for such new growth have to come from reserves held over winter. In most evergreen conifers (but not all, e.g. Norway spruce, *Picea abies*) the older needles from previous years are able to grow sufficient food in early spring to fund the new spring growth and so storage tends to be much more equitable through the year.

Food is not stored just as a starter motor for spring. It is also an insurance against bad times such as losing parts of the canopy to gales, insect defoliation or rot. Trees generally maintain a steady reserve of stored food that is only used in times of great need; depletion of these reserves is a cause of tree death (Chapter 9). The reserves are also a savings account to fund years of abundant fruit production, especially in trees that show masting: see Chapter 5. Palms store impressive amounts of starch in the centre of the trunk to be used in flowering. This has not gone unnoticed by humans. The sago palm of South-East Asia (*Metroxylon sagu*) is felled, the central tissue is scooped out and the starch washed out and used as sago. Or the flower stalks are tapped for their abundant sugary syrup which is used to make sugar and a potent alcoholic drink.

## Sapwood and heartwood

A cut log often has a dark centre (the heartwood) surrounded by a circle of lighter wood (the sapwood; see Figure 3.1). The sapwood contains living tissue (as described above) and the rings that conduct water; the heartwood by contrast contains no living tissue and has materials added (polyphenols, gums, resins, etc.) which contribute to the darker colour (Figure 3.14). In some trees, such as beech (*Fagus sylvatica*), the heartwood is no different in colour from the sapwood, but is still there. Others do not seem to regularly form any heartwood at all; alder (*Alnus glutinosa*), aspen (*Populus tremula* and *P. tremuloides*), a variety of maples and other trees have been found up to 1 m in diameter with living cells and stored starch right to the centre (although some sugars found in, for example, pines, may be a by-product of heartwood formation). Fungal rot can add an extra confusion by producing discoloured wood that has nothing to do with heartwood formation. Sometimes it's not clear: red heartwood in beech is commercially undesirable but is found only in some trees. It was originally thought that the presence of oxygen allowed oxidation of the heartwood, changing its colour, but the concentration of oxygen in red heartwood has been found to be no different from that in normal heartwood. Perhaps the answer lies in fungal rot?

How heartwood is formed is discussed in Chapter 9 but here we'll ask what heartwood does for the tree. Heartwood is usually denser and stronger than sapwood, and, because of the chemicals added to it by the tree, more resistant

**Figure 3.14** Freshly felled elm (Ulmus hollandica 'Vegetata') showing the darker heartwood surrounded by lighter coloured sapwood. Keele University, England.

to rot (it's the part generally used for timber). It can therefore be seen as the strong core which holds the tree up. But this cannot be the entire answer because, as explained in Chapter 9, old hollow trees with little heartwood left may stand up to gales better than solid trees, and it is possible that trees deliberately court heart rot as a way of recycling all the stored minerals. Some might argue that heartwood is a very useful dumping ground for waste products that would otherwise be hard to get rid of. Against this must be set the knowledge that some compounds are expensively produced specifically to be incorporated into heartwood.

Sapwood is widest in vigorously growing trees but there are considerable differences between different species of tree. Northern temperate ring-porous trees, such as oaks, osage orange (*Maclura pomifera*), northern catalpa (*Catalpa speciosa*) and black locust/false acacia (*Robinia pseudoacacia*) normally keep just 1–4 rings of sapwood, amounting to less than 15% of the cross-sectional area, although it can be 20 or more rings wide in fast-growing trees as shown in Figure 3.14. The normally narrow sapwood is generally lost with the bark when sawn for timber. Temperate diffuse-porous trees and conifers, on the other hand, can keep more than 100 sapwood rings, and almost half of the cross-sectional area can be sapwood. In conifers, the cross-sectional area of sapwood at any height up the trunk is normally directly related to the amount of foliage above that point. Tropical diffuse-porous trees such as ebony (*Diospyros ebenum*) may be practically all sapwood, with even a large tree yielding relatively little of the precious heartwood. This all makes sense if you remember that diffuse-porous trees spread their water uptake over a number of rings while ring-porous trees concentrate water movement in the outermost ring (see Growth rings above).

Heartwood does not necessarily follow a clear-cut cone inside the tree following the growth rings. In cross-section the heartwood often meanders across rings, and generally the sapwood is thicker up towards the crown and gets thinner towards the base of the tree. So the top few metres of a tree may contain no heartwood.

Sapwood is expensive stuff to keep alive. It is not just the sugar needed; there is also the problem of supplying it with oxygen. The cambium and phloem get oxygen through the lenticels, but sapwood that is full of water conducts oxygen too slowly for tissue to remain alive. This is primarily solved by dissolved oxygen being carried up with the water from the roots that can contribute 60–100% of the necessary oxygen. Oxygen can also come from the air in cavitated tubes since oxygen can diffuse into wood much more easily through air-filled tubes than it can through those that are water-filled. Wood at 40% gas volume conducts oxygen anywhere between 5 and 1000 times faster than in wood with 15% gas volume.

## Bark

Bark is defined as all those tissues outside of the vascular cambium although, since the cambium comes with bark when it is stripped off, this may pragmatically also be considered part of the bark. (In a perverse way the cambium affects how 'glued on' the bark is: wood cut in summer when the cambium is actively growing tends to lose its bark upon drying, wood cut in winter does not.) Bark is made up of two portions, the inner bark or phloem (which conducts sugary sap around the tree), and the outer bark, which acts as the waterproof skin of the trunk, keeps out diseases and protects the vulnerable living tissue from extremes of temperature, even fire (look back to Figure 3.1).

### Inner bark (phloem)

Movement of the sugars from the leaves, where they are produced, to where they are needed is quite different from the movement of water up through the wood. The inner bark cells where this happens are thin-walled and alive. The active part of the phloem consists of living 'sieve cells' stacked one above the other to form tubes. The end walls are perforated (known as sieve plates) allowing sap to pass along the tube. In hardwoods each sieve cell is closely associated with one or more companion cells. The sieve cells do not have a nucleus, the source of genetic material necessary to keep the cell going (possibly because such a bulky object would impede the flow of sap), and the companion cells provide support to keep the sieve cells functioning. In conifers the sieve tubes are more like the tracheids found in the wood in that they are discrete cells which overlap and communicate with pits rather than a sieve plate, and there are no companion cells. In both hardwoods and conifers, the sieve cells are set in a matrix of strong fibres and general purpose parenchyma cells including rays. The strong 'bast' fibres of the inner bark have been (and are) used to make useful materials: Ugandans make cloth from *Ficus* trees, American First People made ropes and cords (especially from the linden tree, *Tilia americana*) and people of the Pacific islands, China and Korea made writing material from the bark of the paper mulberry (*Broussonetia papyrifera*) long before the pulping of wood.

Sieve tubes are not thickened with lignin like xylem tubes and so remain with thin cellulose walls. It is apparent that whereas the xylem tubes are built to withstand the great suction from above, the sieve tubes would just collapse under such negative pressure. Rather the sieve tubes transport the sap under positive pressure. The exact mechanism is still debated but the hypothesis with most support is one of 'mass flow'. Sugars produced by the leaves are actively pumped into the sieve tubes – this takes energy – which draw water into the tube by osmosis. This creates a high hydraulic pressure forcing the watery sap along the sieve tubes. At places where the sugars are used or stored, they are removed and again water follows by osmosis, lowering the pressure in the tube. With a high pressure at one end (the source) and low pressure at the other (the sink), the contents of the tube will move towards the low pressure. In even a small tree there can be thousands of sources (the leaves) and thousands of sinks (wherever there is a need for sugar) so the flow can be quite complex. To give an idea of speed the sap flow has been measured at 10 cm per hour in European larch (*Larix decidua*) and up to 125 cm per hour in American ash (*Fraxinus americana*).

The flow of sap does not always have to be down the tree, from leaves to roots. In spring, stored food can be transported upwards (along with sugar in the sap; see page 63), and developing fruits at the top of a canopy can draw food upwards from leaves lower in the canopy. One problem with a system under pressure is that any wound could cause the plant to bleed to death. Fortunately, the sudden release in pressure and rapid flow caused by a puncture causes a rapid sealing reaction at the pores of the sieve plates, isolating the damaged element. Flow then continues around the damage.

Sugars can be produced by photosynthesis during daylight more rapidly than they can be exported. The excess sugar is stockpiled as starch and then exported as sugar during the night. The transported sap contains 15–30% dissolved sugar and just traces of other things such as plant hormones, amino acids and minerals such as nitrogen (hence honeydew produced by aphids: the insects want the small quantity of nitrogen and have to filter through a lot of sticky sap to get it; the waste syrup is expelled to drop on whatever is below, usually my car!).

If you look at a piece of bark taken from a tree, you will see that the inner bark (phloem) is comparatively thin (Figure 3.1). There are two reasons for this. Firstly, not much is made: typically 4–14 times more new wood (xylem) is made by the cambium compared to the phloem (although xylem is much more affected by growing conditions than phloem and in bad years the width of phloem and xylem may be almost the same). The second reason is that the phloem is active for normally just 1 or 2 years (5–10 years exceptionally in, for example, limes, *Tilia* spp.). Old phloem on the outside is crushed as the expanding tree meets the constraining outer bark, and is rapidly assimilated into new outer bark. In fact the layer of conducting phloem is usually less than a millimetre thick and has been measured at just 0.2 mm in American ash (*Fraxinus americana*). Incidentally, growth rings appear in the bark in the same way as in wood but are a short-term record only.

## Outer bark

The protective outer bark (officially called the rhytidome) is composed mainly of old inner bark (phloem) that has been replaced by new working phloem from underneath. But cork is added to make the outer bark more weatherproof. To understand the process, let's start at the beginning. In a young stem, be it a seedling or a young twig on an older tree, the epidermis acts as the outer protective coat. Soon, however, the epidermis and green cortex beneath are usually replaced by a corky layer (the periderm; see Figures 3.2 and 3.15). This is produced by a new growing zone (the cork cambium or phellogen) arising in the outermost layers of the cortex (occasionally in the epidermis). This layer grows layers of cork (phellem) on the outside and occasionally a few cells on the inside called the phelloderm. All of these corky layers together are called the periderm (that's the end of the terminology!). As the cork cells mature and die they are filled with air, tannins or waxy material and the walls are impregnated with a fatty substance (suberin) and sometimes other compounds. Suberin accounts for 45% of the weight of cork from a cork oak (*Quercus suber*). It is worth noting that some trees do not bother with producing cork from a cork cambium. Instead, the cortex and epidermis grow throughout the life of the tree to keep it covered, resulting in an effective but very thin bark, as in certain species of holly, maple, acacia, eucalypt and citrus fruits.

When present, the corky layer is water, gas and rot proof: superb as a weatherproof layer but not without its consequences. Firstly, it separates the epidermis from its food and water supply so it dies and scales away, which is a pity because the green cortex can add considerably to the photosynthesis of the whole tree (Chapter 2). However, in some species the phelloderm (the extra cells made on the inner side of the cork cambia) contain chlorophyll,

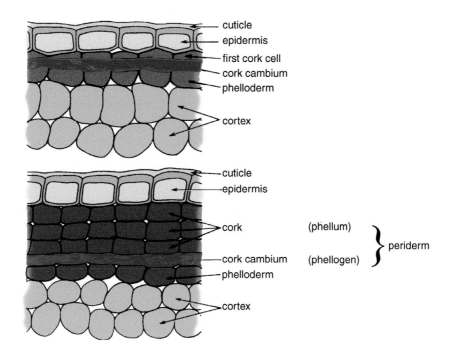

Figure 3.15 Cross-section of a piece of young bark showing the development of a cork cambium and the growth of corky cells. From: Weier, T.E., Barbour, M.G., Stocking, C.R. & Rost, T.L. (1982) *Botany: An Introduction to Plant Biology* (6th edn). Wiley, New York. Reprinted by permission of John Wiley & Sons, Inc.

allowing the corky bark to photosynthesise, especially where the bark is thin (such as in cherries and poplars). The second consequence of having a cork layer is that the exchange of gases across the bark is prevented. The living tissue in the bark, cambium and wood, like all living cells, needs oxygen. As already discussed, some of this oxygen is brought up in the water from the roots, which helps with the oxygen supply for the sapwood. But for the outer living tissues, this problem is primarily solved by having holes in the bark, called lenticels, where the cork layer is ruptured into loose mounds of cells with numerous air spaces to allow gas exchange. These usually form beneath the stomata of the young twig. Lenticels are variously seen on the bark as little pimples through to the horizontal raised lines on birches and cherries, normally covering 2–3% of the bark's surface.

As well as providing a weatherproof skin over the tree, the bark also protects against invasion from anything from bacteria to insects and mammals. In addition to the general toughness of bark, there are a number of chemicals found in some barks which have been shown to reduce attack. Some we find useful: quinine (from the bark of the South American quinine tree, *Cinchona corymbosa*, useful against malaria), the bark of the cinnamon tree native to Sri Lanka (*Cinnamomum verum*), purgatives from buckthorn bark (*Rhamnus* spp.) and tannins for tanning leather from the bark of oak, birch, alder, willow, eucalypts, tropical mangroves (*Rhizophora mangle*) and a good many others.

Most monocots don't grow bark; instead, as in bamboos, the epidermis becomes very hard or, as in many palms, the existing tissue becomes corky beneath the surface. In those monocots with secondary thickening (such as *Yucca*, *Cordyline* and *Dracaena* spp.), a corky layer of sorts appears and may give rise to a very dicotyledon-like bark.

## How does bark cope with the expanding tree?

In most trees, the phloem starts growing early in the spring anything up to 6–8 weeks before the xylem (although it appears more synchronous in some ring-porous trees that start xylem growth particularly early). So, as the new wood starts growing *inside* the bark, the newly formed phloem is immediately put under strain. An immediate solution is for the rays to grow sideways or 'dilate' so that they look like the bell ends of trumpets (Figure 3.16); this is like

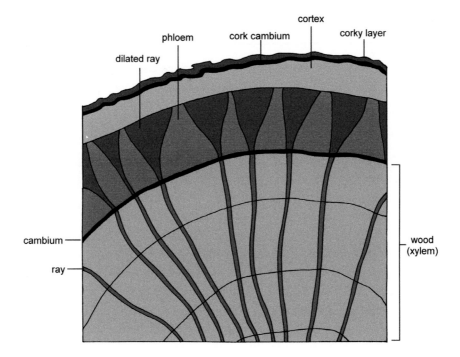

**Figure 3.16** A diagrammatic view of a cross-section of a lime (*Tilia* sp.) showing the expanded rays which 'dilate' to help the inner bark (phloem) to stretch as new wood (xylem) grows under the bark.

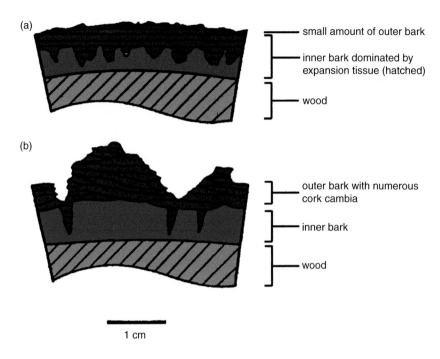

Figure 3.17 Bark cross-sections of (a) beech and (b) oak. Modified from: Whitmore, T.C. (1962) Why do trees have different sorts of bark? *New Scientist*, **312**, 330–331.

sewing a number of extra panels into a shirt that's too tight. Dilation growth is especially common in citrus and lime trees. But what happens as the tree gets fatter each year, and grows in diameter from a few centimetres to sometimes well over a metre, and the original phloem and bark become far too small?

One part of the solution is to have lots of dilation growth, and indeed, in beech (*Fagus* spp.), by the time you get to the outer side of the bark it is almost completely expansion tissue (Figure 3.17a). This is helped by there being only a very small thickness of inner bark produced each year. The second element in this solution is the cork cambium. At its simplest, such as in beech, the cork cambium normally remains active for the life of the tree and keeps up with the expanding girth of the tree by growing sideways. Very little corky tissue is produced and the thin bark is shed in dust-like fragments, and the bark remains smooth. A similar type of growth accounts for the smooth bark of trees such as birches, cherries, hornbeams, some maples and firs, although in these the original cork cambium may be replaced in later life (perhaps after a century) and accounts for the rugged bark found on old cherries, for example.

In most trees the inner bark grows four times or more faster than in beech and there is far more tissue to be dealt with. Here, new cork cambia form successively deeper into the inner bark. These cambial layers usually have the shape of curved shells with their ends pointing outwards (producing plates) or longer lines down the bark, which with the strong bast fibres tends to produce ridges (Figure 3.17b). These cambia may live for several years before being cut off and killed by new cambia below. The dead tissue of the outer bark cannot form expansion tissues so the outer bark splits as the tree increases in girth and changes from the smoothness of youth to the rugged pattern of maturity. The time taken for this change varies from 10 years in Scots pine (*Pinus sylvestris*) to 12 years in lime, 20 years in alder (*Alnus glutinosa*) and 30 years in oaks.

An analogy for these two ways of forming bark is to consider what happens if you eat too much, expand rapidly in girth and your clothes get tight. You have two options if you want to stay covered up. The first is to have thin stretchy clothes similar to beech (but in beech the bark doesn't so much stretch as grow in diameter). The other option is to put on bigger clothes underneath the tight ones (not easy, but ensures continuous coverage) and let the tight ones burst off the outside demonstrating in their remains just how small you used to be. This is similar to how oak copes, where the old layers are left as ridges on the outside.

As an aside, when struck by lightning, trees with rugged bark like oaks are seriously damaged more often than trees with smooth bark like beeches. The reason is very simple. When a tree is struck the charge follows the path of least resistance. With a thoroughly drenched tree this is down the water film on the outside of the trunk (little damage results) but in a dry tree it is through the water in the xylem, which causes the water to explosively boil. The rough bark of oak takes more rainfall to completely wet than does a beech, so oaks are more often at risk.

You can see from the above that bark consists mostly of dead phloem, often with many strengthening fibres, and a small amount of cork produced in thin layers. An exception to this is found in the cork oak (Figure 3.18). Like beech, the cork oak has a single cork cambium but here it is capable of producing a cork layer many centimetres thick over 20–30 years. When this thick, the almost pure cork can be split from the tree by making a vertical slit in the bark and peeling off the cylinder of oak along the cork cambium. This is done in spring when the cork cambium is actively growing and so easily splits. The resulting pink trunk soon turns black and a new cork cambium forms deep in the old inner bark and produces new cork even faster than before, splitting the black skin. Each successive stripping of the oak (every 8–10 years) produces better quality cork until by the third to sixth stripping it is good enough for

Figure 3.18 Cork oak (*Quercus suber*). Birr Castle, Ireland.

wine corks. Remember that cork layers cannot be completely impervious; they have lenticels to allow gases through the bark. Corks for bottles are cut, of course, so that the lenticels (those brown powdery lines) run *across* the cork to prevent spirits or other liquids escaping. Note that most trees die when the bark is stripped off around the tree because it separates from the tree along the *vascular* cambium (see Figure 3.2) and so all the inner bark is lost. In the cork oak the bark splits off through the *cork* cambium leaving the inner bark in place and free to carry on conducting sugary sap.

There is, of course, a current move towards using artificial corks or screw caps for wine bottles, leading to the potential demise of the cork industry (despite proper corks having a lower environmental impact). Cork has fallen into disfavour because it can shrink allowing oxidation of the wine, but the most important problem attributed to the corks is 'cork taint' primarily due to TCA (2,4,6 trichloroanisole), produced in the cork by the fungal action of *Aspergillus* sp. and *Pennicilium* sp., which gives the wine a musty and unpleasant flavour (it is 'corked'). Some of this may be due to the use of chlorine-based pesticides on the oak trees, but some TCA has been produced through the use of chlorine-based bleaches in the wine production. So good husbandry of the trees and good hygiene during wine production may solve the problem and allow real corks a continued role in wine making.

## Bark thickness and bark shedding

Smooth barks can be surprisingly thin, being as little as 0.6 cm (6 mm) thick in a beech 30 cm in diameter (compared with up to several centimetres in a similarly sized oak). Foresters go by the rule of thumb that bark makes up 7% of the volume of a felled beech log, and 18% in oak. This is not just a reflection of how quickly bark grows in oak but also how tenaciously the old bark clings on. In oaks, redwoods and other trees with thick barks, the successive cork cambia stick together so that coastal redwood bark (*Sequoia sempervirens*) can be up to 15 cm thick, a 50 m tall Douglas fir may have a bark around 25 cm thick and giant sequoia bark (*Sequoiadendron giganteum*) can be up to 80 cm or more, representing centuries of old phloem, creating a soft spongy fire-proof blanket around the tree.

On trees that *do* shed their bark, for example plane trees (*Platanus* spp.), the bark separates by tearing through thin-walled cork cells adjacent to the cork cambia, and so the shape and size of the shed pieces reflects the organisation of the successive cambia. Birches and cherries have alternating layers of thin- and thick-walled cork cells which results in the shedding of thin papery sheets. Some 'stringybark' eucalypts grow vertical lines of thin-walled cells in the bark, so it is shed in strings of loosely connected fibrous bark. Shedding of bark is commonest towards the end of the growing season, and especially after hot weather, when shrinkage by desiccation helps pry loosening pieces off the tree. Why shed bark? In some cases it is a defence against things that would live on the bark. Certainly in California, the Pacific madrone (*Arbutus menziesii*), a shrub with colourful orange bark which is regularly shed, is not parasitised by mistletoe as are surrounding trees. The London plane (*Platanus* x *hispanica*) has survived so well in London (and other formerly polluted cities) because it regularly sheds bark, taking with it lenticels blocked by pollution and exposing fresh open lenticels.

Trees grown in shade tend to have thinner bark which explains why tree trunks suddenly exposed to the sun by the felling of neighbours can be damaged on the sunny side by sun-scorch, especially thinner-barked species such as beech, hornbeam, maples, spruce and silver fir (*Abies alba*). Transplanted trees are sometimes protected by being wrapped in paper, raffia or sacking.

## Burrs, buds and coppicing

Abnormal bulges or bumps can be found on the trunks and limbs of nearly every kind of tree. While these burrs or burls are common, we know very little about how many of them form. Insects, bacteria and viruses cause a variety of disruptions to the cambium that can enlarge as new wood is added each year, rather like a cancerous growth. Most burrs, however, are formed where there are a mass of buds buried in the bark which distort the trunk's growth (Figure 3.19). These buds on the trunk (referred to as epicormic or latent buds) are really dormant buds that formed as normal in the axils of leaves on the young shoot (and hence started next to the pith) but which didn't develop any further. Instead they have remained dormant and have grown each year just enough to stay at or just below the surface of the bark (Figure 3.20). Since they often branch as they grow, a whole mass of buds may develop in one spot in the wood causing the familiar burr. If a burr is cut open the trace of the buds can be followed right back to the centre of the tree. You might think that all buds would open in the spring so these dormant buds would be quite rare, but it has been found, for example, that in red oak (*Quercus rubra*) about

Figure 3.19 A giant burr on oak (*Quercus petraea*) that covers most of the trunk. Smaller burrs appear as individual golfball- to football-sized swellings. Cannock Chase, England.

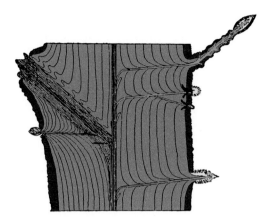

Figure 3.20 Longitudinal section of a tree trunk with epicormic buds arising at the centre of the tree and growing enough each year to stay at the bark's surface, often branching as they go. From: Büsgen, M. & Münch, E. (1929) *The Structure and Life of Forest Trees.* Chapman & Hall, London.

two thirds of buds remain dormant in spring. Many will be subsequently aborted (Chapter 7) but some survive. These dormant buds are the tree's insurance policy; if the crown of the tree is damaged (or we prune it) they will sprout to produce new branches. In trees like common lime (*Tilia* x *europea*) and English elm (*Ulmus procera*) the buds are more exuberant and produce masses of shoots around the trunk's base even in normal healthy trees. Buds may be scattered widely up the trunk as in oak or concentrated at the base as in eucalypts and birch.

Epicormic buds can be 'adventitious', that is, forming afresh from any living cells in the trunk, usually after some sort of injury, especially on older parts of stems. They are often formed on callous tissue: new shoots on cut stumps of poplars, horse chestnuts, beeches and walnuts are often from adventitious buds on the callous tissue rather than from dormant buds. Similarly, most root suckers are from adventitious buds, formed when needed. The main difference between stored epicormic and adventitious buds is that adventitious buds can't be traced back to the centre of the tree because they started life when the trunk or branch was larger. Growth from either type of bud is initially weakly attached and may die and fall off but during each following year new wood grows over the join making it stronger.

Note that a number of tropical trees and the common hazel (*Corylus avellana*) of temperate woods are self-coppicing and will readily grow new shoots from the base to produce the typical mass of shoots, although they tend to be more uneven in age (and thus size) than those resulting from a tree being felled.

Conifers are notoriously poor at growing from epicormic buds (stored or adventitious, but read about the formation of new buds in leaf axils in

Chapter 6) and will die if cut to the base or are otherwise left with no foliage. There are exceptions, of course: the monkey puzzle (*Araucaria araucana*), coastal redwood (*Sequoia sempervirens*) and some pines, including the canary pine (*Pinus canariensis*), pond pine (*P. serotina*), pitch pine (*P. rigida*) and Chihuahua pine (*P. leiophylla*), *will* regrow. Older hardwoods may lose the ability to regrow from epicormic buds. This has caused a problem in the UK with old pollards. Pollards are trees where the trunk is cut, usually at 2 m (6 ft) or so, and left to regrow long straight wood of the sort much used when people lived off the land (coppicing is similar but is done at ground level; pollarding was common in parks where the new regrowth had to be kept out of reach of deer and other grazing animals). Conservation bodies would like to repollard historical trees that were last cut anything up to a century ago but such old branches have few buds left. Fortunately, a large body of research has accumu-lated on how to improve success (see Read 1996).

## Branching and knots

Except for branches arising adventitiously (explained above), all branches can be traced back to the middle of the trunk or branch on which they are growing. As a new branch elongates in the first year, it lays down buds which develop the next year into further young branches (this is examined further in Chapter 7). New layers of wood laid down each year coat the stick and its new side branches in a continuous layer (although rings are thicker on the trunk than the branch, so branch diameter increases more slowly than the trunk) giving the classic pattern seen in trees split along their length, seen in Figure 3.21a, b. The buried portion can no longer increase in diameter so it tapers to the middle of the tree (forming what is often called a spike knot). Alex Shigo, an American forester who has spent his life dissecting trees, put forward an idea for how branches add new wood to increase their strength. He main-tains that new wood grows first along the branch and forms a downwardly turning branch collar (labelled (1) in Figure 3.22). As the trunk tissue begins to form later from above it produces a trunk collar over the top of the branch collar (2). The two collars develop water connections some way above and beneath the branch junction (4) and this interlocking pattern of stem and branch tissue gives it great strength. The resulting collar at the base of a branch can often be seen as in Figure 3.23. As the trunk and branch get fatter, the bark tends to wrinkle in the crotch to form the 'branch bark ridge'. More recent work suggests that this may not be the whole story. Duncan Slater and Claire Harbinson (2010) suggest that the very dense wood inside the ridge plays an important mechanical role in supporting the branch. As noted above, the dense, convoluted wood found here helps constrict flow of water into the

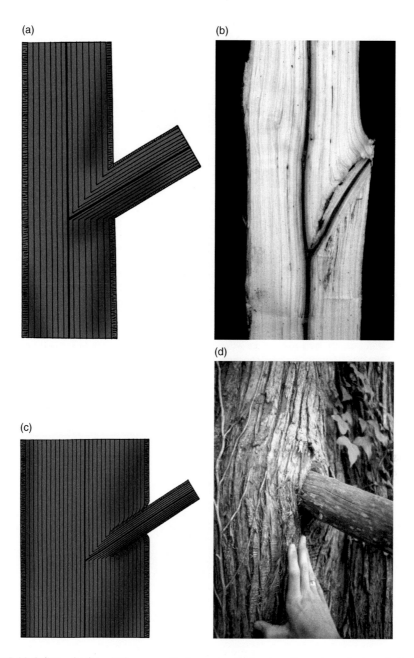

**Figure 3.21** Split trunks show (a) how growth rings cover the trunk and branch producing a 'spike knot' pointing to the tree's centre; a plank cut from this tree perpendicular to the knot will contain an 'intergrown knot': (b) this piece of split poplar shows just how firmly these knots are anchored into the tree. (c) If the branch dies (6 years ago in this example) then planks cut perpendicular to the branch will contain an intergrown knot near the middle of the tree and an 'encased knot' near the bark. Intergrown knots do not fall out of a plank; encased knots can. Encased knots can be surrounded by bark or the bark may fall off before the branch is enclosed by new growth as shown in the western red cedar (*Thuja plicata*) in (d). Photographs from Staffordshire, England.

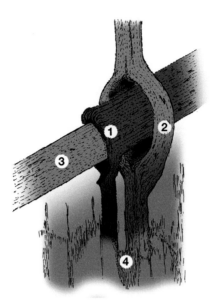

Figure 3.22 Shigo's model of branch attachment to the trunk. The branch collar (1) is grown in spring by the branch (3), followed later by wood growing down the trunk to form the trunk collar (2). The two collars develop water connections some way above and beneath the branch junction (4). From: Shigo, A.L. (1991) *Modern Arboriculture*. Shigo & Trees, Durham, NH.

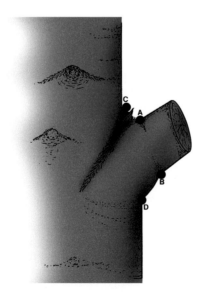

Figure 3.23 Correct pruning. Cuts should be made from A to B leaving the branch bark ridge (C) and the collar (B–D) untouched. From: Shigo, A.L. (1991) *Modern Arboriculture*. Shigo & Trees, Durham, NH.

branch, stopping it getting more than its fair share of water. But this dense saddle-shaped zone may also help to prevent the branch being ripped out of the trees by downward force by 'clipping' the top of the branch into the tree. Whatever the precise role of the wood formed under the branch bark ridge and within the visible collar around the base of the branch, they are certainly both important parts of the tree's mechanical skeleton. Pruning should ensure that these tissues are left intact since this will ensure maximum callous growth and closure over the wound.

When we buy timber we are used to seeing knots in the wood. But why do some seem to be firmly anchored in the wood while others readily fall out? A knot is the portion of a branch buried in the wood as the trunk increases in diameter. If the branch was living as in Figure 3.21a, b, the branch is continuous with the trunk and so the knot is held firmly and cannot fall out: an 'intergrown or tight knot'. If the branch dies, it ceases to increase in diameter and new growth simply surrounds the dead limb stump, so the branch is gradually buried as a cylinder of non-living tissue. When a plank is cut across this, it gives an 'encased or loose knot', often with bark entrapped: these 'black ring' knots are particularly prone to loosen and drop out (Figure 3.21c, d).

## Reaction wood

How do leaning trees straighten themselves? As is often the case, conifers and hardwoods go about it in opposite ways. With a few minor exceptions, conifers push and hardwoods pull. In a leaning conifer, 'compression wood' forms along the underside of the trunk in the new wood that grows after the tree starts leaning. This can be seen in a felled tree as a crescent- or triangular-shaped darker zone of wider rings (Figure 3.24). As it forms, the compression wood expands and pushes the tree towards the upright. It is called compression wood because, obviously, it is in compression, and when cut out of a tree it does expand slightly in length. As you would expect, recovery to the vertical starts at the top (where the tree is thinnest) and gradually proceeds downwards (Figure 3.25). Often the top of the tree has returned to the upright while the lower part is still working at slowly bending the thick trunk, with the result that some years later the top of the tree ends up leaning the opposite way to when it started and it in turn grows more reaction wood to restraighten itself leaving the tree with an S-shaped stem. Compression wood is readily seen in hemlocks (*Tsuga* spp.), which naturally develop drooping new terminal growth each year that by the end of the growing season has straightened itself (or nearly so) using compression wood.

Compression wood is notably different from surrounding normal wood: the tracheids are shorter and rounder with thicker walls that contain fissures, and

Figure 3.24 Section of a young Scots pine (*Pinus sylvestris*) that started to lean to the left 10 years before it was felled. The darker area of wood, with wider rings, on the left is reaction wood (in this case compression wood since it's in a conifer) that would have pushed the tree upright. Wybunbury Moss, England.

there are many spaces between the tracheids, plus they have a higher lignin and lower cellulose content. This all makes the wood denser and harder, but weaker and more brittle than normal with exceptionally high longitudinal shrinkage (ten times normal). Best avoided in woodworking, but bear in mind that compression wood can account for 20–50% of the wood in a Scots pine (*Pinus sylvestris*) grown in a windy area.

Hardwoods, by contrast, form 'tension wood' along the *top* of a leaning stem, although it is not as readily seen since it does not look superficially any different from normal wood. Nevertheless, it is there, contracting as it grows to pull the tree upright. The resulting wood is dominated by thick gelatinous fibres made principally of crystalline cellulose (with a hemicellulose called xyloglucan that creates the tension) but little lignin: the opposite of compression wood. This leaves the wood prone to splitting and cellular collapse on drying. Tension wood also has fewer and smaller vessels than normal.

Reaction wood is not just used to straighten leaning trees. It is permanently present in branches to keep them at the correct angle by resisting bending due

Figure 3.25 Gradual straightening of a young spruce (*Picea jezoensis*) in Japan. Immediately after tilting (a) the tree gradually straightens by the formation of compression wood on the underside of the trunk over (b) 3 days, (c) 1 month (with overcorrection of the tip) and (d) 4 months. From: Yoshizawa, N. (1987) Cambial responses to the stimulus of inclination and structural variations of compression wood tracheids in gymnosperms. *Bulletin of the Utsunomiya University Forests*, 23, 23–141.

to gravity. In some tropical trees there is evidence that tension wood is used to move the crown in an attempt to obtain sufficient light in dense forest. The controlling factor behind reaction wood appears to be plant hormones; auxin and gibberelin regulate tension wood formation, and auxin and ethylene regulate the growth of compression wood.

## 🌿 *Further Reading*

Aloni, R., Alexander, J.D. & Tyree, M.T. (1997) Natural and experimentally altered hydraulic architecture of branch junctions in *Acer saccharum* Marsh. and *Quercus velutina* Lam. trees. *Trees*, 11, 255–264.

Ambrose, A.R., Sillett, S.C. & Dawson, T.E. (2009) Effects of tree height on branch hydraulics, leaf structure and gas exchange in California redwoods. *Plant, Cell and Environment*, 32, 743–757.

Angeles, G., Bond, B., Boyer, J.S., *et al.* (2004) The Cohesion-Tension Theory. *New Phytologist*, 163, 451–452.

Čermák, J., Kučera, J., Bauerle, W.L., Phillips, N. & Hinckley, T.M. (2007) Tree water storage and its diurnal dynamics related to sap flow and changes in stem volume in old-growth Douglas-fir trees. *Tree Physiology*, 27, 181–198.

Cirelli, D., Jagels, R. & Tyree, M.T. (2004) Toward an improved model of maple sap exudation: the location and role of osmotic barriers in sugar maple, butternut and white birch. *Tree Physiology*, 28, 1145–1155.

Cochard, H. (1992) Vulnerability of several conifers to air embolism. *Tree Physiology*, 11, 73–83.

Cohen, Y., Cohen, S., Cantuarias-Aviles, T. & Schiller, G. (2008) Variations in the radial gradient of sap velocity in trunks of forest and fruit trees. *Plant and Soil*, 305, 49–59.

Cramer, M.D. (2012) Unravelling the limits to tree height: a major role for water and nutrient trade-offs. *Oecologia*, 169, 61–72.

Delzon, S., Douthe, C., Sala, A. & Cochard, H. (2010) Mechanisms of water-stress induced cavitation in conifers: bordered pit structure and function support the hypothesis of seal capillary-seeding. *Plant, Cell & Environment*, 33, 2101–2111.

Domec, J.-C., Lachenbruch, B., Meinzer, F.C., Woodruff, D.R., Warren, J.M. & McCulloh, K.A. (2008) Maximum height in a conifer is associated with conflicting requirements for xylem design. *Proceedings of the National Academy of Sciences*, 105, 12069–12074.

Du, S. & Yamamoto, F. (2007) An overview of the biology of reaction wood formation. *Journal of Integrative Plant Biology*, 49, 131–143.

Hacke, U. & Sauter, J.J. (1996) Xylem dysfunction during winter and recovery of hydraulic conductivity in diffuse-porous and ring-porous trees. *Oecologia*, 105, 435–439.

Hacke, U.G. & Sperry, J.S. (2001) Functional and ecological xylem anatomy. *Perspectives in Plant Ecology, Evolution and Systematics*, 4, 97–115.

Jensen, K.H., Liesche, J., Bohr, T. & Schulz, A. (2012) Universality of phloem transport in seed plants. *Plant, Cell & Environment*, 35, 1065–1076.

Koch, G.W., Sillett, S.C., Jennngs, G.M. & Davis, S.D. (2004) The limits to tree height. *Nature*, 428, 851–854.

Milburn, J.A. (1996) Sap ascent in vascular plants: challengers to the cohesion theory ignore the significance of immature xylem and the recycling of Münch water. *Annals of Botany*, 78, 399–407.

Milburn, J.A. & Zimmermann, M.H. (1986) Sapflow in the sugar maple in the leafless state. *Journal of Plant Physiology*, 124, 331–344.

North, G.B. (2004) A long drink of water: how xylem changes with depth. *New Phytologist*, 163, 447–449.

Novaes, E., Kirst, M., Chiang, V., Winter-Sederoff, H. & Sederoff, R. (2010) Lignin and biomass: a negative correlation for wood formation and lignin content in trees. *Plant Physiology*, 154, 555–561.

Offenthaler, I., Hietz, P. & Richter, H. (2001) Wood diameter indicates diurnal and long-term patterns of xylem water potential in Norway spruce. *Trees*, 15, 215–221.

Ohashi, S., Okada, N., Nobuchi, T., Siripatanadilok, S. & Veenin, T. (2009) Detecting invisible growth rings of trees in seasonally dry forests in Thailand: isotopic and wood anatomical approaches. *Trees*, 23, 813–822.

Oldham, A.R., Sillett, S.C., Tomescu, A.M.F. & Koch, G.W. (2010) The hydrostatic gradient, not light availability, drives height-related variation in *Sequoia sempervirens* (Cupressaceae) leaf anatomy. *American Journal of Botany*, 97, 1087–1097.

Panshin, A.J. & de Zeeuw, C. (1980) *Textbook of Wood Technology* (4th edn). McGraw-Hill, New York.

Putz, F.E., Coley, P.D., Lu, K., Montalvo, A. & Aiello, A. (1983) Uprooting and snapping of trees – structural determinants and ecological consequences. *Canadian Journal of Forest Research*, 13, 1011–1020.

Read, H.J. (1996) *Pollard and Veteran Tree Management II*. Richmond Publishing, London.

Rose, D.R. (1987) Lightning damage to trees in Britain. *Arboriculture Research Note* 68/87/PAT.

Russo, S.E., Jenkins, K.L., Wiser, S.K., Uriarte, M., Duncan, R.P. & Coomes, D.A. (2010) Interspecific relationships among growth, mortality and xylem traits of woody species from New Zealand. *Functional Ecology*, 24, 253–262.

Sano, Y. & Fukazawa, Y. (1996) Timing of the occurrence of frost cracks in winter. *Trees*, 11, 47–53.

Slater, D. & Harbinson, C. (2010) Towards a new model of branch attachment. *Arboricultural Journal*, 33, 95–105.

Soliz-Gamboa, C.C., Rozendaal, D.M.A., Ceccantini, G., *et al.* (2011) Evaluating the annual nature of juvenile rings in Bolivian tropical rainforest trees. *Trees*, 25, 17–27.

Sorz, J. & Hietz, P. (2006) Gas diffusion through wood: implications for oxygen supply. *Trees*, 20, 34–41.

Sperry, J.S., Meinzer, F.C. & McCulloh, K.A. (2008) Safety and efficiency conflicts in hydraulic architecture: scaling from tissues to trees. *Plant, Cell and Environment*, 31, 632–645.

Tognetti, R. & Borghetti, M. (1994) Formation and seasonal occurrence of xylem embolism in *Alnus cordata*. *Tree Physiology*, 14, 241–250.

Tyree, M.T. & Sperry, J.S. (1988) Do woody plants operate near the point of catastrophic xylem dysfunction caused by dynamic water stress? Answers from a model. *Plant Physiology*, 88, 574–580.

Wheeler, T.D. & Stroock, A.D. (2008) The transpiration of water at negative pressures in a synthetic tree. *Nature*, 455, 208–212.

Zimmermann, M.H. (1983) *Xylem Structure and the Ascent of Sap*. Springer, Berlin.

Zimmermann, U., Schneider, H., Wegner, L.H. & Haase, A. (2004) Water ascent in tall trees: does evolution of land plants rely on a highly metastable state? *New Phytologist* 162, 575–615.

# Chapter 4: Roots: the hidden tree

A common view of tree roots is that they plunge deep into the ground producing almost a mirror image of the canopy. Yet in reality a tree usually looks more like a wine glass with the roots forming a wide but shallow base (Figure 4.1). Most trees fail to root deeply because it is physically difficult and unnecessary. The two main functions of roots are to take up water and minerals, and to hold the tree up. In normal situations, water is most abundant near the soil surface (from rain), and this is also where the bulk of dead matter accumulates and decomposes releasing minerals (nitrogen, potassium, etc.). It should not be surprising, therefore, to find that the majority of tree roots are near the soil surface. The flat 'root plate' also serves very well for holding up the tree; deep roots are not needed (see Chapter 9).

Roots have other functions as well. They store food for later use (see Chapter 3) and they play an important role in determining the size of the tree. Roots normally account for 20–30% of a tree's mass (although it varies from as little as 15% in some rainforest trees up to 50% in arid climates). However, if the trunk is ignored (which forms 40–60% of the total mass), the canopy and the roots come out at roughly the same mass. This helps put into perspective the relative value of the roots and the leaves to each other. Too few roots and the canopy suffers from lack of water. Too few leaves and the roots get insufficient food. There has to be a balance. The roots 'control' the canopy partly through water supply but also by the production of hormones. If you doubt this control, think of fruit trees that are kept small by being grafted onto dwarfing root stock.

Water and dissolved minerals move only slowly through soil so roots have to go to the water rather than waiting for the water to come to them. Although roots respond in a general way to gravity, they do not grow towards anything or in any particular direction, rather they are opportunistic in following cracks, worm runs, old root channels, and in proliferating where they find water or in areas of nutrient-rich soil. Since soil is often very variable over short distances, root systems tend to be more variable in shape than shoot systems, and it is often hard to see any pattern in their growth. Nevertheless, the roots of most trees can be divided into a central root plate and an outer portion.

(a)

(b)

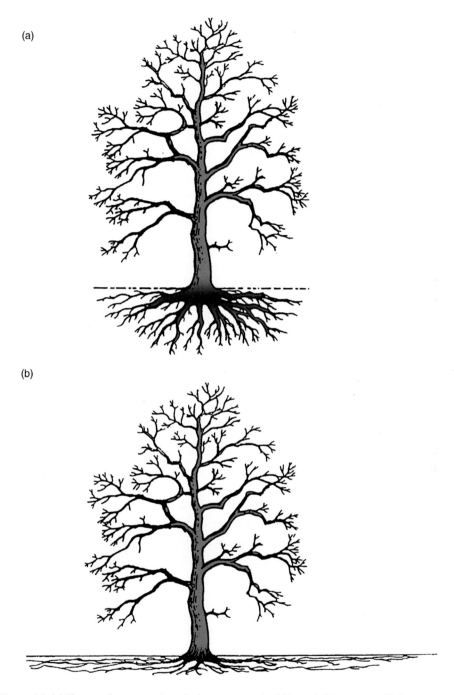

**Figure 4.1** (a) The popular conception of what tree roots look like and (b) a more realistic representation. Based on: Helliwell, D.R. (1989) Tree roots and the stability of trees. *Arboricultural Journal*, 13, 243–248.

## The root plate

Newly germinated seedlings produce a single root (see Chapter 8) which grows down as a young tap root. In some species (generally those with small seeds such as spruces, limes, willows, poplars and birches) the tap root is small, easily deflected and seldom plays a major role; in a few species, such as willows, the tap root is so rudimentary that they are often described as having fibrous root systems. In a small number of species, including pines, oaks, walnuts and hickories, the tap root grows more vigorously and initially dominates the root system and may play an important role in anchoring young trees during storms, acting like a stake stuck into the ground. In oaks it may reach half a metre in the first season. But even here the tap root is not essential: trees from open-ground sowings usually lose their tap roots when transplanted and do not suffer although they may well regrow one or more tap roots from where they have broken off. In all but a handful of species the importance of the tap root is diminished as a number of major 'framework roots' (commonly 3–11) grow out sideways from the top of the tap root (Figure 4.2). These lateral roots may stay very close to the surface (as in spruce, fir and beech; Figure 4.3a) or, as in oaks, they may descend obliquely down several tens of centimetres before growing horizontally (or rather, parallel to the soil surface since roots can grow 'uphill'). In a few trees, such as birches, larches, limes and Norway maple (*Acer platanoides*), these laterals continue growing down at an oblique angle to produce a 'heart-root' system, more akin to a root ball, but which works in a similar way to a thinner root plate (Figure 4.3b).

Although the lateral roots can be well over 30 cm diameter at the base of the tree (helping to cause the marked swelling at the base of the trunk, the root collar or root flare), they quickly taper over the first metre to often less than 10 cm diameter, and down to just 2–5 cm over the first 1–4 m from the trunk becoming less rigid and more rope-like. As these lateral roots grow out from the tree like the spokes of a wheel, they fork, branch and overlap. Within this criss-crossing framework the rigid large roots cannot move away from each other. As they grow fatter, and so are pushed hard against each other, the internal tissues fuse together (aided sometimes by the swaying of the trunk wearing away the bark of crossing roots), creating a solid physical connection between the roots (Figure 4.2b). This produces a solid root plate, the same width as the canopy or slightly wider, which moves as a unit as the tree sways and provides much of the anchoring and support for the tree. The root plate is made more solid and heavier still by the inclusion of rocks trapped between expanding, grafted roots. Darwin noted on the voyage of the *Beagle* that the inhabitants of the Radack Archipelago in the Pacific sharpened their tools using stones prised from the roots of trees washed up on their beaches.

(a)

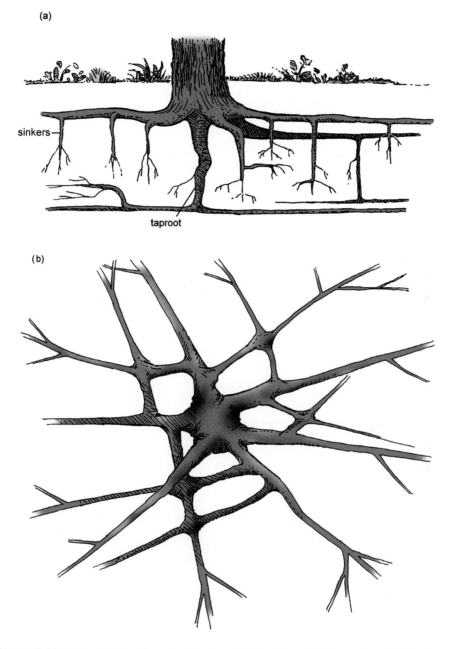

Figure 4.2 Diagrammatic view of a root system from (a) the side and (b) the underneath, showing a range of possible features.

Lateral woody roots tend to grow longer, and to a certain extent thicker, and be more numerous heading into the direction of prevailing wind. Roots on the leeward side also tend to be plentiful and thick to better resist the bending at the root plate. In response to wind, the shallow, horizontal roots develop an

(a)

(b)

Figure 4.3 (a) The very thin root plate of a fallen beech tree (*Fagus sylvatica*) lifted out of the ground (Keele University, England) (b) a fallen maple (*Acer* sp.) showing a heart-root system made up of a number of laterals growing obliquely downwards (Abbotsbury Subtropical Gardens, Dorset, England).

I-beam or T-beam in cross-section to counter the bending caused by the trunk rocking, especially near the stump.

## Deeper roots

In temperate trees, 80–90% of roots are in the top 60 cm of the soil and 90–99% are in the top 1 m. In deep soils, however, trees gain extra support and access to deep water supplies from a few deeper roots, normally in the form of sinker roots (also called striker or dropper roots) which normally grow down from the lateral roots within a metre or two of the trunk (creating a 'dimorphic' root system; Figure 4.2a). Sinker roots can equal the length and thickness of the tap root, both of which grow down until they meet either an obstruction, the water table or low oxygen concentrations. In temperate and tropical soils this leads to roots penetrating normally 1–2 m, and up to 3–7 m on well-drained soils (which debunks the myth that all tropical trees have shallow roots). For example, beech (*Fagus sylvatica*) in Germany will normally reach no more than 1.4–1.7 m down, but on very dry sandy soils, with more oxygen lower down, roots can go down 3.5 m. sessile oak (*Quercus petraea*) goes somewhat deeper and is less sensitive than beech to flooding or soil compaction (Leuschner *et al.* 2001). In temperate areas, the deepest roots may be 4–5 m down in deciduous trees and 7–8 m in Mediterranean conifers (Körner 2005), but these are exceptional figures. Upon reaching their lowest depth the roots either grow horizontally to form a lower layer of horizontal roots (but generally not as vigorous or as branched as those near the surface) or they repeatedly branch to produce a bushy end like a broom (Figure 4.2a). To make the pattern more complicated still, in red oak (*Quercus rubra*) horizontal roots may grow obliquely upwards part or all the way back to the surface before again growing horizontally. These lower roots provide a safety-net in helping to catch and recycle nutrients washed down below the normal level of roots that would otherwise be lost to the trees and other plants. Thus on nutrient-poor heathland in the UK, birch trees help to keep the soil less acidic by catching nutrients with their deeper roots and dropping these nutrients back onto the soil surface in their leaf litter.

These sinker roots are also important in helping the tree to stand up, since on the windward side of a tree they help pin the root plate to the ground. In some cases in trees having a large tap root and numerous sinkers from the lateral roots, this results in a rigid 'root cage' rather than a root plate. This holds large amounts of soil and rocks which are guyed by long horizontal lateral roots near the surface. This is seen in the European maritime pine (*Pinus pinaster*), where the cage comprises 70% of the root volume.

Jiri Hruska and colleagues (1999) mapped the large roots of a stand of 50-year-old, 18 m high oaks (*Quercus petraea*) in the Czech Republic. They worked within a 6 by 6 m area, detecting the roots using ground-penetrating radar that could identify roots more than 3–4 cm diameter. They found on average 6.5 m of these thick roots below each square metre of ground. Figure 4.4a shows how interwoven these roots were, and Figure 4.4b shows the root system of just one of the trees, a large oak 100 years old, 17.6 m tall with a trunk 35 cm in diameter and with a timber volume of 0.939 m³. This tree had 82 m of large roots that were between 0.2 and 2.2 m deep (this tree is on a dry ridge with deep soil which explains why these roots go deeper than is normal for oak). The root density was 2.32 m below each square metre. The root plate radius was 3.35 m (compared to the canopy radius of 2.7 m (that is 125% of the crown radius), covering a total of 35.3 m², compared to the crown area of 22.5 m² (157% of crown cover). The estimated volume of soil containing these large roots was 70.6 m³ compared to a crown volume of 117 m³ (60% of the crown volume).

Some trees are more capable than others of modifying their inherent growth pattern to suit soil conditions. For example, pines, which tend to have a long tap root and good laterals, can still grow on a range of shallow soils. Similarly, some willows and spruces, which have fibrous root systems and are normally at home on shallow or waterlogged soil, can root deeply on dry soils. In contrast, other trees including silver fir (*Abies alba*), sycamore (*Acer pseudoplatanus*) and oaks depend strongly on a deep tap root from which laterals are produced and do not do well on shallow soils. That is not to say that they are not capable of any modifications since droughted sycamore seedlings will grow deeper roots, but their plasticity is rather limited.

Although most trees do not root more deeply than a metre or two, there are many examples around the world of dry habitats where roots have been found to go much deeper to reach water: 12 m in an acacia unearthed in the excavation of the Suez Canal; up to 13 m for the tap root of cork oak (*Quercus suber*), 15 m in pines on the western slopes of the Sierras, USA, and 10–30 m in eucalypts. Deeper still were the roots of a mesquite bush (*Prosopis* spp.) found 53 m down in a gravel bed of an open-pit mine near Tucson, Arizona. At first the roots were thought to be fossilised and only after they were carbon dated were they found to be less than 6 years old! In the deep sands of the Kalahari in Botswana, roots of a local tree, *Boscia albitrunca*, were found more than 68 m down a borehole (with the water table at 141 m). And fig roots in South Africa have been reported at 120 m depth (400 ft). In these cases the roots grow deeply simply because in the easily penetrated, dry soils they physically can and this is where they can find water. In such extreme cases, where water is available primarily from deep water tables, the sinker and tap roots may replace the laterals as the main roots, producing the mythical rooting pattern

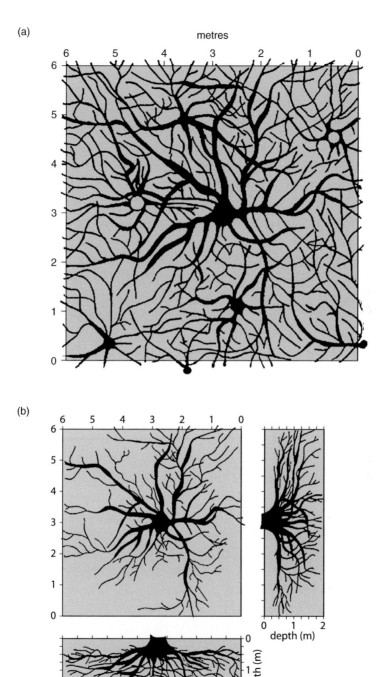

**Figure 4.4** Depth and spread of roots of an oak (*Quercus petraea*) in the Czech Republic under a 6 × 6 m square; (a) shows all the roots in the square while (b) depicts just the roots of the central oak tree. Redrawn from: Hruska *et al.* (1999) Mapping tree root systems with ground-penetrating radar. *Tree Physiology*, 19, 125–130, by permission of Oxford University Press.

expected in most trees: these trees are referred to as 'phreatophytes', plants using very long roots to reach the water table. The dependence of 'normal' trees on the water table is discussed in Box 4.2.

## Beyond the root plate

After a wind storm, leaning or fallen trees can be seen with the root plate lifted from the ground (Figure 4.3). This has undoubtedly led to the impression in many people's minds that tree roots spread only as far as the branches. Yet the lateral roots do not end at the edge of the root plate. Rather, these thin, and thus easily snapped, roots will explore the soil around the tree often for considerable distances. They appear unimportant: they are thin and rope-like (2–5 cm in diameter) with little taper but they can represent over half the mass and surface area of all the roots. Their role is not so much to hold the tree up (although they do act as important guy ropes helping to hold down the root plate in high winds), but to search for water. At the edge of the central root system, the lateral roots are already typically 1–2 m apart. To help fill the void between them as they spread out like the spokes of a wheel, they repeatedly fork. Forking seems to be primarily in response to injury or when the tip is forced to grow around an object: roots meeting a rock commonly fork either side of it. This may seem a very chancy process but is apparently reliable enough to work.

### How far do roots spread?

In temperate trees the total spread away from the trunk is usually 2–3 times the radius of the canopy, and even up to four times the radius on dry sandy soils. Alternatively, root spread can be estimated as 1½-2 times the height of the tree. To put this into meaningful terms, the roots of an oak, poplar or red maple (*Acer rubrum*) can reach up to 30 m away from the trunk (you should try pacing this out to appreciate how far it is). Underlying this is an inherent difference between trees. Some species such as European ash (*Fraxinus excelsior*) and the American southern magnolia (*Magnolia grandiflora*) have long, fast-growing, moderately branched 'pioneer' roots designed to exploit a large volume of soil; roots may be found up to four times the canopy radius away from the trunk. Others such as the European beech (*Fagus sylvatica*) and the American green ash (*Fraxinus pennsylvanica*) have short, slow-growing roots with many branches to utilise a small volume more effectively. This may explain why beech suffers in droughts; it uses up the available water and cannot exploit new areas of soil quickly enough. But beech can be very competitive in gaining water in non-drought conditions. In a southern German woodland, it was found that oak and beech had roughly the same amount of fine roots when grown on their own, but in mixed stands, oak produced far fewer roots and beech had 4–5 times more fine roots than the oak even though both trees had similar stem densities and leaf areas. Also, the beech fine roots were more common outside of its own canopy where nutrient-rich litter had built up while the oak roots were equally abundant in all directions from the trunk (see Leuschner *et al.* 2001 for more detail). Such large variations in root abundance through competition have also been seen between pine and eucalyptus, and maples and pines, and doubtless occur in many mixed stands of trees.

## …and how much damage can roots do?

Given that roots spread so far, and knowing that they can cleave rocks, should we be growing trees close to our homes? Fortunately, roots are only comparatively rarely implicated in damaging buildings. Direct damage to buildings can occur by the expanding girth of the trunk and thick bases of the lateral roots. But once away from the trunk, roots do not thicken appreciably and tend to grow around obstacles given the option. These roots can sometimes still lift lighter structures such as paths and driveways but this is most likely by those trees with an abundance of surface roots such as ash, cherry, birch and pine.

Roots are not very good at penetrating compact or hard substrates and are therefore easily stopped unless they can follow a line of weakness such as an old root channel or a crack in your house foundations. Once in, though, they can exert considerable sideways force as the roots become woody. This can be enough to split rocks along bedding planes and displace masonry in an impressive show of force. The take-home message is that providing your house foundations are sound, tree roots should cause you few problems. Unless your house is on clay, that is.

A common problem is building subsidence due to shrinkage of clay soils caused by trees extracting water. In the UK this is particularly bad on the very shrinkable London clay and, to a lesser extent, on the medium shrinkage Oxford clays. In a survey in southeast England, Cutler and Richardson (1981) found that on London clays oaks and poplars could damage buildings 30 m away from the trees, and willows 40 m away (Box 4.1). Oaks and poplars were particularly menacing because they caused damage out of all proportion to their abundance. Other large trees such as ash, elm, lime, maple and horse chestnut were found to have damaged buildings more than 20 m distant. Even *Prunus* and *Sorbus* species and fruit trees (all in the family of Rosaceae) could cause subsidence over 10 m away. The good news, if you live on similar clay soils, is that the roots of the increasingly common cypresses, including Leyland cypress (X *Cupressocyparis leylandii*), spread a long way (20+ m) but 90% of the recorded damage was within 5 m of the tree. Felling a large tree on clay

**Box 4.1** Distances over which trees have been seen to cause damage to buildings by subsidence on predominantly clay soils in southeast England

| Common name | Scientific name | Maximum tree-to-damage distance (m) | Distance within which 90% of damage cases were found (m) | No. of trees |
|---|---|---|---|---|
| Willow | *Salix* | 40 | 18 | 124 |
| Oak | *Quercus* | 30 | 18 | 293 |
| Poplar | *Populus* | 30 | 20 | 191 |
| Elm | *Ulmus* | 25 | 19 | 70 |
| Horse chestnut | *Aesculus* | 23 | 15 | 63 |
| Ash | *Fraxinus* | 21 | 13 | 145 |
| Lime | *Tilia* | 20 | 11 | 238 |
| Maple | *Acer* | 20 | 12 | 135 |
| Cypresses | *Cupressus & Chamaecyparis* | 20 | 5 | 31 |
| Hornbeam | *Carpinus* | 17 | – | 8 |
| Plane | *Platanus* | 15 | 10 | 327 |
| Beech | *Fagus* | 15 | 11 | 23 |
| False acacia | *Robinia* | 12 | 11 | 20 |
| Hawthorn | *Crataegus* | 12 | 9 | 65 |
| Rowan + Whitebeam | *Sorbus* | 11 | 10 | 32 |
| Cherries, etc. | *Prunus* | 11 | 8 | 144 |
| Birch | *Betula* | 10 | 8 | 35 |
| Elder | *Sambucus nigra* | 8 | – | 13 |
| Walnut | *Juglans regia* | 8 | – | 3 |

| Box 4.1 (cont) | | | | |
|---|---|---|---|---|
| Common name | Scientific name | Maximum tree-to-damage distance (m) | Distance within which 90% of damage cases were found (m) | No. of trees |
| Laburnum | *Laburnum* | 7 | – | 7 |
| Fig | *Ficus carica* | 5 | – | 3 |
| Lilac | *Syringia vulgaris* | 4 | – | 9 |

These are the results of a survey from 1971 to 1979 conducted by the Royal Botanical Gardens, Kew (Cutler & Richardson 1981).
Dashes indicate that data are not available.

soils can be equally damaging as the water previously used by the tree causes the clay to expand causing heave.

Subsidence problems can be complicated by the pattern of water uptake by roots during drought. Glenda Jones and colleagues (Jones *et al.* 2009) at Keele University looked at water uptake from the soil around a willow (*Salix* sp.) and an oak (*Quercus robur*) growing on highly shrinkable London clay in southeast England. They found that the high density of roots up to twice the width of the canopy tended to keep the soil fairly dry, so changes in ground level were restricted to 1.0–1.5 cm. Movement was most pronounced at the edge of the canopy along the 'drip line' presumably due to a higher concentration of roots using the water running off the outside of the canopy. Nevertheless, shrinkage tended to be fairly slow and consistent over large areas; the whole area went gently up and down, more or less as a single lump. Outside of this, however, up to around three times the width of the canopy, was a zone that filled rapidly with water during winter and which was equally rapidly emptied by root extension as a response to drought, especially at the extreme limit of their spread. This resulted in 3–6 cm of vertical movement of the soil over short distances of 3–5 m, which would have potentially disastrous effects on a building. So, paradoxically, problems of shrinkage were more acute 15 m or more away from the tree and less acute near to the canopy.

Perhaps the most widespread problem caused by tree roots is the penetration and clogging of drains and sewers, which, being warm and wet, must be root paradise! And they can be big: one champion willow root taken from a storm sewer in Utah by RotoRooter Corporation was 35 m long (Figure 4.5). Roots will proliferate around pipes that have condensation on the outside or are leaking,

Figure 4.5 An illustration of how long roots can grow in a storm sewer. This willow root (35 m long) was shown in the RotoRooter Corporation magazine in 1954.

and where they find an opening or crack they waste no time in taking advantage of it. Poplar, willow, horse chestnut and sycamore tend to be particularly invasive.

The paradox of these strong yet weak roots is often seen in planted trees. Smooth compacted sides in a dug hole, especially in clay soils, cause the new lateral roots to grow around the inside of the hole producing 'girdling roots' that wrap themselves around and around the trunk (especially prominent in maples). The roots may never escape from the planting hole (with implications for stability and water uptake) and if the tree survives long enough, the girdling roots can restrict the expansion of the trunk and seriously hinder growth. The same problem can be caused by roots growing around the inside of a container; once they start they tend to continue even when planted out. Root pruning can cause similar problems when new lateral branches (at right angles to the original root) grow tangentially across the trunk. In commercial conifer growing, the small containers used have vertical ridges to break the pattern, or they can be coated with copper sulphate; when the roots reach it they stop growing. Other innovations include Air-Pots® (Figure 4.6) which have a textured surface like an egg carton so there are no flat surfaces to deflect roots, and the pot surface is punctured with holes so the outermost roots tend to dehydrate and 'air prune'. It has also been seen that there are fewer girdling roots when using white-coloured pots because the high light intensity inside the pot stops the roots growing as they reach the edge of the pot (Figure 4.7).

(a)

(b)

Figure 4.6 Air-Pots®. Trees grown in these do not suffer from girdling roots that grow around the inside of a normal pot. (a) Shows a dawn redwood (*Metasequoia glyptostroboides*) removed from an Air-Pot and (b) shows the open plastic bottom that also prevents roots building up at the base. When the tree is planted out, the root ball should start growing roots directly out through the soil. Photographs by (a) Andy Hirons, (b) Natt Hazlewood and Vicki Southern.

Figure 4.7 A White Pot™ used by Barcham Tree Nursery, Cambridgeshire to prevent root girdling. The piping above the pot is for irrigation.

Tree roots can in theory be stopped by various permeable barriers made of tightly woven plastic. When buried in the soil, between, say, a tree and a driveway, these barriers allow through water but not roots, although roots are known to grow down and under such barriers.

## Fine roots

The large woody roots of the root plate and outer root systems dictate the overall size and shape of the whole root system, but it is the plethora of increasingly fine roots that provide the really intimate contact with the soil. These fine roots are often called 'feeding' roots but it is better to avoid this term. Tree 'food' is manufactured by the leaves as sugar; roots take up water and minerals which are used by the tree in conjunction with these sugars to live and grow.

From the radiating lateral roots, small-diameter woody side branches grow outwards and upwards, often reaching several metres long, branching four or more times to end in fans of short, fine non-woody roots. In red oak (*Quercus rubra*) there may be as many as 10–20 of these branches along a metre of lateral root within the root plate area, dropping to one branch every 1–5 m in the outer root system. Side roots tend to stay thin, seldom over 4–6 mm diameter in red oak. They are there to hold the finest roots, not to hold the tree up.

The finest roots are typically 1–2 mm long and 0.2–1.0 mm in diameter but can be down to 0.07 mm in diameter. Since they are easily broken with a trowel they usually go largely unnoticed. The painstaking work of Walter Lyford (1980) sheds light on the impact of such fine roots. He dissected 1 cm³ cubes of soil taken a few centimetres down into the floor of a red oak stand in Harvard Forest, Massachusetts. In these small cubes he found an average of 1000 root tips, more than 2.5 m of root with a surface area of 6 cm² (six times the area of the top of the cube) not counting mycorrhizas (see below) or root hairs. Although this may sound as if the soil must have been solid root, most of the roots were so fine that they actually filled only 3% of the cube volume. Scale this up and a mature red oak may have 500 million live root tips. Reynolds (1975) estimated that in Douglas fir, half the total length of all the roots were less than 0.5 mm thick and 95% less than 1 mm.

We have already seen that tree roots as a whole tend to be near the surface of the soil. In typical clay-loam soils, up to 99% of roots are in the top 1 m and most are less than 20–30 cm below the surface. But if we consider just fine roots, the shallowness is even more impressive. In a North Carolina oak forest 90% of the mass of roots less than 2.5 mm diameter were in the top 10–13 cm of soil. Similarly, in an acidic beech forest of Germany it has been calculated that 67% of roots less than 2 mm in diameter were in the top 5 cm and 74% in the top 7 cm. In tropical trees the pattern is less clear but fine roots are probably even more strictly limited to the upper 5 cm of soil and often form a mat of roots over the soil surface. In Costa Rica, for example, this root mat can contain a third to a half of all the fine roots.

In a woodland situation these very fine roots are actually growing high up in the leaf litter, often in the moist compacted litter below the last leaf fall. Here the very small roots ease their way between the layers of leaves (sometimes helped by dissolving a path), branching and spreading in the same plane as the laminated leaf litter. It follows, therefore, that in woodlands, fields and gardens the finest roots are amongst the grass and herb roots, competing side by side for water and minerals. Thus the common view of tree roots being large and deep needs to be replaced with a picture of a stout central framework with radiating rope-like roots complemented by an incredibly delicate system of small woody and non-woody roots very close to the soil surface.

Having said that fine roots are very shallow, it is worth noting that some will always be found deeper, associated with the structural roots. To use the Costa Rica example again, around 5% of the fine root mass and 13% of all roots can be found 85–185 cm deep into the soil. The fact that there are some deep fine roots has been utilised in agroforestry where trees and crops are grown together with the assumption that the crops are using water near the soil surface and the trees are using deeper water, maybe 20–30 cm down. It is certainly true that in dry conditions, trees can often utilise water deeper in the soil than other plants. For example, in the Sierra Nevada Mountains of California, Rose *et al.* (2003) found that during the dry summer period Jeffrey pine (*Pinus jeffreyi*) used water over 3 m below the surface, deeper than the shrubs it was growing with. However, the assumption that trees *just* use deeper water is undoubtedly not true since even deep rooted trees will also have some shallow roots, and while the trees may exploit a greater volume of soils this includes the surface layers where the crops are rooted. Many trees get between 30% and 40% of their yearly water needs from the top 0.3 m of soil, using progressively deeper water once the surface is fairly dry. Certainly, crop yield is normally lower near the trees. This can be overcome to a degree by root pruning the trees by digging a trench between the trees and crops.

## Fine roots and tree health

The fact that fine roots are close to the soil surface has several implications for looking after trees. When fertilising trees there is no real need to dig pits or use 'tree spikes'; research has found no difference in the response of trees to fertiliser placed in holes versus that scattered on the surface. The roots of a tree and grass are competing in the same space for the same things, which explains why grass and weed control around newly planted trees is so important (a grass-free circle 2 m in diameter can dramatically increase root development in a sapling). In the same way, when we spread herbicides on our lawns or on paths and driveways (where it seeps into joints and cracks) we can also be killing tree roots. This may not seem too important but Figure 4.8 shows the roots spreading from a field maple 10.5 m tall and 40 cm diameter growing in a 1 m wide lawn with a tarmac road to one side and a path and lawn on the other. There were very few roots that extended below the tarmac of the road, and most of the structural and fine roots were in the front lawn because that's where most water and nutrients were available. Herbicide on that lawn could be devastating to the tree. How many gardeners would worry about killing an oak tree 15 m away? And yet, as we have seen, the roots can spread over 30 m. For a similar reason, digging in the garden is another common cause of tree root death. You might ask at this point how lone trees can then survive in the middle of a ploughed field. The answer is that with repeated disturbance the roots proliferate below the level reached by the plough or spade. The problem comes when an area is dug or ploughed for the first time damaging a high proportion of the roots.

Figure 4.8 raises several other potential problems for urban trees. As can be seen, the spread of roots can be seriously affected by the impervious surfaces of roads and courtyards. Open soil can be no better when it is compacted by feet and pigeons are a prime offender since their small feet exert more pressure per square centimetre than heavy machinery. Compaction seals the surface

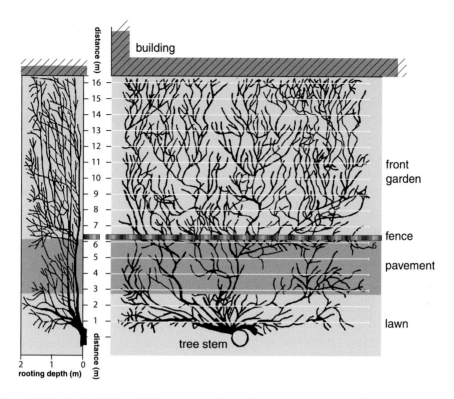

**Figure 4.8** Roots of a field maple (*Acer campestre*) growing on a grassy verge with a road on one side and a pavement and garden on the other. Taken from a tree in the Czech Republic. Redrawn from: Čermák, J., Hruška, J., Martinková, M. & Prax, A. (2000) Urban tree root systems and their survival near houses analyzed using ground-penetrating radar and sap flow techniques. *Plant and Soil*, 219, 103–116. With kind permission from Springer Science and Business Media.

preventing gas exchange and water penetration, and makes root penetration impossible. Fortunately, many trees can brave the low oxygen and water content beneath these barriers, pushing roots through to more favourable areas (trees such as cherries are known for producing root suckers on opposite sides of roads) or following the cracks between paving stones. Others, exemplified by the Japanese katsura tree (*Cercidiphyllum japonicum*), cannot apparently tolerate a paved surface over their roots, presumably because this must restrict root spread with obvious consequences in a drought or high wind. Perhaps the biggest problem comes when the area around a previously free-growing tree is paved: many of the existing roots die as oxygen is consumed in the soil, leading to an uncertain period when the life of the tree is in the balance before new roots can grow into new areas rich in oxygen and moisture.

Similarly, if the soil level is raised around established trees either as part of landscaping (or naturally by the addition of silt by flooding) ill health and death can follow. Sinker and deep horizontal roots are likely to be starved of oxygen but, perhaps more significantly, the fine roots end up below the zone where most water, minerals and oxygen are available unless the added material is very porous. If you like to mulch your garden you'll be pleased to know that even 50 cm of wood chips is sufficiently porous to cause little problem for tree roots. As is usually the case, some trees will have a solution to such problems: willows (*Salix* spp.) can cope with extra soil or silt by producing new roots from the buried stem, just below the new surface, within days of burial.

Underground utilities can be devastating for tree health. Cables and pipes are usually buried deeper than the roots so the trouble comes in putting them in and repairing them. A trench dug along the pavement in Figure 4.8 would neatly sever most of the main roots of the tree making it much more likely to die of drought or be blown over. So why don't we see rows of dead trees along streets if trenching is that bad? The answer is that sick urban trees are usually removed branch by branch or are felled long before they get to the stage of being an obvious danger. We usually notice the problem when we realise that the trees are missing. But there is plenty of advice available to help our beleaguered urban trees by sympathetic handling as you can see in Figure 4.9. This is increasingly reinforced by codes of practice such as the British Standard on *Trees in Relation to Construction – Recommendations* (BS5837) in the UK.

## Root loss and death

Root loss raises the question of how many roots can be lost before a tree suffers. This obviously depends upon several factors but as a general rule many practitioners would consider the risk to be small if roots are cut off beyond the edge of one side of the canopy. A tangential straight-line cut along the edge of the canopy would cut off about 15% of roots. If the straight-line cut is made midway between the edge of the canopy and the trunk then around 30% of roots will be severed and trees of reasonable health, with roots previously unhindered in any direction, should be able to survive even this. In practice, 50% of roots can sometimes be removed with little problem provided there are vigorous roots elsewhere. A general rule of thumb for root protection is to allow no disturbance (such as adding or removing soil, or soil compaction from heavy machinery) under the canopy of the tree (sometimes referred to as not going beyond the 'drip line' of the canopy). This is why tree wells (empty areas left around the trunk when soil is added for landscaping or building) that are less than a metre wide around the trunk are useless. More specifically, the British Standard mentioned above recommends that the root protection zone under a tree should be a circle with a radius 12 times the diameter of the tree (measured at 1.5 m above ground) for single stems and 10 times for a tree with more than one stem from near the base. This is up to a maximum of 15 m

**1**

**THE LIFE OF THE TREE IS IN YOUR HANDS**
Even the biggest trees have nearly all their roots just below the surface: they're the trees' life support system. If the roots get chopped, the tree will be damaged and may **die!**

**2**

**PLAN WELL AHEAD**
Street trees need special treatment- cable trenching should be planned well ahead.

**3**

**NO MACHINERY**
Mechanical diggers, power drills and slab cutters all damage roots without you noticing. Don't use machines where there are branches overhead.

**4**

**DIG CAREFULLY BY HAND**
Take care with roots thicker than your thumb. Don't break or scuff the roots. Don't let the roots dry out: cover them with damp sacking or spray with water.

**5**

**SLIDE THE DUCT UNDERNEATH THE ROOTS**
Shallow cables may need to be buried deeper than normal.

**6**

**BACKFILL CAREFULLY**
Use sand as backfill around the roots. Hand tamp around and between the roots. Don't use a whacker, except on the very top layer of tarmac. If you're not sure what to do, ask the boss!

**Figure 4.9** A guide to cable trenching along streets with trees, produced by the Black Country Urban Forestry Unit, West Midlands, England.

radius. So a single tree 30 cm in diameter should have a root protection zone of 3.6 m radius (equal to an area of 41 m$^2$). And indeed this normally takes the root protection zone outside of the canopy. The British Standard does allow the circle around the tree to be a different shape if protecting a lopsided canopy or a particular rooting area, providing it's the same equivalent area.

(a)

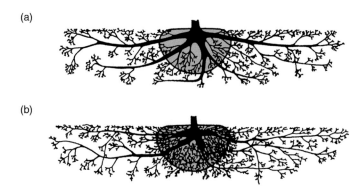

(b)

Figure 4.10 Root pruning and its effect on the production of new roots: (a) an unpruned tree with few roots in the root ball, and (b) a pruned tree that has now grown many fine roots within the root ball. From: Watson G.W. (1986) Cultural practices can influence root development for better transplanting success. *Journal of Environmental Horticulture*, 4, 32–34.

Inevitably a large amount of root loss will temporarily slow canopy growth and even lead to some dieback. Youth plays its part: younger trees can stand more loss. Seedlings can (and do) lose as many as 90–98% of roots when transplanted as bare-rooted stock (i.e. with only the larger woody roots intact) and still survive. A simple rule of thumb is that when transplanting, a tree should have a 12 inch (30 cm) diameter root ball for each 1 inch (2.5 cm) of trunk diameter.

The odds of a transplanted tree surviving are increased if the nurseryman has encouraged a dense bushy root ball by root pruning: cutting through far-reaching roots a short way from the tree 6–10 weeks before the tree is moved. Once a main woody root has been cut, new branches are produced just behind the damage over the next few months. A root 3–5 cm in diameter can produce up to 30 new tips (Figure 4.10). This is not without its problems. The majority of the new roots which grow from cut root-ends can be deeper than most roots would normally grow and can cause a 'deep root system', leading to poor growth or death in transplanted trees (Watson 2012). When the root ball is dug up it should ideally be slightly bigger than the root-pruned ball to take as many of the new roots as possible. However good the transplanting, long-term survival is dependent upon growth of new fine roots once the tree is replanted. Temperate trees are best planted in the autumn and early winter to stimulate new fine root development without a great demand for water from the shoot. Since evergreen trees are usually good at controlling water loss (see Chapter 2) they can better tolerate spring planting.

## How much soil is needed?

Trees planted in holes are often just as much in a pot as are trees planted into containers, either because of the compacted hole sides (see above) or because of surrounding buildings and roads. Just how much soil do we need to give them? A general rule of thumb in temperate areas (Europe and the USA) is that a tree requires about 1/2–3/4 cubic metres of soil per square metre crown projection (the shadow cast by an overhead sun). This translates into around $5\,m^3$ of soil for a medium-sized tree and up to $60$–$150\,m^3$ for a big tree. In effect the smaller the amount of soil, the more likely that the tree will suffer drought. Such a tree will grow more slowly than an open planted sibling because the roots are restricted (as discussed at the beginning of the chapter). But it will grow and as the canopy gets bigger and needs more water, and the roots can't enlarge, it will grow closer and closer to the edge of drought and disaster.

## Is the water table needed?

Trees will grow in restricted soil volumes but tend to pay the price of slow growth and early death. Moreover, as we saw in Chapter 2, trees can use prodigious quantities of water: hundreds of litres per day. These facts usually lead people to think that, for trees to grow really big, they need access to the abundant supplies of the water table. Some do: in arid areas, the water table is an important source of water and trees here tend to have a dual root system with shallow roots to soak up rain and a deeper set of roots to utilise the water table. This includes the tamarisks (*Tamarix* spp.) around the world which grow in salty areas. But these trees are the exception: most trees are dependent upon just rainfall for their water needs.

The empirical evidence can be seen in trees growing at the edge of tall cliffs where the water table is completely out of reach. Moreover, trees will suffer in dry years just like any other plant but seldom suffer if the water table is dropped as is all too common through borehole extraction of drinking water. A more calculated analysis that the amount of water held in soil is sufficient for most trees is given in Box 4.2.

### *Hydraulic lifting*

Although most trees do not utilise water from the water table, they do have access to deeper supplies of water than most herbaceous plants, as discussed above under agroforestry. But it's more than just having access, it's what the trees do with the water that's interesting. At least 60 different trees and shrubs, both temperate and tropical, absorb water from around their deeper roots which is passed into the shallow roots and even out into the soil around them. This happens mostly at night (but sometimes during overcast days) when there is little suction of water up through the trunk (Figure 4.11a). The roots appear to act like passive wicks helping water seep from wet to dry soil through the roots' internal plumbing. That the process is passive is demonstrated by it

> **Box 4.2** Can a tree in temperate areas get enough water from the soil without using the water table?
>
> Assume that a modest temperate tree (12 m high with a canopy of 4 m radius) is using 40 000 litres of water per year. Over a growing season of 6 months/180 days this means 220 litres per day.
>
> Root spread is likely to be something over 8 m in diameter. Let's choose 8.4 m to keep the sums easy! A circle of 8.4 m radius gives a rooting area of 220 m$^2$ ($\pi r^2$).
>
> In a day the tree therefore needs 220 l from 220 m$^2$ of soil or 1 l from 1 m$^2$
> 1 l is 1000 cm$^3$ of water and 1 m$^2$ is 10 000 cm$^2$
> So 1 l from 1 m$^2$ becomes:
> 1000 cm$^3$ from 10 000 cm$^2$
> = 0.1 cm$^3$ from each cm$^2$ of soil surface.
>
> This is equal to 0.1 cm (=1 mm) height of water from each square centimetre of soil each day during the growing season. Over the 180-day growing season the tree needs 180 mm of water.
>
> This is the right order of magnitude. A whole forest uses 2–3 mm per day and individual temperate trees in the open (which will use most water) normally use 1–5 mm per day. (For comparison, a really thirsty tree like a young eucalypt in Australia would need maybe 700 mm per year.)
>
> The average rainfall in Britain is 600 mm of rain. Not all of this will be available to the roots (some is intercepted and evaporated from the canopy and other plants will root in the same area) but there is more than enough to supply the tree.
>
> A good loam can hold 130–195 mm of rain in 1 m depth (enough water for the entire growing season) and sandy/gravel soil may hold nearer 50 mm (nearly 2 months' worth for the tree if we can assume that this is all available to the tree). We undoubtedly can't assume that but trees can have access to water in even deeper soil layers. Accepting these figures even within an order of magnitude does illustrate that temperate trees do not normally need the water table.

not happening when the soil near the surface is wet from rain. In reality this should be called hydraulic 'redistribution' rather than lifting since water can go down as well as up. When the soil near the surface is wet from rain, water can be carried down through the roots to recharge the deeper layers (Figure 4.11c). In a study using various eucalyptus species in Western Australia

(a)    (b)    (c)

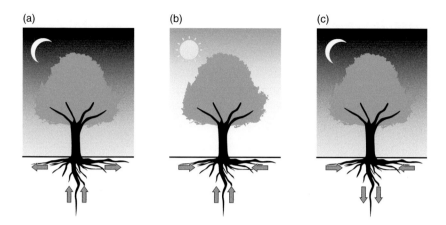

Figure 4.11 Hydraulic lifting in a hypothetical tree. In dry times (a) water is lifted from moist areas deep in the soil at night and it escapes into the dry shallow soil. During the next day (b) the tree takes up water from the remoistened shallow soil and from deep areas. Following rain (c) water moves in the opposite direction at night from wet surface layers to deep in the soil. Based on Lee, J.-E., Oliveira, R.S., Dawson, T.E. & Fung, I. (2005) Root functioning modifies seasonal climate. *PNAS*, 102, 17576–17581.

the tap root readily carried water down to 2.6 m below the saturated surface. Water can also be moved from roots on one side of the tree to those on the other side, passing through the bottom few metres of the stem.

The benefits are obvious; water is brought up during the night ready for use the next day without the extra suction needed to bring it from deep in the soil at the time when it's needed (Figure 4.11b). Water movement can amount to 80–100 l per night. Depending on the size of the tree and how much water it is transpiring in a day, the hydraulically lifted water can be anything from 15% to 80% or more of the tree's daily water needs, especially in dry conditions. This can make a huge difference to tree survival during hot dry summers and periods of drought. It may also be beneficial in helping keep fine roots and their valuable mycorrhizas (see below) alive in a drought. In turn the mycorrhizas may help transfer hydraulically lifted water to vulnerable seedlings helping them survive. There is some evidence, however, that hydraulic lifting is not always beneficial. For example, in the parts of the Amazon forest that experience drought, the ready movement of water from deeper soil layers during the early stages of a drought can use up available water quickly and cause problems later. This is rather like having a larder full of food that needs to last a month but eating everything within the first week because it's easily done, and then starving, and possibly dying. Nevertheless, this may not be true of temperate trees which, as discussed under agroforestry above, tend to use water intensively in the top 0.3 m of the soil (aided by hydraulic lifting) until this is too dry to supply more, and then to use progressively deeper supplies.

# Further increasing water and nutrient uptake

There is a limit to how many roots a tree can produce because of the cost of producing and maintaining them. The way around this is to grow roots where they can absorb most and to make better use of the ones already there. Trees, in common with other plants, have come up with several ways of increasing the surface area of roots.

## Canopy roots

Rainforest trees (both tropical and temperate) are often festooned with thick mats of epiphytes (orchids, bromeliads, mosses, etc., which hitch a ride on the tree but take nothing from it), which are very good at extracting nutrients from the atmosphere and from water running down the host trunk. Some of the festooned trees produce 'canopy roots' from the trunk and branches to exploit the high nutrient content of the humus produced in these mats and caught in branch forks and hollows. One tree of the Australian rainforest (*Ceratopetalum virchowii*) has canopy roots in clumps or enveloping the trunk like a fibrous coat but which do not root in the soil. These roots are able to recover nutrients from water running down the trunk that have been leached out higher up the canopy. The red mangrove (*Rhizophora mangle*) does the same sort of thing with sponges that grow on its prop roots: it grows a mass of fibrous roots into the live sponges. But in this case both partners benefit. The mangrove gets nitrogen from the sponges and they in return are given carbohydrate from the roots. Canopy roots are also found in a number of temperate trees as well, including the European beech (*Fagus sylvatica*). These are rooted into pockets of humus that accumulate in branch junctions or rotting stems but, in this case, probably don't produce that much extra nutrition for the tree.

## Root hairs

Root hairs are outgrowths of root cells behind the apex which greatly increase the root's intimate contact with the soil (Figure 4.12) and anchor the end of the root helping it to grow forward through the soil. In red oak (*Quercus rubra*) they are just 0.25 mm long but can vary from 0.1 mm in apple (*Malus* spp.) to 1.0 mm in blackcurrant (*Ribes* spp.). Most live for only a few hours, days or weeks, being replaced by new hairs as the growing tip elongates. Some trees, including honey-locust (*Gleditsia triacanthos*) and the Kentucky coffee-tree (*Gymnocladus dioica*), retain root hairs for months to years but such persistent root hairs appear relatively inefficient in absorption. On the other hand, despite their usefulness, root hairs are comparatively lacking in many trees and may be totally absent from some conifers and at times from pecan (*Carya illinoensis*) and avocado (*Persea americana*). This is tied up with mycorrhizas.

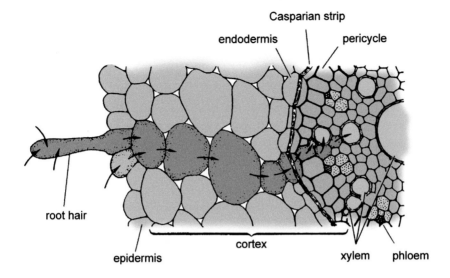

Figure 4.12 A root cross-section showing the passage of water and dissolved minerals from a root hair through the Casparian strip (a waterproof cylinder) into the xylem. From: Esau, K. (1960) *Plant Anatomy*. Wiley, New York.

## *Mycorrhizas*

A mycorrhiza is an association between a root and a fungus which works to the benefit of both (symbiosis). The fungi can also be referred to as endophytic fungi as they are partly inside the plant; *endo* within, *phyto* plant. Usually the tree gets extra water and hard-to-get nutrients (particularly phosphorus but also nitrogen and other nutrients), and perhaps some protection from fungal diseases and soil toxins such as heavy metal pollution. In return the fungus gets carbohydrates and other products from the tree. The fungus appears to work not by having access to different supplies of nutrients but by permeating more widely than the roots and by having a more intimate connection to soil particles than root hairs can achieve. Many trees, including beeches, oaks and pines, require the association to prosper, for others such as maples and birches it is not essential, while yet others, notably members of the Proteaceae (one of the most prominent families of the southern hemisphere encompassing proteas, banksias, grevilleas and the macadamia nut, *Macadamia integrifolia*), rarely, if ever, form mycorrhizas. But it's a truism that mycorrhizas are usually only found on trees that need them: specimens on rich soil or with adequate sources of nutrients tend not to have mycorrhizas. This makes good economic sense; they cost the tree in carbohydrate and many mycorrhizal fungi can become parasitic if the tree is too weak to provide the necessary sugars.

Mycorrhizas are found in around 80% of the world's vascular plants. Two main types are found in trees. Ectomycorrhizas (ECM) are found in only 3% of flowering plants and are confined almost entirely to woody plants, primarily of the cool temperate regions of the northern hemisphere (Box 4.3). Arbuscular mycorrhizas (AM; what used to be called Endomycorrhizas) are far more common in flowering plants and are found in a wide scattering of different trees. Ericoid mycorrhizas are also found in trees and shrubs of heathlands around the world, including heathers in the heath family (Ericaceae) but are not considered further here. While the occurrence of the two main mycorrhizal types on different trees seems very clear-cut, there are recorded instances where normally ectomycorrhizal trees can form other types of mycorrhiza,

| Box 4.3 Types of mycorrhiza associated with trees | |
|---|---|
| **Ectomycorrhiza (ECM)** | Found in:<br><br>• 90% of temperate trees of the northern hemisphere including most conifers<br>• Some southern hemisphere trees such as southern beeches (*Nothofagus*) and eucalypts (*Eucalyptus*)<br>• Less commonly in various families of the tropics including the dipterocarps (Dipterocarpaceae) and in species-poor tropical forests<br><br>At least 6000 species of fungi are involved, usually basidiomycetes, rarely ascomycetes |
| **Arbuscular mycorrhizal (AM)** (used to be called **Endomycorrhiza**) | Common in species-rich tropical forests and found in a wide variety of families:<br><br>• Sapindaceae (maples)<br>• Betulaceae (birches & alders)<br>• Salicaceae (willows & poplars)<br>• Fabaceae (most acacias)<br>• Juglandaceae (walnuts & hickories)<br>• Ulmaceae (elms)<br>• Oleaceae (olives & ashes)<br>• Magnoliaceae (magnolias & the tulip tree)<br>• Hamamelidaceae (sweet gums & witch hazels)<br>• Cupressaceae (cypresses & junipers plus redwoods & swamp cypress)<br>• Araucariaceae (monkey puzzles & araucarias)<br>• Taxaceae (yews)<br>• Ginkgoaceae (the ginkgo)<br><br>The fungi are usually zygomycetes |

sometimes both at once or one after the other (e.g. poplars, willows, junipers, eucalypts and tulip trees, *Liriodendron* spp.). Within the same mycorrhizal type, trees may be infected with several species of fungi at the same time, and show a succession of different species as the tree ages.

Ectomycorrhizas appear on fine roots as a smooth fungal sheath coating side branches less than 0.5 cm long. Inside the sheath, the fungus penetrates between the cells of the root to form a complex system of branching hyphae (the individual threads of the fungal growth) giving a large and intimate area for the exchange of materials. Outside of the fungal sheath the fungus permeates out through the soil, often in well organised strands, over large distances. The basic advantage to the tree is therefore to increase the effective size of the root system. Estimates have been made for various trees as to how much this adds: for willows there may be 100–300 cm of fungal hyphae for every centimetre of root, while in pines this may be 1000–8000 cm per cm of root (Lipson & Näsholm 2001). Once infected, roots stop growing and lose their root hairs; they remain in one place and delegate exploration of the soil to the radiating hyphae. By contrast, fast-growing large-diameter woody roots on the same tree generally have root hairs and are non-mycorrhizal. Fungi that form ectomycorrhizas may only be able to invade slow-growing roots; perhaps the rarity of this type of mycorrhiza in the tropics may be due to continuous and rapid growth of fine roots, aided by the relative scarcity of basidiomycete fungi.

Arbuscular mycorrhizal roots look more normal. They keep their root hairs, and on the outside of the root the only usual sign of the fungus is a few wispy hyphae. Most of the active part of the fungus grows into the cells of the root (not between cells as in ectomycorrhizas) producing highly branched structures (arbuscules) for the exchange of materials, hence the name.

These two sorts of mycorrhiza work in different ways although they produce comparable amounts of nutrients to their hosts. Litter from trees with ectomycorrhizas is harder and slower to decompose but the fungi can directly unlock these nutrients for their host by digesting this litter and transferring nutrients straight to the tree. This is not cheap: up to 20% of the carbon fixed by the trees may be used by the large amounts of fungus involved. But on the cold, nutrient-limiting soils typical of ectomycorrhizal trees the benefits outweigh the cost.

By contrast, the rapidly decomposing litter of arbuscular mycorrhizal trees releases nutrients directly into the soil. The fungi are largely unable to preferentially access nutrients from organic matter so their beneficial role is to expand the volume of soil from which nutrients can be extracted. This may be just a matter of millimetres or centimetres from the root but enough to make a difference. The lesser role of the fungus is correspondingly cheaper for the tree: 2–15% of carbon fixed.

Arbuscular mycorrhizal fungi have low host specificity and completely unrelated plants can be linked into a common mycorrhizal-mycelium system (mycelium being the collective name for all the hyphae). This may help to reduce competitive dominance and promote coexistence and species diversity. It may also help partially explain why AM communities (e.g. tropical forest) are species rich. Experiments have shown that phosphorous is preferentially transferred from dying roots of one plant to neighbours with the same mycorrhizal type. Ectomycorrhizal fungi tend to be more host specific and are less likely to form widespread links but even here the joining of different individuals may be very important. For example, a study of Douglas fir seedlings (*Pseudotsuga menziesii*) in Oregan found that seedlings were rooting into a hyphal mat produced by two species of ectomycorrhizal fungus which covered 28% of the forest floor. Since Douglas fir is shade-intolerant, the energy for survival when seedlings are in deep shade must come in large part from the mature trees. Seedlings tapping into the fungal network grow quicker and are heavier than those that do not. This peaks at 2.5–5.0 m from the mature trees (any nearer and there is probably too much competition from the mature tree). There is evidence that some pines (ectomycorrhizal) may in fact inhibit the establishment of arbuscular mycorrhizal trees and herbs. But it is not always that simple: under oak trees in Britain (ectomycorrhizal) the commonest seedlings can be ash and sycamore (both AM).

## Other aids to nutrition

Although the atmosphere is 78% nitrogen, it is not in a form that is useable by plants. Trees in the pea family (Fabaceae; what used to be the Leguminosae), along with its herbaceous members, have swellings or nodules on the roots which contain bacteria (one of several *Rhizobium* species) that can 'fix' atmospheric nitrogen into an organic form that is useable by the bacteria and the plant (Figure 4.13c). Again it is a two-way relationship; the tree gains nitrogen and the bacteria get a protected home and a supply of carbon. The benefit to the tree is exemplified by the false acacia/black locust (*Robinia pseudoacacia*) which is excellent at invading disturbed areas and is a fast-growing tree: up to 12 m in 10 years has been recorded. As in mycorrhizas, the formation of nodules and nitrogen fixation decreases as the level of soil nitrogen increases; fertilise the soil and the tree needs less from the bacteria.

Nodules can also be found outside the Fabaceae in a range of (primarily) temperate hardwood trees from a range of habitats: wetlands to deserts (Box 4.4). In these the nitrogen-fixing organisms are Actinomycetes (*Frankia* spp., related to filamentous bacteria). The nodules are not so much swellings as a series of short densely branched roots, giving a coral-like appearance.

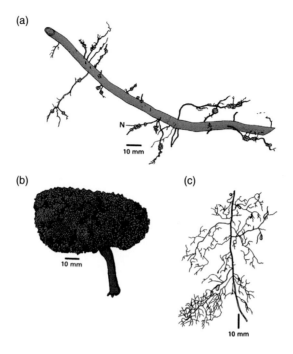

Figure 4.13 (a) Nitrogen-fixing nodules (N) on the roots of alder (*Alnus glutinosa*), (b) a single large nodule on the alder, and (c) small nodules on the roots of an acacia (*Acacia pravassima*) in the pea family. From: Bell, A.D. (1998) *An Illustrated Guide to Flowering Plant Morphology*. Oxford University Press, Oxford. Reprinted by permission of Oxford University Press.

In alders (*Alnus* spp.) the nodules can reach the size of a tennis ball and live for up to a decade (Figure 4.13a, b). Perhaps most unusually, in the primitive Australian cycad *Macrozamia riedlei*, a photosynthetic bluegreen alga (an *Anabaena* sp.) invades the roots which then grow up near or even above the soil surface, allowing the alga to photosynthesise and fix nitrogen.

Several nutritional aids can be used together: red alder (*Alnus rubra*) has nodules and is commonly ectomycorrhizal, and its nitrogen fixation rates (up to 300 kg per hectare per year) are as high as published figures for the Fabaceae.

Roots are well known for their exudates released in response to nutrient deficiency, which are rich in a variety of organic compounds including sugars, amino acids, organic acids and phenolics. Although these represent only a small part of the tree's total energy (1–5% of net carbon produced by photosynthesis) these exudates play an important role in helping with nutrient uptake. They influence the solubility of nutrients (by changing the soil pH) and their uptake by roots both directly and via their effects on soil microbes. This effect may only extend a few millimetres or centimetres into the soil (technically called the rhizosphere) but that's all that is needed. These chemicals can also help the plant to gain a competitive advantage, an effect called allelopathy, described in more detail in Chapter 9.

| Box 4.4 Trees that fix nitrogen in root nodules | | | | |
|---|---|---|---|---|
| Scientific name | Common name | Family | No. of species nodulated/ total no. of species in the genus | Notes |
| – | Leguminous trees and shrubs | Fabaceae | – | Including acacia, laburnums and pagoda tree; widespread |
| *Coriaria* | Coriarias | Coriariaceae | 13/15 | Warm temperate shrubs |
| *Dryas* | Avens | Rosaceae | 3/4 | Cool temperate dwarf shrubs |
| *Purshia* | – | Rosaceae | 2/2 | Western N America |
| *Cercocarpus* | – | Rosaceae | 4/20 | Western USA & Mexico |
| *Casuarina* | She-oaks | Casuarinaceae | 24/25 | Australia & Pacific coast |
| *Myrica* | Myrtles | Myricaceae | 26/35 | Almost cosmopolitan |
| *Comptonia* | Sweet fern | Myricaceae | 1/1 | N America |
| *Alnus* | Alders | Betulaceae | 33/35 | Mostly north temperate |
| *Elaeagnus* | Oleasters | Elaeagnaceae | 16/45 | Europe, Asia, N America & Australia |
| *Hippophaë* | Sea buckthorn | Elaeagnaceae | 1/3 | Europe & Asia |

| Scientific name | Common name | Family | No. of species nodulated/ total no. of species in the genus | Notes |
|---|---|---|---|---|
| *Shepherdia* | Soapberries | Elaeagnaceae | 3/3 | N America |
| *Ceanothus* | Californian lilacs | Rhamnaceae | 31/55 | N America |
| *Discaria* | - | Rhamnaceae | 2/10 | Australasia & S America |
| *Colletia* | - | Rhamnaceae | 1/17 | S America |

Based on: Torrey, J.G. (1978) Nitrogen fixation by actinomycete-nodulated angiosperms. *Bioscience*, 28, 586–592; Becking, J.H. (1975) Root nodules in non-legumes. In: *The Development and Function of Roots* (edited by J.G. Torrey & D.T. Clarkson). Academic Press, London, pp. 507–566.

Parasitic trees, which steal their nutrition and water from other plants, are very rare, but one does exist. *Parasitaxus ustus* (Podocarpaceae) is the only parasitic conifer, native to New Caledonia. This island broke free from Gondwanaland before flowering plants had arrived and so the conifers diversified and took on this parasitic role.

## Root grafting

As mentioned when describing the root plate, the big roots produced by a tree near its base are commonly fused together into a solid network. What is equally interesting when thinking of tree nutrition is that roots of different trees can also graft together. In the relative stillness of the soil, roots readily intermingle without the same 'shyness' found between tree crowns (see Chapter 7). For example, in a mixed hardwood forest it is possible to get the roots of 4–7 trees below the same square metre of soil surface (Figure 4.14) with most of the woody roots of different species no more than 10–20 cm apart. When roots meet and push against each other, they can graft together, joining their vascular tissue. Grafts between roots of different individuals of the same species are common in both conifers and hardwoods. Grafts between different species are possible but rare and many are 'false grafts' where the bark may fuse but the vascular tissue of the roots remains separate. The big framework roots

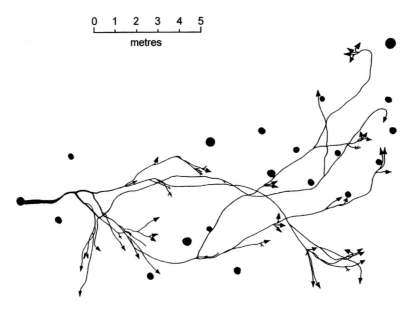

**Figure 4.14** Plan view of a single lateral root of a red maple (*Acer rubrum*) about 60 years old. Circles show the location of other trees in the stand. Arrows indicate that the root tips were not found; these roots therefore continued somewhat further than is shown. From: Lyford, W.H. & Wilson, B.F. (1964) Development of the root system of *Acer rubrum* L. *Harvard Forest Paper*, 10.

are most likely to graft because they are the roots that grow largest in diameter and so are most likely to press together. Indeed, it is unlikely that fine non-woody roots ever touch despite their abundance, and thus they do not graft. (There are exceptions: the aerial roots of the strangler fig *Ficus globosa* have short hairs which will fuse together on contact, holding the bigger roots together, aiding grafting.) Since framework roots are usually involved, the frequency of root grafts between individuals depends on the closeness of trees. For example, in American studies almost all American elms (*Ulmus americana*) were root grafted together if less than 2 m apart but this fell to 43% for elms up to 5 m apart and 29% for those up to 8 m apart.

Once grafted together there is considerable potential for transfer of materials between roots. Sugars and hormones flowing in the phloem move readily across grafts (although water and minerals in the xylem are less likely to exchange because they tend to flow mainly within the grain of one root). You might think that root grafting is another way for the big to become bigger at the expense of others, but it seems that the general movement is from the dominant individual to the underdog, although probably not enough to upset the dominance pattern. This is the natural outcome of the sugars in phloem sap flowing from the source to the sink where they are needed, as discussed above; in this case the sink being a needy tree joined via a root graft. This is nicely illustrated when conifers are felled. Most conifers do not regrow once cut and the stump usually lives for no more than one growing season. But there are cases where stumps have been kept alive via root grafts for one or two decades, producing new rings and callous growth over the cut top. Epstein (1978) recounts the case in which 22 years after a forest of Douglas firs was selectively thinned, 23% of stumps were still alive, kept so by root grafts to the remaining trees. Root grafts between different species can have other bizarre consequences. For example, the wood of a fig species that was root grafted to an Indian tree, *Vateria indica*, that yields aromatic gum, had the smell of the *Vateria*, which in turn exuded a milky latex like that of the fig.

Root grafts between individuals are not always beneficial. Dutch elm disease and other diseases can be transmitted between trees by them. Herbicides applied to one tree can affect others by 'backflash' through root grafts. In this case, however, caution is needed; herbicides can move between trees not grafted together either by being exuded by roots, and subsequently being taken up by other trees, or via mycorrhizal fungi and other organisms living around the root.

## Food storage

Many herbaceous plants store food underground in modified stems and roots. This is a safe option because there are comparatively few large herbivores feeding underground; there are no underground cows. For a tree, however, the tough woody skeleton provides almost as much protection above ground as do woody roots below. Consequently, food is stored throughout the woody skeleton, above and below ground. In the event of the trunk being removed by, for

Figure 4.15 Lignotuber of a eucalypt in eastern Australia.

example, fire or coppicing, the food stored below ground is used towards new growth. In extreme cases, storage problems have resulted in the lignotuber, a large swollen root collar in the soil covered in a mass of buds (Figure 4.15). Lignotubers are found in many tropical trees in seasonally dry savanna woodlands prone to frequent fires, but especially in the 'mallee' species of eucalyptus in Western Australia but also the strawberry tree, *Arbutus unedo* and the ginkgo, *Ginkgo biloba*. The largest lignotuber on record was 10 m across out of which branched 301 stems. The 'burrs' of shrubby heathers, such as *Erica arborea* from which smoking pipes are made, are similar structures.

## Development and growth of roots

Roots and shoots grow in a similar way (as described in Chapter 1): they get longer by new growth at the end (primary growth) and some (in the case of roots) also get fatter (secondary growth).

### *Root elongation: primary growth*

As in herbaceous plants, a root lengthens from a small group of dividing cells (the meristem) just behind the root tip. The new growth is protected as it pushes through the soil by a root cap: a shield of tissue (which can be quite loose and easily wobbled with a fingernail) continually added to at the back and often lubricated by mucilage. Root caps are best developed in strongly growing roots and may be completely absent on short side roots. Behind the growing tip, the new root differentiates into separate tissues. The centre of the root is dominated by the xylem (see Chapter 3) which radiates out into a number of star-like arms (technically called arches). In roots, and unlike stems, there is no central pith (see Figure 3.2). Nestling in the gaps between the arms are strands of phloem. Around all this vascular tissue is a layer of cells (the pericycle) that gives rise to branch roots (Figure 4.12). This central 'stele' is surrounded by a one cell thick cylinder (the endodermis) which in turn is surrounded by the cortex and outer skin (the epidermis). Water and dissolved minerals can pass relatively easily between the cells of the cortex but in the endodermis, the cells are tightly touching, and where they touch they are impregnated with suberin (a waterproof compound also found in bark) to form the Casparian strip; in effect this strip ensures that all water and minerals that reach the internal plumbing have to pass *through* a cell and cannot sneak *between* cells.

You might ask how water is taken up by a root. At first thought it might seem that roots merely act like a sponge passively soaking up water. In actuality it is a more controlled process. Minerals in the water are actively 'pumped' through the Casparian strip and the water then follows passively by osmosis.

Thus, the movement of water and minerals into the root is subject to the regulatory activity of the living cells which gives the root more control over what is taken up and by how much. But, as mentioned below, there can be cracks in old roots that allow some water to get straight into the stele, bypassing the normal regulation.

Side roots arise some way back from the root tip from the outer layer of the stele (specifically, the pericycle) usually opposite the points of the xylem star. These emerge through the outer tissues of the parent root either by digesting a path or by sheer mechanical force. The new side roots form vascular connections with the stele of the main root and are then able to take part in the water uptake business.

## Root thickening: secondary growth

For secondary growth to occur (i.e. thickening of the roots) there needs to be cambium present (again, see Chapter 3). This starts to develop before the end of the first year in the gaps between the xylem arms and soon joins together to form a complete cylinder between the xylem and phloem (see Figure 3.2). Each year, new xylem and phloem are produced (xylem on the inside and phloem on the outside) in annual rings, just as in the trunk, and the root gets thicker. As the roots thicken, a different type of cambium forms in the increasingly stretched pericycle: cork cambium (again, just as in the trunk). This produces corky cells, heavily laden with waterproof waxy substance (suberin) to form bark similar to that above ground complete with breathing holes (lenticels) (see Chapter 3). The cortex surrounding the root, increasingly cut off from the food supply of the stele, is starved, stretched and finally ruptured to be shed with the endodermis. Since the cortex often accounts for two thirds the width of a new root, the young corky root may be somewhat thinner than the succulent tip. As a root grows longer, only the growing tip remains unsuberised. In North Carolina, USA, both the loblolly pine (*Pinus taeda*) and tulip tree (or yellow-poplar, *Liriodendron tulipifera*) were found by Kramer & Bullock (1966) to have usually less than 1% of the surface area of the roots unsuberised. It is analogous to a race: in fast-growing root systems large lengths may be unsuberised but as growth slows down, suberisation catches up. Although the endodermis is shed, this does not necessarily mean free entry of water because the developing suberised bark has taken its place. There was a long debate in the second half of the 1900s about whether suberised portions of root can absorb water and the evidence points to at least some water getting into roots through holes and cracks in the bark such as at the ends of dead roots.

A cross-section of a root with secondary thickening increasingly resembles that of a branch (Figure 3.2). Upon looking closer, however, the wood of a root is commonly less dense than trunkwood (especially in hardwoods) with fewer fibres and a less clear distinction between growth rings. Moreover, ring-porous woods tend to become diffuse-porous with increasing distance along the roots from the trunk. In consequence, root wood often looks nothing like the wood in the trunk and requires specialised skill in its identification (see, for example, Cutler *et al.* 1987). Such structural differences appear to be due to the growing conditions because roots of hardwoods exposed to light and air assume most of the characteristics of trunkwood.

Thickening in roots is concentrated at the base of the big lateral roots and in particular on their upper side, leading to eccentric roots that often bulge above ground, producing in extreme cases the buttress roots well known in tropical trees. Just why thickening should be so localised is still not completely understood. It may be because of the direct continuity of the upper side with the phloem in the trunk and better food supply. Also, however, portions of root exposed to air are known to grow better: they can develop chlorophyll in the inner bark which provides some food but more importantly leads to the production of growth hormones and local mobilisation of food reserves. It is also possible that the general thickening at the root collar may be due to stimulation of the cambium because it is under greater mechanical stress at this point (see Chapter 7).

Away from the trunk in the outer root system, secondary thickening is much less pronounced resulting in the long rope-like roots that thicken slowly, if at all. The production of wood is much more irregular than in stems and growth rings often vary greatly in thickness around and along a root, appearing in some parts and not others.

## Speed of growth

Roots, like shoots, exhibit apical dominance; the main woody roots grow most rapidly and for longer periods compared to side roots. Small non-woody roots of red oak (*Quercus rubra*) elongate at an average of 2–3 mm per day, whereas large-diameter lateral roots grow at 5–20 mm per day, and up to 50 and 56 mm per day in hybrid black poplar (*Populus* × *euramericana*) and false acacia/black locust (*Robinia pseudoacacia*), respectively. As noted above, mycorrhizal root tips may not grow at all for the year or so they remain alive. At the other extreme are the delicate pink aerial roots of the liana *Cissus adnata* in the New World tropics which may be no more than 1 mm thick but over 8 m long; they have been recorded as growing at more than 3 cm per hour after rain. The closely related *C. sicyoides* can grow at a constant 4 mm per hour day and night, a rate of almost a metre in a day! Most tree roots grow faster during the night.

## Control of growth

Growth of the tops of trees, even tropical ones, is carefully regulated to be in tune with the changing seasons. Below ground, however, life is more anarchic, with roots growing whenever conditions are suitable. Temperature is most often the limiting factor. Roots normally grow best between 20 and 30°C but in many trees will continue down to 5–6°C (this minimum may be as low as 0°C for some northern trees or as high as 13°C for citrus species). Thus in the tropics the roots of rubber trees (*Hevea* spp.) grow continuously in contrast to the rhythmic growth of their shoots. In temperate trees, roots may continue growing in winter long after the shoot system has become dormant, and in mild winters roots may grow right through to spring. The roots of red maple (*Acer rubrum*) in eastern N America are normally dormant between November and April but if they are kept warm at 20°C the fine roots continue to grow at 5–10 mm per day even though the stem is dormant. As temperate soils warm in the spring, root growth (if it has stopped) usually resumes before the shoots. Lack of oxygen (for example by compaction of the soil or leaks from gas pipelines) can reduce or stop root growth but they will grow quite happily down to 10% oxygen (air contains 21%) and will only stop at around 3%. Water also has relatively little influence on root growth simply because roots are the last to feel water stress working down from the leaves, and are the first to recover. But temperate trees usually show a mid-summer lag as the soil dries to be followed by a new peak of root growth in late summer-autumn when shoots are growing very little. Root growth in hardwoods tends to peak in early summer whereas that of conifers is more uniform through the season. This might be tied in with competition for resources within the tree; when the shoot is growing strongly it is better at attracting more than its fair share of food so slowing down root growth.

There has been a great deal of debate as to whether root growth is governed purely by soil conditions or whether hormonal control, so evident in the shoot, extends down into the roots. On the one hand, root tips cut from a tree will grow quite happily in a moist dish, and roots of felled conifer stumps begin growing at the same time as those of neighbouring intact trees. On the other hand, roots still attached to the tree show cyclic periods of growth even when kept under uniform environmental conditions. On the whole, the shoot does have a definite influence on roots via supply of food and hormones. Part of the reason why this is not always obvious lies in the capability of the roots to store necessary food for use when the stem is dormant and in the fact that hormones in roots work at such low concentrations they are often difficult to detect. Root elongation is controlled by one set of hormones (auxins) arising in the stem while the formation of new cells and lateral roots depends on a subtle balance between auxins and another set (cytokinins) produced by the root tips.

As mentioned before, root growth must inevitably be determined, at least in part, by the shoots to ensure an equitable balance between the amount of roots and shoots. Too few roots leads to lack of water in the canopy, whereas too many roots is a waste of resources that could be put towards canopy growth and reproduction. This balance does vary. For example, during the early stages of a seedling's growth there is a priority for root growth over shoot growth, especially in dry areas where survival is dependent on the roots reaching a reliable water supply. Conversely, fertilising a tree commonly leads to a disproportionately greater increase in shoot growth. We normally think that by fertilising a tree it will grow big healthy leaves, make more sugar and so be able to grow more roots. But this assumes the root growth is limited by lack of carbohydrates and not by soil conditions or grass competition. Fertilisation may not always help.

## Longevity of roots

In many ways the main structural roots are like the trunk and branches – long-lived – and the fine roots are like the leaves, flowers and fruits, regularly shed and regrown. The root system of a tree can be thought of as a persistent woody frame with disposable fine roots. Fine roots may live for just 1 week (as in some varieties of apple), one summer, or persist for 3–4 years (e.g. Norway spruce, *Picea abies*) growing year round. In temperate zones, many small roots, especially those close to the surface, die in the winter; walnuts (*Juglans regia*) may lose more than 90% of their absorbing roots in winter, though others such as the tea plant (*Camellia sinensis*) may lose less than 10%. The shedding of fine roots is as ecologically important as above-ground leaf fall, and the two are about equal in mass.

There is uncertainty about what precisely causes fine roots to die. Some people consider root death to be a normal physiological process governed by the tree's internal clock, just like leaf fall. Their arguments are based on the cost of keeping roots alive. For example, when soil is dry (or 'physiologically dry' when soil and roots are too cold to work) the fine roots do nothing except consume food and could consume their own dry mass in reserves in 1 week. It therefore makes good economic sense to lose these wasteful roots and retain only a skeleton of woody roots from which new fine roots can grow when conditions improve.

Others argue that root death is not due to the tree's actions but is a direct effect of unfavourable soil conditions and various pests and diseases. This fits better with the idea that roots are opportunists, growing even during the winter when conditions are suitable. Fine roots are very susceptible to dry conditions and may shrivel and die after only a few minutes of exposure to dry air (take note when planting trees!), and all remains can be gone within a few days. Nor have roots developed the extreme cold tolerance of shoots; temperatures below −4 to −7 °C can kill. Thus, the fine roots in woodland litter are easily exposed to lethal temperatures in temperate and northern climates. This explains why potted trees and shrubs (exposed to low temperatures from several sides) are less hardy and often need special protection in winter. Fine roots can also be killed by movement of the root plate as the tree sways, and a whole range of subterranean predators from nematodes to small mammals are capable of damaging and cutting small roots. Fungal and insect attacks on the canopy can also strongly affect root mortality by reducing food and hormone export to the roots. In balsam fir (*Abies balsamea*) attacked by spruce budworm in eastern N America, 'rootlet' death rises from less than 15% in healthy trees to more than 30% in trees with 70% defoliation, to more than 75% in completely defoliated trees.

## Roots in wet soils

Oxygen in flooded soils is quickly used up by roots and microorganisms. Some oxygen diffuses down through the waterlogged soil but this is usually only enough to keep the top few centimetres of soil oxygenated. Below that, the soil is devoid of oxygen (anaerobic). Woody roots can survive such conditions for some time when in a dormant state but when active, fine roots die quickly, leading to poor water and mineral uptake, wilting of leaves and reduced photosynthesis. As well as the direct effect of oxygen starvation, there are problems of toxic compounds produced by the soil (e.g. hydrogen sulphide) and also directly by the tree. The roots of many *Prunus* species (including cherries, peach, apricot and almond, but not plum) contain cyanogenic glucosides as a defence against being eaten, which break down to release cyanide gas when oxygen is limited. Excessive flooding can also leach valuable nutrients from the soil.

Trees vary tremendously in their ability to tolerate flooding. To some degree this is dependent upon the state of the individual tree (well-grown and dormant trees are generally more tolerant), but there is an underlying inherent difference between species (see Box 4.5). There are a number of trees such as the swamp tupelo (*Nyssa aquatica*), swamp cypress (*Taxodium distichum*) and willows that can survive flooding for several months or even permanently. Others such as alders (*Alnus* spp.) have *some* of their roots permanently in water. And these are not isolated examples. Swamps and floodplains occupied by flood-tolerant trees can cover extensive areas; for example, some 2% of the Amazon Basin (more than 100 000 km$^2$) is flooded by 2–3 m of water for 4–7 months every year. Saplings may be completely submerged for 7–10 months or

| Box 4.5 Flood tolerance of cultivated trees and shrubs exposed to a summer flood on the lower Fraser River Valley, British Columbia, Canada | | |
|---|---|---|
| Very susceptible to death and injury | Holly | *Ilex aquifolium* |
| | Hazel | *Corylus avellana* |
| | Lilac | *Syringia vulgaris* |
| | Mock orange | *Philadelphus gordonianus* |
| | Cotoneaster | *Cotoneaster* spp. |
| | Cherry | *Prunus* spp. |
| | Cherry laurel | *P. laurocerasus* |
| | Rowan | *Sorbus aucuparia* |
| | Japanese red cedar | *Cryptomeria japonica* |
| Less susceptible but often killed or severely injured | Hawthorn | *Crataegus laevigata* |
| | Box | *Buxus sempervirens* |
| | Blackberry | *Rubus procera* |
| Some damage with leaf yellowing | Evergreen blackberry | *Rubus laciniatus* |
| | Pears | *Pyrus* spp. |
| | Roses | *Rosa* spp. |
| | Grapes | *Vitis* spp. |
| | Rhododendrons | *Rhododendron* spp. |
| | False acacia | *Robinia pseudoacacia* |
| No obvious injury | Apples | *Malus* spp. |
| | Walnuts | *Juglans* spp. |
| | Manitoba maple | *Acer negundo* |
| | Chestnuts | *Castanea* spp. |

Based on Brink, V.C. (1954) Survival of plants under flood in the lower Fraser River Valley, British Columbia. *Ecology*, 35, 94–95.

longer each year. Yet these trees retain their leaves and grow even when completely submerged, although height growth usually stops within 1–3 weeks. The flooding results in low oxygen levels around the roots and stems (especially in the warm water full of respiring organisms), the leaves and roots being covered by sediment (by up to a metre a year) and growth being slow due to low light levels caused by the turbid water, and there is also the mechanical strain. Most of these plants eventually go into a quiescent phase while flooded. Leaves that are flooded normally quickly close their stomata, but Amazonian flooded species do not. The thick leaves with a thick cuticle maintain a thin gas layer over the surface which prevents the entrance of water (with associated bacteria) and so they don't rot and can start working again as soon as they are in the air again. It also allows some photosynthesis to go on while conditions are suitable.

What is visually most striking about many wetland trees, especially in the tropics, is the production of aerial roots (really 'adventitious' roots: roots appearing 'out of the usual place'), including stilt, peg and knee roots. Mangroves growing in tidal muds of the Old and New World tropics (e.g. *Rhizophora* spp.) and a wide range of other trees in wet areas, produce a hula skirt of stilt roots from the trunk (often branched) which, once rooted, graft together into a rigid three-dimensional latticework (Figure 4.16a shows dry-land stilt roots). The mangrove may be held almost clear of the mud with the stilt roots taking the place of the sparse, short-lived normal roots. Other mangroves (e.g. *Avicennia* spp.) and several species in freshwater marshes (including the bay willow, *Salix pentandra*, in temperate areas, and various palms) have a shallow underground root system which produces short pencil-shaped peg roots which stick out of the ground (Figure 4.16b). These soft and spongy roots are rarely more than 1 cm in diameter but may be up to 2 m high and a single tree may produce as many as 10 000 of them. Both stilt and peg roots act as snorkels (and, certainly with stilt roots, provide stability in loose mud). Above ground the roots have large lenticels that feed into wide air passages (officially called aerenchyma) connected with the spongy air-filled underground roots. The stilt roots of *Rhizophora* have 5% gas space above ground, increasing to c. 50% after penetration into the mud. Oxygen is thus able to diffuse down into the underground root, and toxic gases produced by incomplete aerobic metabolism can escape (although many flood-tolerant trees have biochemical adaptations to reduce the production, accumulation and impact of such toxins). The oxygen leaks out into the mud forming an oxygenated envelope around the root, allowing it to function. This mechanism also works in the trunks of trees without modified roots (such as the swamp tupelo *Nyssa sylvatica*, alders and ashes). The air flows down through the bark and air-filled sections of wood

(a)

(b)

**Figure 4.16** Modified roots. (a) Stilt roots of the Seychelles stilt palm (*Verschaffeltia splendida*) growing in Singapore; (b) peg roots of the grey mangrove (*Avicennia marina*) in Brisbane, Australia, which act as snorkels allowing oxygen to reach flooded roots; (c) knee roots of the swamp cypress (*Taxodium distichum*) in Florida, USA; (d) buttress roots of the kapok tree (*Ceiba pentandra*) in Singapore; and (e) pillar roots as seen on figs, in this case the Moreton Bay fig (*Ficus macrophylla*) in Sydney, Australia.

(c)

(d)

Figure 4.16 (cont)

(e)

Figure 4.16 (cont)

(xylem); the cambium was long thought to be impervious to gas, but it is now evident that this is not true of wetland trees.

Some flood-tolerant species, including those in the flooded Amazon, produce new (adventitious) roots on the submerged part of the stem up near the water surface where there is more oxygen or may even emerge into the air. Such roots are found in a diverse range of trees from eucalypts and bottlebrushes (*Melaleuca* spp.) in the southern hemisphere to the swamp tupelo, elms, willows and ashes of the north.

A number of unrelated tropical hardwood species, including mangroves (*Avicennia* spp.), and the important genus *Terminalia*, plus a palm or two in the genus *Phoenix*, can produce knee roots (properly termed pneumatophores) in wet areas. Here, a loop of root arches up into the air to form a distinct knee. Again, knees act as aerating organs for the underground roots. The only conifer that can produce knees (and is also the only non-tropical tree), is also probably the most famous: the bald or swamp cypress (*Taxodium distichum*) of seasonally flooded freshwater marshes in southeast USA (Figure 4.16c). These knees are also produced in a different way, growing as a thickening of the upper surface of a horizontal root, producing the conical swelling that comes above the water. The knees are most frequent in regularly flooded sites and can grow to almost 4 m high, depending upon the normal annual high waters. However, these knees do not appear to be involved in aerating submerged roots. Robert Lamborn published a wonderful paper in 1890 suggesting that the knees were primarily for mechanical support, utilising the mass of roots that spreads from each knee as ballast; to quote Lamborn, 'I have never found a healthy cypress that had fallen before the fierce hurricanes that sweep through the southern forest-lands'. And, indeed this does appear to be their main function. He also suggested that by catching floating debris the knees may help the tree's nutrition.

## Buttresses, pillars and stranglers

The aerial roots discussed above have been primarily on tropical wetland species, but they do also occur in temperate and dry-land trees (although they are not as common, large or as variable). Stilt roots, for example, are well developed in the screw pines (*Pandanus* spp.) of the Old World tropics and a range of palms (Figure 4.16a). These monocot trees do not have secondary thickening and therefore use the roots to hold up the increasingly tall and wide canopy.

### *Buttresses*

In temperate trees, the root collar may flare a little way up the tree to form small buttresses. These flanges, which are really part root and part stem, are common in oaks, elms, limes and poplars, especially where the soil keeps the roots very shallow. For spectacular buttresses, however, we need to look in the tropics where each buttress can be many centimetres thick, extending out 1–2 m from the tree and rising 2–3 m, and sometimes 10 m, up the tree (Figure 4.16d). Some buttresses are thin and flat enough to be used to make walls of buildings; others are twisted and fused together to make a series of tank-like hollows. Buttresses are usually associated with tall trees that do not develop a tap root, and in some trees the stem reduces to about a third of its full diameter or all but disappears as it approaches ground level so that it is literally held up by the buttresses. That vertical buttresses close to the stem encourage mechanical stability seems beyond question; they help brace the roots to the trunk like angle brackets reducing the stress on the root collar when the tree sways in the wind. Certainly buttresses in the tropics are most common on trees that grow tall and emerge above the canopy, and so meet wind. Sinker roots from the bottom of the buttresses anchor the whole structure in place, especially on wet, flooded soils. However, buttresses may also aid the nutrition of the tree, especially those that are not very tall but spread a long way from the tree, by allowing fine roots to be much closer to the soil surface and therefore more competitive on poor soils. The buttresses also trap litter and aid soil structure, and stimulate the release of nitrogen from the litter, to the benefit of the tree. The big question is not what use they have but just why they are rare outside the tropics. It may be that the large surface area makes them prone to damage by the temperature fluctuations of temperate or cold climates, or fire.

## Pillar roots

The weeping fig (*Ficus benjamina*), the banyan (*F. benghalensis*) and a number of other figs use pillar roots beautifully. Slender free-hanging roots, which may be no more than 2 mm in diameter, grow down from the branches at up to 1 cm per day. Once anchored into the soil they form tension wood which has the effect of contracting the root, so much that they can lift large flower pots from the ground. This contraction has the effect of straightening the root so that as the root thickens it forms a straight pillar able to bear the weight of the sideways spreading branches (Figure 4.16e). In this way, the trees can grow outwards to form groves whose 'trunks' are actually pillar roots. Indeed, banyans may cover the largest area of any living plant; for example one tree planted in 1782 in the Royal Botanic Garden of Calcutta is 412 m (1350 ft) in

Figure 4.17 'Chichi' growing on an old ginkgo (*Ginkgo biloba*) at the Tsurugaoka Hachimangu Shrine, Kamakura, Japan. This venerable tree, photographed in 1990, which stood beside the main stairway of the shrine, unfortunately collapsed in March 2010. Traditionally, sacred places are marked off with special ropes (shimenawa) and strips of white paper (gohei).

(a)

(b)

(c)

(d)

**Figure 4.18** Development of strangler figs. The fig germinates in the canopy and sends roots down to the ground (a). Once a water supply is secured, more roots grow and fuse together (b) which prevent the host tree growing outwards and so strangles it (c). The host eventually dies leaving a hollow tube of roots as the new strangler fig 'trunk' (d). Photographs (a)–(c) *Ficus aurea* in Florida, USA, (d) the weeping fig, *Ficus benjamina* in Singapore (widely grown as a non-strangling houseplant).

circumference and covers an area of 1.3 ha (3 acres) with 2800 'trunks' under which a suggested 20 000 people could shelter.

In temperate areas, the ginkgo/maidenhair tree (*Ginkgo biloba*) produces downward-growing woody knobs called 'chichi' (Japanese for nipples) from the underside of large old branches (Figure 4.17). These knobs, which look like aerial roots, bear buds and are in reality leafless branches, which grow down to the ground where they take root and produce new shoots. The largest ones measured, in Tokyo, are 2.2 m long and 30 cm in diameter.

## *Strangling trees*

A variant of pillar roots is seen in strangling trees. These are mostly figs (*Ficus* spp.), but are also found in the unrelated genera of *Schefflera* (Araliaceae), *Clusia* (Clusiaceae), *Griselinia* (Cornaceae) and *Meterosideros* (Myrtaceae) including the northern rata (*Meterosideros robusta*) of New Zealand. In the classic picture of the strangler fig, the seed germinates in the canopy of a tree and the roots rapidly grow down to the ground. Once there, the roots thicken, branch and graft (there may also be an element of vertical contraction) to form a network of roots, which can be more than 30 m high, encasing the trunk of the host (Figure 4.18). Note that the 'trunk' of the strangler is really root (sometimes referred to as a pseudotrunk). The strangler may then slowly kill the host by preventing it from increasing in girth: a slow version of the boa constrictor! *Ficus leprieuri* has been seen to kill a large host tree in 30 years. The strangling part has probably been overdramatised and it is likely that competition with the host for water and light are probably more significant causes of host decline. Typically, however, less than 10% of infected hosts are killed, and many figs put down just a single harmless pillar root to ensure a water supply without having to grow their own trunk. Most stranglers (including the common rubber trees *Ficus elastica* and *F. benjamina)* will grow as normal small trees if they germinate in the soil, with or without pillar roots, as many keepers of house plants can testify.

## 🍃 *Further Reading*

Aloni, R., Aloni, E., Langhans, M. & Ullrich, C.I. (2006) Role of cytokinin and auxin in shaping root architecture: regulating vascular differentiation, lateral root initiation, root apical dominance and root gravitropism. *Annals of Botany*, 97, 883–893.

Alvarez-Uria, P. & Körner, Ch. (2007) Low temperature limits of root growth in deciduous and evergreen temperate tree species. *Functional Ecology*, 21, 211–218.

Burgess, S.S.O., Adams, M.A., Turner, N.C., White, D.A. & Ong, C.K. (2001) Tree roots: conduits for deep recharge of soil water. *Oecologia*, 126, 158–165.

Burgess, S.S.O. & Bleby, T.M. (2006) Redistribution of soil water by lateral roots mediated by stem tissues. *Journal of Experimental Botany*, 57, 3283–3291.

Crow, P. (2005) *The Influence of Soils and Species on Tree Root Depth. Information Note 78.* Forestry Commission, Edinburgh.

Cutler, D.F. & Richardson, I.B.K. (1981) *Tree Roots and Buildings.* Construction Press (Longman), London.

Cutler, D.F., Rudall, P.J., Gasson, P.E. & Gale, R.M.O. (1987) *Root Identification Manual of Trees and Shrubs.* Chapman & Hall, London.

Dawson, T.E. (1996) Determining water use by trees and forests from isotopic, energy balance and transpiration analyses: the roles of tree size and hydraulic lift. *Tree Physiology*, 16, 263–272.

Ellison, A.M., Farnsworth, E.J. & Twilley, R.R. (1996) Facultative mutualism between red mangroves and root-fouling sponges in Belizean mangal. *Ecology*, 77, 2431–2444.

Emerman, S.E. & Dawson, T.E. (1996) Hydraulic lift and its influence on the water content of the rhizosphere: an example from sugar maple, *Acer saccharum. Oecologia*, 108, 273–278.

Epstein, A.H. (1978) Root graft transmission of tree pathogens. *Annual Review of Phytopathology*, 16, 181–192.

Gill, A.M. & Tomlinson, P.B. (1975) Aerial roots: an array of forms and functions. In: *The Development and Function of Roots* (edited by J.G. Torrey & D.T. Clarkson). Academic Press, London, pp. 237–260.

Gilman, E.F. (1988) Predicting root spread from trunk diameter and branch spread. *Journal of Arboriculture*, 14, 85–89.

Griffiths, R.P., Castellano, M.A. & Caldwell, B.A. (1991) Hyphal mats formed by two ectomycorrhizal fungi and their association with Douglas-fir seedlings: a case study. *Plant and Soil*, 134, 255–259.

Helliwell, D.R. (1993) *Water Tables and Trees. Arboricultural Note No 110.* Arboricultural Advisory & Information Service, Farnham, Surrey.

Hruska, J., Cermák, J. & Sustek, S. (1999) Mapping tree root systems with ground-penetrating radar. *Tree Physiology*, 19, 125–130.

Jones, G.M., Cassidy, N.J., Thomas, P.A., Plante, S. & Pringle, J.K. (2009) Imaging and monitoring tree-induced subsidence using electrical resistivity imaging. *Near Surface Geophysics*, 7, 191–206.

Khuder, H., Stokes, A., Danjon, F., Gouskou, K. & Lagane, F. (2007) Is it possible to manipulate root anchorage in young trees? *Plant & Soil*, 294, 87–102.

Körner, Ch. (2005) An introduction to the functional diversity of temperate forest trees. In: *Forest Diversity and Function* (edited by M. Scherer-Lorenzen, Ch. Körner & E.-D. Schulze). Ecological Studies 176. Springer, Berlin, pp. 13–37.

Kramer, P.J. & Bullock, H.C. (1966) Seasonal variations in the proportions of suberized and unsuberized roots of trees in relation to the absorption of water. *American Journal of Botany*, 53, 200–204.

Kurz-Besson, C., Otieno, D., Lobo do Vale, R., *et al.* (2006) Hydraulic lift in cork oak trees in a savannah-type Mediterranean ecosystem and its contribution to the local water balance. *Plant and Soil*, 282, 361–378.

Lamborn, R.H. (1890) The knees of the *Taxodium distichum. The American Naturalist*, 24, 333–340.

Lehto, T. & Zwiazek, J.J. (2011) Ectomycorrhizas and water relations of trees: a review. *Mycorrhiza*, 21, 71–90.

Leuschner, C., Hertel, D., Coners, H. & Büttner, V. (2001) Root competition between beech and oak: a hypothesis. *Oecologia*, 126, 276–284.

Lindsey, P. & Bassuk, N. (1991) Specifying soil volumes to meet the water needs of mature urban street trees and trees in containers. *Journal of Arboriculture*, 17, 141–149.

Lipson, D. & Näsholm, T. (2001) The unexpected versatility of plants: organic nitrogen use and availability in terrestrial ecosystems. *Oecologia*, 128, 305–316.

Lyford, W.H. (1980) *Development of the Root System of Northern Red Oak (Quercus rubra L.). Harvard Forest Papers, No. 21*. Harvard University, Massachusetts.

Nadkarni, N.M. (1981) Canopy roots: convergent evolution in rainforest nutrient cycles. *Science*, 214, 1023–1024.

Neumann, R.B. & Cardon, Z.G. (2012) The magnitude of hydraulic redistribution by plant roots: a review and synthesis of empirical and modeling studies. *New Phytologist*, 194, 337–352.

Newbery, D.M., Schwan, S., Chuyong, G.B. & van der Burgt, X.M. (2009) Buttress form of the central African rain forest tree *Microberlinia bisulcata*, and its possible role in nutrient acquisition. *Trees*, 23, 219–234.

Pandey, C.B., Singh, L. & Singh, S.K. (2011) Buttresses induced habitat heterogeneity increases nitrogen availability in tropical rainforests. *Forest Ecology and Management*, 262, 1679–1685.

Parolin, P. (2009) Submerged in darkness: adaptations to prolonged submergence by woody species of the Amazonian floodplains. *Annals of Botany*, 103, 359–376.

Prieto, I., Armas, C. & Pugnaire, F.I. (2012) Water release through plant roots: new insights into its consequences at the plant and ecosystem level. *New Phytologist*, 193, 830–841.

Redmond, D.R. (1959) Mortality of rootlets in balsam fir defoliated by the spruce budworm. *Forest Science*, 5, 64–69.

Reynolds, E.R.C. (1975) Tree rootlets and their distribution. In: *The Development and Function of Roots* (edited by J.G. Torrey & D.T. Clarkson). Academic Press, London, pp. 163–177.

Rolf, K. & Stål, Ö. (1994) Tree roots in sewer systems in Malmo, Sweden. *Journal of Arboriculture*, 20, 329–335.

Rose, K.L., Graham, R.C. & Parker, D.R. (2003) Water source utilization by *Pinus jeffreyi* and *Arctostaphylos patula* on thin soils over bedrock. *Oecologia*, 134, 46–54.

Rowe, R.N. & Catlin, P.B. (1971) Differential sensitivity to water logging and cyanogenesis by peach, apricot, and plum roots. *Journal of the American Society for Horticultural Science*, 96, 305–308.

Torrey, J.G. (1978) Nitrogen fixation by actinomycete-nodulated angiosperms. *BioScience*, 28, 586–592.

Wagar, J.A. & Barker, P.A. (1993) Effectiveness of three barrier materials for stopping regenerating roots of established trees. *Journal of Arboriculture*, 19, 332–339.

Wang, G., Alo, C., Mei, R. & Sun, S. (2011) Droughts, hydraulic redistribution, and their impact on vegetation composition in the Amazon forest. *Plant Ecology*, 212, 663–673.

Warren, J.M., Brooks, J.R., Meinzer, F.C. & Eberhart, J.L. (2008) Hydraulic redistribution of water from *Pinus ponderosa* trees to seedlings: evidence for an ectomycorrhizal pathway. *New Phytologist*, 178, 382–394.

Watson, G. (2012) Fifteen years of urban tree planting and establishment research. In: *Trees, People and the Built Environment* (edited by M. Johnston & G.Percival). Research Report. Forestry Commission, Edinburgh, pp. 63–72.

# Chapter 5: Towards the next generation: flowers, fruits and seeds

Like other plants, trees have to engage in sex by proxy, using the wind, water or an animal as an intermediary to get pollen from one tree to another (see Box 5.1). Unlike many other plants, the sheer size of trees raises extra problems of pollination, and eventually seed dispersal, which are solved in ingenious ways. The original trees, the conifers, were (and still are) wind-pollinated. The flowering plants (angiosperms), which includes hardwood trees, evolved hand in hand with insects to be, not surprisingly, primarily insect-pollinated. Yet some have reverted to the old way of wind pollination, and for very good reasons. These are linked to geography: most trees in high latitudes are wind-pollinated, but animal pollination (insects, birds and mammals) becomes more important the closer one gets to the tropics, reaching 95% of trees around the equator. Figure 5.1 gives an overview of general flower structure.

## Animal pollination

Animal pollination is primarily the world of the insect; in the wettest Costa Rican forests, for example, 90% of trees are insect pollinated. But within insect pollination there are different strategies. Some trees, like magnolias, apples, rowan (*Sorbus aucuparia*), European spindle (*Euonymus europaea*), some maples, hawthorns (*Crataegus* spp.) and a long list of others, go for quantity. They are generalists that spread the pollen on a wide range of flies and beetles in the hope that some will arrive on another flower of the same species. Common features are open flowers, often facing upwards, a drab colour, many stamens, easily reached nectar and a strong scent, especially at night (see Box 5.2 and Figure 5.2).

Other trees opt for quality, catering to a more limited number of specific pollinators that are more likely to go straight to another tree of the same kind. Bees, butterflies and moths fall into this category. Darwin noticed that certain pollinators are attracted to flowers of certain colours. Box 5.2 shows that each animal group tends to be attracted to different types of flowers. Notice that day-flying moths and butterflies, and bees can be attracted to similar coloured

154

**Box 5.1** Flower and fruit characteristics of native and common introduced trees and shrubs of the British Isles (the flower and fruit types are explained in the text)

| Common name | Scientific name | Flowering date (months) | Pollinator | Flower type | Fruit/cone type |
|---|---|---|---|---|---|
| **Hardwoods** | | | | | |
| Alder | *Alnus glutinosa* | 2–3 | Wind | Monoecious | Woody cone containing nuts |
| Alder buckthorn | *Frangula alnus* | 5–6 (–9) | Insects (esp. bees) | Hermaphrodite | Drupe (2–3 stones) |
| Apple | *Malus* spp. | 5 | Insects | Hermaphrodite | Pome |
| Ash | *Fraxinus excelsior* | 4–5 | Wind | Dioecious or mixed | Samara |
| Beech | *Fagus sylvatica* | 4–5 | Wind | Monoecious | Nut |
| Birch | *Betula* spp. | 4–5 | Wind | Monoecious | Samara |
| Blackthorn/Sloe | *Prunus spinosa* | 3–5 | Insects | Hermaphrodite | Drupe |
| Broom | *Cytisus scoparius* | 5–6 | Insects (large bees) | Hermaphrodite | Exploding legume plus seeds with elaisosomes |
| Butterfly bush | *Buddleia davidii* | 6–10 | Insects (butterflies) | Hermaphrodite | Capsule |
| Box | *Buxus sempervirens* | 4–5 | Insects (bees and flies) | Monoecious | Capsule and seeds with elaisosomes |
| Cherry | *Prunus* spp. | 4–5 | Insects | Hermaphrodite | Drupe |

155

**Box 5.1** (cont)

| Common name | Scientific name | Flowering date (months) | Pollinator | Flower type | Fruit/cone type |
|---|---|---|---|---|---|
| Currant | *Ribes* spp. | 3–5 | Insects | Hermaphrodite (rarely Dioecious) | Berry |
| Elder | *Sambucus nigra* | 6–7 | Insects (esp. small flies) | Hermaphrodite | Drupe |
| Elm | *Ulmus* spp. | 2–3 | Wind | Hermaphrodite | Samara |
| Hazel | *Corylus avellana* | 1–4 | Wind | Monoecious | Nut |
| Holly | *Ilex aquifolium* | 5–8 | Insects (honey bees) | Dioecious, rarely Hermaphrodite | Drupe (3+ stones) |
| Hornbeam | *Carpinus betulus* | 4–5 | Wind | Monoecious | Nut with bract |
| Horse chestnut | *Aesculus hippocastanum* | 5–6 | Insects (bees) | Hermaphrodite with some male flowers | Capsule |
| Lime | *Tilia* spp. | 6–7 | Insects (bees) | Hermaphrodite | Nut with bracts |
| Maple | *Acer* spp. | 4–7 | Wind and small insects | Hermaphrodite or mixed | Samara |
| Oak | *Quercus* spp. | 4–5 | Wind | Monoecious | Nut |
| Poplar | *Populus* spp. | 2–4 | Wind | Dioecious rarely Monoecious | Capsule with plumed seeds |

| | | | | | |
|---|---|---|---|---|---|
| Rhododendron | *Rhododendron ponticum* | 5–6 | Insects | Hermaphrodite | Capsule with small winged seeds |
| Rose | *Rosa* spp. | 5–7 | Insects | Hermaphrodite | Achenes enclosed in a fleshy hip |
| Spindle | *Euonymus europaeus* | 5–6 | Small insects | Hermaphrodite or mixed | Capsule and seeds with arils |
| Strawberry tree | *Arbutus unedo* | 9–12 | Insects (?) | Hermaphrodite | Warty berry |
| Sweet chestnut | *Castanea sativa* | 7 | Insects | Monoecious | Nut |
| Viburnum | *Viburnum* spp. | 6–7 | Insects | Hermaphrodite | Drupe |
| Whitebeam | *Sorbus* spp. | 5–6 | Insects | Hermaphrodite | Pome |
| Willow | *Salix* spp. | 2–5 | Insects and birds | Dioecious rarely Monoecious | Capsule with winged seeds |
| **Conifers** | | | | | |
| Juniper | *Juniperus communis* | 5–6 | Wind | Dioecious | Fleshy cone |
| Scots pine | *Pinus sylvestris* | 5–6 | Wind | Monoecious | Cone |
| Yew | *Taxus baccata* | 3–4 | Wind | Dioecious | Seed with aril |

Based on information from Clapham, A.R., Tutin, T.G. & Moore, D.M. (1987) *Flora of the British Isles*. Cambridge University Press, Cambridge; Snow, B. & Snow, D. (1988) *Birds and Berries*. Poyser, Calton, Staffordshire; and Sedgley, M. & Griffon, A.R. (1988) *Sexual Reproduction of Tree Crops*. Academic Press, London.

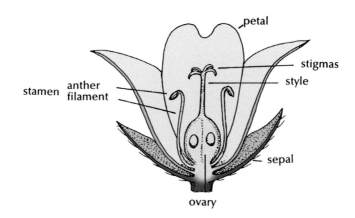

Figure 5.1 The make-up of a flower. The petals collectively form the corolla. The sepals collectively form the calyx. Together they form the perianth. The male stamens contain the pollen-producing anthers. The central carpel is composed of an ovary (containing the ovules which will become the seeds) with a stigma for catching pollen mounted on a style. From: Hayward, J. (1987) *A New Key to Wild Flowers*. Cambridge University Press.

| Box 5.2 Types of flowers associated with different animal pollinators | | | | |
|---|---|---|---|---|
| Pollinator type | Flower | | | |
|  | Type | Colour | Smell | Reward |
| Beetles | Upward facing bowl | Brown, white | Strong | Pollen, nectar |
| Flies | Upward facing bowl | Pale, dull | Little | Nectar |
| Bees | Often asymmetrical, strong, semi-closed | Yellow, blue | Fairly strong | Nectar |
| Butterflies, moths | Horizontal or hanging | Red, yellow, blue (day); white or faintly coloured (night) | Strong, sweet | Nectar |
| Birds | Hanging or tubular, copious nectar | Vivid red | Absent | Nectar |
| Bats | Large, strong, single flowers or brush-like | Greenish, cream, purple | Strong at night | Nectar, pollen |

flowers: butterflies will visit predominantly bee-flowers such as privet (*Ligustrum* spp.), alder buckthorn (*Frangula alnus*) and lime (*Tilia* spp.) in Europe.

Colour can also be used to discourage insects from visiting flowers that are already pollinated. For example, in the horse chestnut (*Aesculus*

Figure 5.2 Magnolia flower (*Magnolia wilsonii*) which appeals to a wide range of insects by offering an open flower, a bright colour and many stamens. Keele University, England.

*hippocastanum*), markings on the petals change from yellow to red in pollinated flowers. At the same time, nectar production is reduced. To bees the red appears black and these unattractive flowers are avoided. Both the tree and bee gain; precious time is not wasted on flowers that do not need pollen and which contain no reward. Why doesn't the tree simply shed the petals from the pollinated flowers? In experiments with Brazilian *Lantana* shrubs, which repel butterflies in a similar way, the insects homed in even on flower heads with a large proportion of individually unattractive purple flowers. It seems the pollinated flowers of *Lantana* and horse chestnut are retained because they are bright and conspicuous and attract insects from a distance, and the flower colour directs the insect to individual flowers once it has arrived. For the horse chestnut this is particularly useful in its dry native habitat of Greece and Albania since trees are fairly widely scattered and the large signal of intact but pollinated flowers is useful for attracting pollinators from over a long distance. In a similar way, woody plants such as dogwoods (*Cornus* spp.) and bougainvillea (*Bougainvillea spectabilis*) use large petal-like bracts (really modified leaves) to help make small, inconspicuous flowers more attractive to pollinators; and the European guelder rose (*Viburnum opulus*) uses large, sterile, white flowers around the outside of the flower head to advertise the smaller and less showy fertile flowers in the middle.

Birds are also important pollinators of trees, especially in the tropics with hummingbirds and honeycreepers in the New World, and honeyeaters and lorikeets in the Old World. Bird-pollinated flowers tend to conform to a pattern (Box 5.2), exemplified by the fuschias commonly grown in gardens: tubular flowers (to hide the copious nectar from the wrong pollinators), brightly

Figure 5.3 Flowers of a eucalypt (bell-fruited Mallee, *Eucalyptus preissiana*) from SE Western Australia.

coloured orange or red (although many Australian bird flowers such as eucalypts are yellow or white) but odourless (birds have a poor sense of smell). However, in the dry climate of Australia the bottlebrush trees (*Banksia* spp.) and other related members of the Proteaceae are typical bird-pollinated flowers with masses of protruding stamens to dust the birds' feathers as they suck the copious nectar. The eucalypts take this further; the petals have been lost and the calyx is modified into a cap over the flower (the operculum) which is pushed off as the bud opens; the great tuft of stamens is the real attractant to birds (Figure 5.3).

Although commonest around the tropics, bird pollination is occasionally found in temperate areas. In Europe small warblers and blue tits have been seen taking nectar from gooseberry, cherry, *Mahonia* and almond. A most striking example is the furry catkin of willows (look ahead to Figure 5.14), which for many years were thought to be wind- and insect-pollinated. The stamens and stigmas are exposed for wind pollination but each flower in a catkin also has a large nectary producing a large and easily seen glistening drop of nectar. Moths are well known for using these willows for food in early spring but more importantly there are also many records of blue tits feeding on this nectar which is readily accessible to such a small-beaked bird (see Kay 1985 for further details). Pollen is clearly seen liberally dusting the face and chest feathers as they fly from bush to bush. Despite being large and warm-blooded, and so needing a lot of energy, it is calculated that blue tits in Britain can get their total energy needs for a day in less than 4 hours of feeding on willows (although they may need to balance their diet with insects). This demonstrates how rich and abundant nectar in bird-pollinated flowers can be. Indeed the nectar from Australian bottle-brush trees (*Callistemon* spp.) is gathered for food by the Aboriginal peoples. Bird pollination is obviously very expensive to the plant; is it worth it? Where there is a comparative dearth of highly developed flower-visiting insects, such as in the tropics and Australia, or where the tree is hedging its bets, as in the willows, the answer appears to be yes. Certainly birds are very effective pollinators and will move pollen further than bees. They also show considerable constancy to flowers of a single species and can visit many thousands of flowers a day.

Mammals also make good pollinators, especially the flying ones. Bats pollinate a number of tropical trees including the kapok tree (*Ceiba pentandra*), the fluffy seeds of which has been used to stuff furnishings (again look forward to Figure 5.19), balsa (*Ochroma lagopus*) and the infamous durian (*Durio zibelthinus*) whose fruits smell like sewers and old socks but are a delicacy in Southeast Asia (from experience I can say that they are an acquired taste but worth persisting with!). Bat-flowers tend to be large dull things (bats are colourblind) which open at night, producing copious nectar and a sour or musty smell. Many species flower when they are leafless, while others produce flowers away from the leaves such as on the trunk or edge of the canopy to give plenty of room for manoeuvre. Some bats take nectar while hovering, others land (requiring a strong flower to take the weight). The baobab (*Adansonia digitata*) is an example of the latter (Figure 5.4); the bat hangs onto the tufts of purple stamens, from which they lap up the nectar.

Non-flying mammals tend to be poor cross-pollinators. Nevertheless, a variety of monkeys and possums are implicated in the pollination of trees. Some proteas (*Protea* spp.) of South Africa are pollinated by rodents, and the traveller's palm in Madagascar (*Ravenala madagascariensis*; Figure 2.1b) by a lemur. In Australia the bird-pollinated banksias and eucalypts are also visited by marsupial mice and honey possums, equipped with long tongues for feeding on nectar, and fur which picks up pollen. Perhaps most bizarre is the knobthorn acacia (*Acacia nigrescens*) in semi-arid savannahs of Africa that appears to be pollinated by giraffes that come to eat the flowers at the end of the dry season! Although the trees lose flowers (they make up to 40% of the giraffe's annual diet) the long-legged animals can travel over 10 miles a day between stands of trees, ensuring pollen is well spread.

## Treating your guests right

Flowers are not always passive dining halls: they can be quite abusive to their guests. Broom (*Cytisus scoparius*), a European shrub of heaths and open woods,

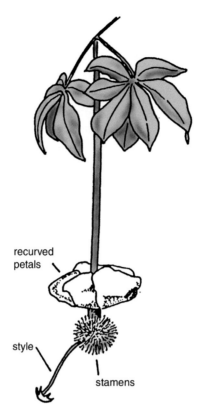

recurved
petals

style

stamens

Figure 5.4 Bat pollinated flower of the Baobab tree (*Adansonia digitata*). From: Proctor, M.C.F. & Yeo, P. (1973) *The Pollination of Flowers.* Collins, London.

has explosive pea-like flowers. As the bee lands, the two bottom 'keel' petals are pushed apart by the insect's weight releasing the sprung stamens and style. Five short stamens hit the bee's underside and five long ones and the style strike the bee on the back of the abdomen (Figure 5.5). Like a firework, each flower is used just once, delivering and collecting pollen in one explosion to an insect of the correct weight. The mountain laurels (*Kalmia* spp.) of N America are also explosive. Here the 10 stamens arch back with the tip of each held in a little cavity in the petal (Figure 5.6). When an insect lands on the flat flower and pushes the petals down, the stamens are released with a jerk, dusting the unsuspecting insect. Other shrubs have stamens that are not captive but 'irritable'. The closely related barberrys (*Berberis* spp.) and Oregon grapes (*Mahonia* spp.) have six stamens pressed against the petals. When an insect pushes against the base of the stamens they spring inwards within 45 thousandths of a second to dust the insects with pollen. Unlike the one-off mechanisms in broom and mountain laurels, the barberry stamens slowly bend back to their original position over several minutes although *Mahonia* only returns to an erect position.

(a)　　　　　　　　　　　　　　　　(b)

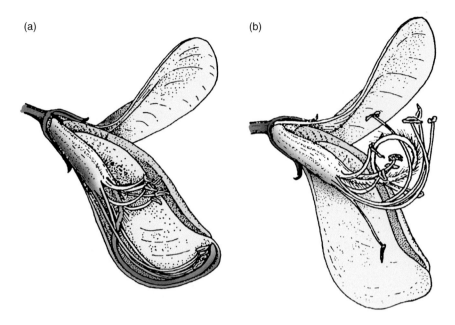

Figure 5.5  Broom (*Cytisus scoparius*). (a) A newly opened flower with half the calyx and corolla removed to show the stamens and style held under tension. (b) The flower 'exploded' following an insect visit. From: Proctor, M.C.F. & Yeo, P. (1973) *The Pollination of Flowers*. Collins, London.

Figure 5.6  Flowers of mountain laurel (*Kalmia latifolia*) from eastern USA showing the captive stamens. Photograph by pfly (originally posted to Flickr as Mountain Laurel), reproduced under Creative Commons Attribution-Share Alike 2.0 Generic, courtesy of Wikimedia Commons.

Luring the right pollinator is a matter of providing the right attractions and the right rewards. The attractants are usually fairly obvious, various colours and smells which we often find quite aesthetically attractive ourselves. Plant fragrances contain a huge range of organic compounds; over 1700 have been identified through all plants. Some uses of fragrance may be less obvious. For example, the male and female cones of burrawang cycads (*Macrozamia*, spp.) native to Australia attract pollen-feeding thrips by smell. To stop the thrips being too comfortable on the male cones and never leaving, the cones periodically heat up and produce a very high concentration of smell to drive the thrips away with hot, toxic odours. The female cones maintain a gentle smell, a thousand times weaker than the male cones, to attract the evicted thrips, and thus to complete pollination. This is like being attracted by a woman's perfume but being repelled by the concentrated smell of the same perfume in the duty free shop of a hot airport (or is that just me?). Terry *et al.* (2007) give more details of this wonderful mechanism.

In terms of rewards, nectar is mostly sugars with (usually) small quantities of amino acids to supplement the diet of those animals which largely depend upon it for food. The precise nectar composition is tailored to the preferences of the main pollinator. Pollen is also used as a food by many animals, including bees, some butterflies and bats. This may also be tailored to the pollinator; in some New World trees, at least, the pollen contains amino acids which are apparently useless to the plant but essential for its bat pollinators.

Trees may provide other rewards. The cocoa tree (*Theobroma cacao*) – from which we derive chocolate – is pollinated by biting midges that breed in the decaying pods. Other tropical trees 'pay' moths one leaf crop which is devoured by the caterpillars and in return the adults do the pollinating. Perhaps the ultimate relationship between plant and pollinator is found in the figs, many of which are large trees or stranglers (Chapter 4). Here the 'flower' looks more like a gourd-like green fruit because the base of the flower head has grown up around the flowers to create a hollow sphere lined on the inside with male and female flowers (called a syconium; Figure 5.7i). Pollination of different fig species is by specific small female fig wasps. She is attracted to the right species of fig purely by the smell of the fig and then has to squeeze in through a small hole, losing her wings and antennae as she does so (Figure 5.7ii). Once inside, the female wasp sets about laying her several hundred eggs into the flowers (oviposition). At this stage the female flowers are ripe and are pollinated as she wanders around with the pollen she has brought with her. The eggs hatch and the grubs feed and develop inside the flowers, their presence stimulating the tree not to drop that fruit (Figure 5.7iii). Not all is lost for the fig because it has two layers of ovaries; the upper are infested with grubs but the lower are too deep to have received eggs and quietly get on with growing seeds. Several weeks later (Figure 5.7iv) tiny wingless males hatch out, locate female-containing flowers, chew their way in and mate. Their last job is to tunnel out of the fig, but since they are wingless they are doomed to go nowhere. Once the fig is punctured, the accumulated

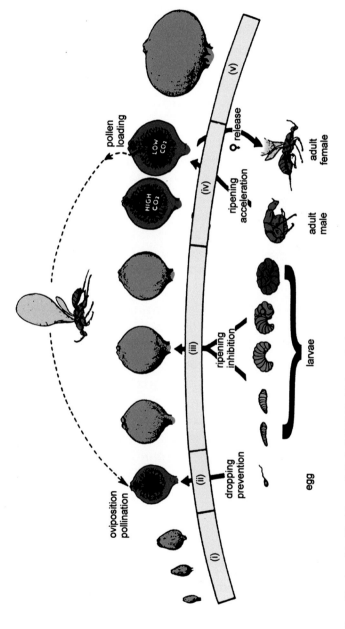

Figure 5.7 Fig pollination. Modified from: Galil, J. (1975) Fig biology. *Endeavour, New Series*, **1**, 52–56. With permission from Elsevier Science.

high levels of carbon dioxide leak out. This is the signal for the male flowers to ripen and for the female wasps to emerge. On their way out of the male-made exit holes the females collect pollen. With some wasps, like that of the common fig, this is a matter of getting coated on the way out, but in others the females quite literally stop and stuff special pockets with pollen using their legs. Then the cycle starts again with the female hunting by smell for another flower of the same species, leaving the fig of its birth to ripen, change colour and hopefully be eaten to spread the seeds. In case this puts you off eating figs ever again, relax; figs for consumption come from *Ficus carica* which is dioecious, and just female trees are grown which are parthenocarpic (see below), developing fruit without seeds or fertilisation!

Keeping the wrong animals from stealing pollen or nectar without offering pollination services is often a matter of hiding away the pollen and nectar so only the right animals get the reward. An obvious example is putting nectar at the end of a long spur so that only long tongued butterflies or moths will bother visiting. Under this category can be classed 'buzz pollination'. Some tropical trees, exemplified by the Indian laburnums (*Cassia* spp.) in the New and Old World have anthers which do not split open to reveal the pollen, as is normal, but have small openings in the end of the anther. The only way to get the pollen out is to shake the anther at the right frequency. An alighting bee sits over the stamen and rapidly vibrates its flight muscles (but not the wings) causing the pollen to spray out of the pores and catch on the hairy bee. This is quite an event and can be heard up to 5 m away.

Pollinators, of course, are not dumb servants and have ways of cheating the system. *Vespula* wasps in Europe, for example, steal nectar from heathers by boring holes in the side of the flower. Numerous similar tropical examples involving insects and birds could have been chosen. This entrepreneurial spirit explains why exotic trees not known for self-pollination can sometimes set seed even when its natural pollinator is absent. But this is not always so: oil production from the oil palm (*Elais quineensis*), originally from W Africa, was greatly improved in Malaysia after the introduction of its natural pollinator, a weevil.

## Wind pollination

Wind pollination is often seen as being primitive and wasteful in costly pollen and yet it is surprisingly common, especially in higher latitudes. There must be a good reason for this. It is tempting at first to say that this is explained by there being fewer insects and higher wind speeds in northern forests. But the whole answer is a little more subtle. Wind is very good at moving pollen a long way. Pollen can be blown for hundreds of kilometres; only birds can get pollen anywhere near as far. The drawback is that wind is obviously unspecific as to where it takes the pollen. It is like trying to get a letter to a friend at the other end of the village by climbing onto the roof and throwing an armful of letters into the air and hoping that one will end up in their garden. For the relatively few dominant tree species that make up temperate forests, where there are many individuals of the same species within pollen range, this is quite a safe gamble. If all my friends around me were throwing letters off roofs, I'd be bound to get one. Indeed, wind pollination is found in other groups with large frequent clones such as grasses, sedges and rushes. By contrast, in the tropics, where each tree species has few, widely scattered individuals, the chance of wind blowing pollen to another individual is sufficiently slim that animals are a safer bet. Even tall trees in the tropics emergent above the surrounding tree canopy are usually not wind-pollinated despite being in windy conditions. In a similar way, trees in temperate forests that *are* insect pollinated, such as whitebeams (*Sorbus* spp.), hawthorns (*Crataegus* spp.) and apples, tend to grow as solitary, widely spread individuals.

## Modifying the flower for wind

Since wind-pollinated flowers have no need to attract insects or other animals, they have dispensed with bright petals, nectar and scent. These are at best a waste and at worst an impediment to the transfer of pollen in the air. The result is insignificant-looking flowers and catkins.

Despite looking so nondescript, wind-pollinated flowers and catkins are based on normal flower structure. The odd-looking mass of elm flowers (*Ulmus* spp.) that appear in early spring before the leaves (Figure 5.8) are perfect little flowers with four stamens and a central ovary with two styles, all stretched out into the wind. The petals have been reduced to a four- or five-lobed fringe around the base. In many ways, oaks are similar except here the male and female parts are borne in separate flowers (the arrangement of sexes is further discussed below) both appearing with the opening leaves. The male flowers are very simple; the remains of petals (a 4–7 lobed perianth; see Figure 5.1) are dwarfed by the 4–12 stamens offering copious pollen straight to the wind (Figure 5.9). To make them even more accessible to the wind the male flowers are loosely grouped together on hanging catkins up to 10 cm long. The female flowers are surrounded at their base by a series of overlapping bracts (modified leaves) which will go on to form the cup of the acorn (Figure 5.9b, c). Just poking out above these bracts is all that is left of the petals and sepals: a toothed border surrounding the three styles. Other wind-pollinated trees have their flowers in even more pronounced catkins ('little cat' after the resemblance to a kitten's tail). Some put just the male flowers in catkins (including

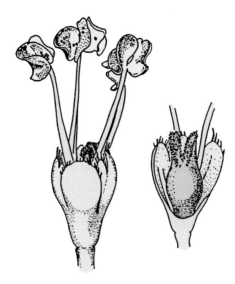

Figure 5.8 One elm flower from a mass, showing the four stamens and the ovary with stigmas in the cut-away section. From: Proctor, M.C.F. & Yeo, P. (1973) *The Pollination of Flowers*. Collins, London.

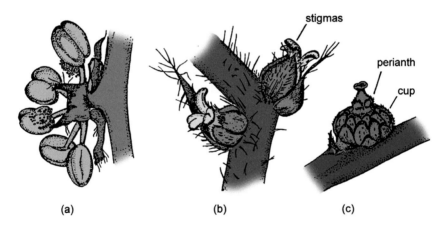

Figure 5.9 Oak flowers (*Quercus robur*). (a) Side view of a single male flower on a catkin; (b) two female flowers; and (c) a young developing acorn. From: Proctor, M.C.F. & Yeo, P. (1973) *The Pollination of Flowers*. Collins, London.

hazel, *Corylus avellana*, and alders); others use them for both sexes (hornbeam, birches, poplars and walnuts). Even though they look so strange the catkins still contain 'normal' flowers. For example, in birch (Figure 5.10) the male catkin is made up of a series of scales or bracts below which nestle three flowers, each with two deeply divided stamens. The remains of the perianth are there but very small and insignificant. The female catkin is made up in a very similar way. A catkin is really a branch with leaves (bracts) and flowers

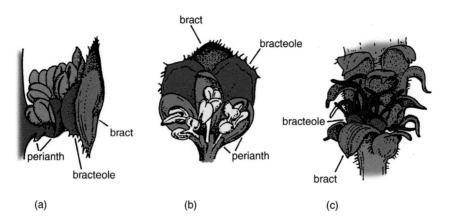

**Figure 5.10** Birch flowers (*Betula pendula*). (a) Side view of a group of male flowers borne on a single catkin scale; (b) a similar group seen from below; and (c) part of a female catkin. From: Proctor, M.C.F. & Yeo, P. (1973) *The Pollination of Flowers*. Collins, London.

that has been condensed down into a small flexible dangling branch: a perfectly adapted way of getting pollen into the wind.

Conifer cones (botanically called strobili) are built in a similar way to catkins. A central stem or axis bears the (usually) spirally arranged scales (really modified branches), and it is these scales that carry the male and female parts. In most conifers the female cones have two ovules on each scale which will become seeds. In the ephemeral male cones (which drop off once they have shed their pollen), the papery scales bear pollen sacs. The ovules in conifers are 'naked', unprotected by an ovary and so with no stigma and no style.

## Sending pollen by air

Wind pollination does, of course, require a lot of pollen. Birches and hazel can produce 5.5 and 4 million pollen grains *per catkin*, respectively (pollen production is affected by many factors such as weather, and some trees like ash and elm follow 3-year cycles of maximum pollen production). There are various adaptations to help as much of the pollen as possible go as far as possible. Most deciduous wind-pollinated trees produce their pollen while the canopy is bare of leaves to reduce the surrounding surfaces that 'compete' with the stigmas for pollen. Evergreen conifers have less to gain from spring flowering and indeed some flower in the autumn (true cedars, *Cedrus* spp.) or winter (northern incense cedars, *Calocedrus* spp. and the coastal redwood *Sequoia sempervirens*). They improve pollen dispersal by placing the cones on the ends of branches.

Pollen produced higher in the canopy is likely to go further; it is windier (and gustier) and the pollen can be blown further before hitting the ground (however, conifers tend to keep male cones lower to help prevent self-pollination; see below). Moreover, in dangling catkins like hazel, the scales touch and so hold the pollen in until the wind is strong enough to bend the catkin, ensuring that pollen is only shed into the air when the wind is blowing hard. Weather is also important. Pollen is shed primarily when the air is dry to prevent too much sticking to wet surfaces or being knocked out of the air by rain. Despite these adaptations, much of the pollen fails to leave the canopy and only between 0.5 and 40% gets more than 100 m away from the parent. But once this far, significant quantities can go a kilometre or more. Indeed, pollen can travel many thousands of kilometres at high altitude although this is undoubtedly of such low density that it has little value in pollination. Since these are all floating around in the air, is it any wonder that wind-pollinated trees are a major source of hay fever.

Once the pollen has been snatched by the wind, the fate of the pollen is obviously up to the vagaries of the wind, but not everything is left to chance. Wind-borne pollen is dry, rounded, smooth and generally smaller than in insect-pollinated plants to help it fly through the air (20–30 μm diameter[1] in wind-pollinated hardwoods, 50–150 μm in conifers compared to 10–300 μm in insect-pollinated plants). But size is a two-edged sword. Small grains may be blown further but they are also more prone to be whisked round the waiting stigma because smaller particles tend to stay entrained in the streamlines that flow around the stigma. But stigmas create turbulence and this 'snow fence' scenario may help pollen stick in the low air speed of the turbulence. Moreover, pollen grains start with a negative charge on the anther but acquire a strong positive charge as they fly through the air, and female flowers have a negative charge which may also help pluck pollen electrostatically from the streamlining air.

Conifers don't have sophisticated stigmas to catch pollen, but most do produce a sticky 'pollination drop' at the tip of each scale (Figure 5.11). Pollen sticks to the drop (probably aided by turbulence) and is pulled down towards the young unfertilised seed as the drop dries. This may be less passive than it seems because if pollen is added to a pollination drop it disappears within 10 minutes while others remain unchanged for a few days to a few weeks. Incidentally, the pollen grains of pines and some other conifers have two air-filled bladders on the sides which are said to help the flotation of the relatively large grains in the air. But settling rates for pollen in still air are the same for conifers and hardwoods (3–12 cm per second on average) and the sacks are reported to

---

[1] 1 μm is 0.001 mm.

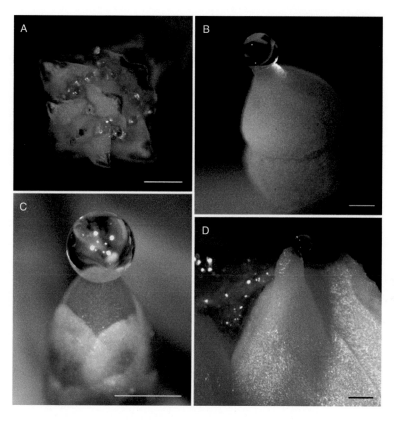

Figure 5.11 Pollination drops in conifers. (a) Lawson cypress (*Chamaecyparis lawsoniana*), (b) ginkgo or maidenhair tree (*Ginkgo biloba*), (c) a hybrid yew (*Taxus* x *media*), and (d) Douglas fir (*Pseudotsuga menziesii*). Each drop arises from one ovule. Scale bars are 1 mm long. From: Coulter, A., Poulis, B.A.D. & von Aderkas, P. (2012) Pollination drops as dynamic apoplastic secretions. *Flora*, 207, 482–490.

shrink in dry air anyway. It may be that the sacks are more important in orientating the pollen grain on the pollination drop. Although pollination drops are found in most conifers (including primitive conifer-relatives such as cycads, the ginkgo (*Ginkgo biloba*) and Gnetales; see Chapter 1) they are not found in true firs (*Abies*), cedars (*Cedrus*), larches (*Larix*), Douglas firs (*Pseudotsuga*), hemlocks (*Tsuga*), the southern kauris (*Agathis*) or monkey puzzle relatives (*Araucaria*). Instead, they rely on a variety of shaped cone scales to effectively trap pollen as it drifts past. Douglas fir, for example, has a slippery slope leading right down to the ovule, sending any falling pollen on a helter-skelter ride down into the depths of the cone away from any wind that might try to tear the pollen away.

## Blurring between animal and wind pollination

Although the distinction between wind and animal pollination, with their very different sorts of flowers, appears to be clear cut, it is likely that in many trees there is a balance between the two. Pollen from all the best known European wind-pollinated trees (oak, beech, ash, birch, hazel) has been found on honey bees. And even trees normally thought of as being insect-pollinated and which produce abundant nectar (limes, maples, sweet chestnut and euca-lypts) release appreciable quantities of pollen into the air. These cases can be thought of as 'accidental leaking' of pollen to the insects or wind but undoubt-edly really represent an adaptation to avoid putting all the tree's eggs in one basket. This is not a static situation: as environmental conditions change so the balance of advantages of wind versus animal pollination can change, such as we have seen in willow on the cusp of wind, insect and bird pollination. Nor is this blurring restricted to hardwood trees. The primitive conifer-like cycads and welwitschia (*Welwitschia mirabilis*; Figure 2.1d) have a pollination drop and appear to be wind-pollinated but it is possible that the pollination drop acts as nectar to attract insects to the female cones while pollen attracts them to the male cones.

## The problem of being large

A large animal-pollinated tree in full flower faces distinct problems because of its size. It needs to produce enough reward in its flowers to attract the pollin-ators but runs the risk of producing so much within one canopy that the pollinators linger rather than go on to the next tree. Moreover, if all other individuals of a species are in flower (which they would need to be to ensure cross-pollination) there may not be enough pollinators to go around the huge number of flowers produced.

One solution to these problems is to flower when other species are not so that you have the pollinators to yourself. In temperate regions this is con-strained by having to work around winter, but flowering of insect-pollinated trees is indeed spread over a large part of the summer. But this still does not solve the problem of encouraging pollinators to leave the abundant food of one tree to go to another. Some large tropical trees get around this by having some parts of the tree in flower, others in bud and others in fruit, giving pollinators the impression of a series of small trees. A more extreme example, used by other tropical trees, is to produce just a few flowers at a time over a long period, possibly all year. Hawkmoths, hummingbirds, bats and, especially, solitary euglossid bees exploit these extended blooming trees by 'traplining'. Here they fly over complex feeding routes repeatedly visiting widely spread flowers, just like a trapper in the Arctic visiting his spread out line of traps. These bees can travel more than 20 km during these daily excursions (by comparison, temperate bees move pollen comparatively shorter distances: honeybees routinely travel 600–800 m and bumblebees 1.5 km). This appears to benefit the pollinators by saving them time looking for nectar, and helps the trees by moving pollen longer distances than randomly searching bees. The benefits to the bees may not be entirely for food; it has been suggested that the bees show territorial behaviour involving scents taken from plants.

At the other extreme, however, are the hundreds of tropical tree species in many different families which are 'mass blooming', showing synchronous opening of huge numbers of flowers across a large area. This is especially seen in the aseasonal (no regular dry period) tropical evergreen forests of Borneo and Malaysia where the tallest canopy trees (mostly species of dipterocarp) may all flower and fruit synchronously at intervals of 2–7 years. 'Over a period lasting a few weeks to a few months, nearly all dipterocarps and up to 88% of all canopy species can flower after years of little or no reproductive activity... The region over which such a mass-flowering event occurs can be as small as a single river valley or as large as northeastern Borneo or peninsular Malaysia' (Ashton *et al.* 1988). With individual dipterocarp trees presenting up to 4 million flowers, this mass flowering seems to ignore the above problems. But as is often the case, the solutions are subtle and not immediately obvious. Firstly, the dipterocarps flower sequentially, in the same order each time (Figure 5.12), thereby reducing competition for pollinators. Secondly, although an individual tree can flower over 2–3 weeks, individual flowers may last only a day. The main pollinators are small thrips, attracted by the overpowering scent of an opening flower to spend the night tramping around inside, eating it and the pollen. In the morning the flower falls, complete with its happy thrips (leaving the ovary on the tree to develop into a fruit). As the next wave of flowers opens the following evening the thrips fly up for their next feed, blown by even light winds to, hopefully, land in a flower of a different tree to the night before. Once there they deliver their load of pollen as they wander around feeding. This works only because, unlike most tropical trees, the dipterocarps tend to grow clumped together (the heavy winged seeds of dipterocarps spin like maple 'helicopters' but do not go far). Where do all the necessary thrips come from? They persist at low levels between mass flowerings and explode in numbers as the trees come into flower. Lastly, cross-pollination is encouraged by many of the dipterocarps being self-sterile or even having separate male and female trees (see below).

In general, mass blooming trees use relatively unspecialised pollinators, which gives them a greater number of pollinators to go round. The strong

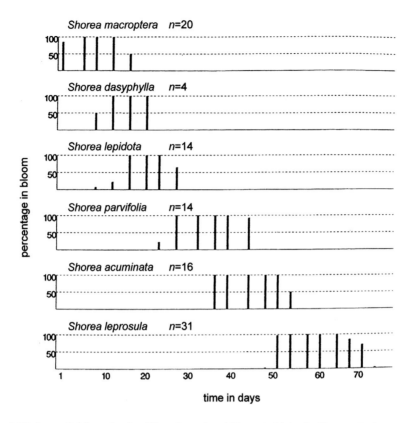

**Figure 5.12** Sequential flowering by different species of *Shorea* in Malaysia. The graph shows percentage of trees in bloom on a given day after 14 March 1976 for six species (*n* is the number of trees looked at in each species). From: LaFrankie, J.V. & Chan, H.T. (1991) Confirmation of sequential flowering in *Shorea* (Dipterocarpaceae). *Biotropica, 23,* 200–203.

visual image created by mass flowering can attract pollinators in large numbers from over great distances. Up to 70 species of bee have been observed visiting a single tropical tree; such high rivalry can work to the benefit of the tree because in the ensuing battle for flowers some bees can end up being ejected and going to the next tree. But there may be even more subtle ways of encouraging insects to move from one tree to the next. It has been suggested that trees are a continually changing source of quantity and quality of nectar and bees regularly sample nectar from different trees to find the best current source. Why go to this trouble? The answer seems to be that mass blooming and fruiting can be an important mechanism for reducing the predation of flowers and seeds (see page 189).

Wind-pollinated trees do not entirely escape the problems associated with mass flowering. Wind dispersal of pollen is not affected by the number of flowers but the big danger is swamping a tree's flowers with its own pollen.

Even if self-sterile, the stigmas could get so coated with the tree's own pollen that other pollen will physically not be able to reach the stigma. How a tree solves this problem is discussed below. Even in those trees where self-pollination is possible, too much pollen may be counterproductive: it has been experimentally shown in walnuts that adding extra pollen to flowers resulted in fewer fruits and seeds: too much of a good thing!

## What controls when flowers appear and then die?

Most plants can be divided into short-day, long-day or day-neutral plants. Short-day plants start to flower when the day length is shorter than a critical shortness and so flower in early spring and autumn, and long-day plants when days are longer than a critical length and so flower towards mid-summer. Day-neutral plants are, as the name suggests, not fussy about day length. In reality it is not the length of the day that plants measure but the length of the night. This is done by a set of pigments called phytochromes that, in effect, allow the plant to track how long the nights are. Having said all this, woody plants in general are less sensitive to day length than most herbaceous plants, but there are a few trees, especially near the tropics, that are stimulated to flower by short days, including tea (*Camellia sinensis*), coffee (*Coffea* spp.), bougainvillea (*Bougainvillea* spp.) and poinsettia (*Euphorbia pulcherrima*).

In some trees the monitoring of day length is very precise, as seen in the synchronous flowering of many tropical species. Here day length variation through the year is minimal yet a large number of species flower near the spring equinox, others near the autumn equinox, and a few at both times. It seems that the trees are still measuring changes in day length between the equinoxes even though this is less than a minute per day over the year. These trees tend to flower around the time that sunrise or sunset changes are fastest in spring and autumn. Evidence from the rubber tree (*Hevea brasiliensis*) suggests that the small increases in amount of bright sunlight at the equinoxes might also be part of the trigger. For others, such as the dipterocarps of Southeast Asia, the trigger for mass flowering appears to be irregular periods of drought providing it has been long enough since the last mass flowering year for sufficient food reserves to accumulate. For the majority of trees, flowering is triggered not by day length but by other features of the environment, particularly temperature.

How many flowers are produced each year is determined partly by weather (since most flower buds start forming the autumn before they come out; see Chapter 6) and partly by internal food stores (this is explored further under masting below). The need for abundant sugar supplies to support flowering explains why girdling of the trunk (cutting through part or all of the bark around a tree) can lead to increased flower and seed production by stopping sugars going down to the roots. In a study using 42-year-old Japanese larch trees (*Larix leptolepis*), 20 trees were double girdled, that is two cuts were made into the cambium around 60% of the circumference from opposite sides, one above the other 20cm apart. The girdled trees produced an average of 398 cones and other trees left untouched produced an average of 2!

Unlike leaf death, which is usually dependent upon environmental changes (light, temperature; Chapter 2), the death of flowers is wholly controlled internally. Thus, the life span of a flower is ultimately genetically controlled, so each tree species will tend to keep flowers for a set period if everything else is equal. The main factor altering this is pollination: once pollinated, the flower usually dies very quickly. This makes good economic sense since the flower is costly to maintain and it is a waste of energy and often water to keep it any longer than necessary. Also, a flower is a good point of entry for pathogens.

## Self- and cross-pollination

It has been assumed so far that cross-pollination is better than self-pollination. This perhaps needs to be justified. In the long-term it is better for trees and other plants to be cross-pollinated (i.e. pollinated with pollen from a different individual) to prevent inbreeding and the expression of harmful genes. Studies have demonstrated the many effects of 'inbreeding depression' such as poor germination, reduced survival of seedlings, chlorosis of seedlings (lack of chlorophyll) and reduced height growth. Indeed 'hybrid vigour', where a hybrid is more vigorous than either of its parents, may partly be an expression of this outcrossing in comparison to the parents which may be inbred to an unknown degree. The other side of this coin is that cross-pollination helps mix up genes within a group of trees so that each new tree is as genetically different from others as possible; this has important implications for helping a group of trees cope with attacks by insects and fungi (see Chapter 9). Having said this, self-pollination is a useful short-term way of producing some seeds if no other pollen is available (although trees do this less well than other plants probably because being long-lived, missing a year or two of seeds is less important). The common solution is to aim for cross-pollination but to fall back on self-pollination if all else fails. The mechanics of doing this are varied, as discussed below.

### *Self-incompatibility*

Some trees are completely 'self-incompatible' and so cannot pollinate themselves. This is particularly common in wind-pollinated trees where lots of 'self' pollen is likely to be blowing around. Thus in beech (*Fagus sylvatica*), self-pollination only leads to empty nuts, and well-spaced trees produce fewer nuts than those in woodlands. But self-incompatibility is also found in insect-pollinated trees such as apples. Most apples are cloned by grafting and so an orchard of one variety is, in effect, all the same tree; the solution is to plant two or more other varieties. This explains why a lone apple tree may be barren (unless it has several varieties grafted onto one stem, while a single Victoria plum (self-fertile) fruits well. The majority of trees, however, have at least some capacity for self-pollination as an insurance policy. The European mountain ash (*Sorbus aucuparia*), which as the name suggests can live in the hostile uplands, has widely spread stamens in warm weather, exposing the abundant nectar to attract pollinators but in dull weather, when few insects are around, the stamens converge to touch the stigma leading to self-pollination. Others, by a variety of physical and chemical means, favour pollen from another tree but will often accept their own pollen if no other is available. One chemical mechanism is for a tree's own pollen to grow more slowly through the style; for example, the pollen of Timor mountain gum (*Eucalyptus urophylla* – from East Timor) takes 120 hours to grow through its own style while pollen from a different tree takes 96 hours, and in rose gum (*E. grandis*) it is 96 hours versus 72 hours. This means that 'foreign' pollen will fertilise the seeds if it is available, and if not, its own pollen will eventually do the job.

## Producing the sexes at different times

A common mechanism in trees to reduce or prevent self-pollination is to mature the male and female parts of the flower at different times (called dichogamy). Predominant in wind-pollinated plants (but also found in, for example, the magnolias, pollinated by insects) are flowers where the female stigma is receptive to pollen before the stamens start producing pollen (proto-gyny). The advantage is obvious: the stigma can catch 'foreign' pollen before being besieged by its own abundant pollen, but can be self-pollinated if foreign pollen is absent. The opposite situation, where the male stamens produce their pollen and wither before the stigma becomes receptive (protandry), is more common in animal-pollinated flowers. Self-pollination here is trickier but can be achieved by such mechanisms as the bending/curling of the stigma to touch parts of the flower still dusted in its own pollen, or having a late-maturing set of anthers to do the job. Another variation is to have some individuals in a population that are protogynous and some that are protan-drous (officially called heterodichogamy). This is found in many walnut rela-tives (*Juglans* spp.) and around half of maple species. It does work: up to 10% self-pollination has been found in the Manitoba maple (*Acer negundo*).

## Male and female parts in separate flowers on the same tree: monoecy

Where the separation of the sexes in time becomes so great that either the male or female parts occur too late in the flower's life to work very well, it leads to a functional separation of the sexes into different flowers. Thus we move from having 'perfect' flowers (hermaphrodite flowers with the two sexes in one flower, as in limes, elms and horse chestnuts) to 'imperfect' flowers holding just one sex. Such separation (called dicliny) is particularly common in wind-pollinated trees and in some with relatively unspecialised insect pollination. For most animal-pollinated trees such separation is too expensive since pollinators will only be doing half the fertilisation process in any one flower – collecting or delivering pollen – effectively doubling the costs of attracting animals.

Although the male and female flowers are separate, they can be very close. In chestnut (*Castanea*) species, for example, the sexes are normally in separate catkins but female flowers can occur at the base of an otherwise male catkin. Normally, however, there are male and female flower heads which may be on the same branch or in different parts of the tree. This is taken to an extreme in conifers where the female cones tend to be high up in the crown and the males low down (though it is not always so clear cut). Self-pollination is reduced since pollen seldom moves directly upwards but turbulence carries pollen upwards into the canopy of surrounding trees helping cross-pollination. And the heavy seeds, being high up in the canopy, are likely to spread further in the wind once released.

Male and female flowers are not always produced in different parts of the canopy purely for reasons of effective pollination and seed dispersal. In tropical trees especially, flowers and fruits can spring directly from large leafless branches or the trunk (a condition called cauliflory). This is common in trees with large fruits that need a solid support, such as the cocoa tree (*Theobroma cacao*), durian (*Durio zibethinus*) and some figs (Figure 5.13). The Judas tree (*Cercis siliquatrum*) is one of the few cauliflorous species outside the tropical forest; here the inflorescence springs from an old leaf scar or from very dwarf shoots just on the bark.

Unisexual flowers can be either on the same tree (monoecious: both sexes in 'one home') as described above, or different trees (dioecious: the sexes in 'two homes' i.e. two trees).

## Why be dioecious?

A moment's reflection shows that dioecious trees (having separate male and female trees; Figure 5.14) are apparently at a disadvantage. True, it ensures cross-pollination but only roughly half the trees (the ones with female flowers) can produce seeds. Moreover, if the individuals are widely spaced the chances

Figure 5.13 A cauliflorous tree bearing fruits directly on the trunk. Roxburgh fig (*Ficus auriculata*) in Brisbane Botanic Gardens, Australia.

Figure 5.14 Dioecious flowers of goat willow (*Salix caprea*). The branch on the left is from a female tree and that on the right, a male tree. Keele University, England.

of breeding are much reduced: for example, there is no seed production in relict populations of juniper (*Juniperus communis*) in Wales where the remaining males and females are widely separated. Poor seed set by the dwarf shrub, cloudberry (*Rubus chamaemorus*) at the southern limit of its growth in the Pennines of England is considered to be due to the presence of large single-sex clones and a greater frequency of male plants. Dioecy can also make pollination more difficult. Since only half the trees produce pollen, nectar needs to be the main reward, cutting down the potential number of pollinators (and, as noted above under monoecy, animal pollination is more expensive with unisex flowers). Female willow catkins contain three times as much nectar as males and the bird pollinator, blue tits, visit mainly female catkins; presumably the odd visit to a male catkin and most visits to female plants is the most efficient use of the pollinator.

Despite these apparent disadvantages dioecy is particularly common amongst trees. Only about 6% of all flowering plants (angiosperms) in the British flora, and the world, are dioecious but in the temperate forests of North Carolina dioecy is found in 12% of trees (26% if shrubs are included), rising to about 20% (32% with shrubs) in Costa Rica, and 40% for Nigeria. Dioecy is scattered through many families but only a few are entirely dioecious, notably the Salicaceae: willows, aspens and poplars. Dioecy is found in the ginkgo (*Ginkgo biloba*), cycads, yews, many junipers, the kauri of New Zealand (*Agathis australis*), the yellow-woods (*Podocarpus* spp.), about 15% of maples, the temperate holly (*Ilex aquifolium*), tree of heaven (*Ailanthus altissima*), the shrubby butcher's broom (*Ruscus aculeatus*) and many tropical trees. But dioecy is not found in the pines, the monkey puzzle relatives (*Araucaria* spp.), cypresses, or any members of what used to be the Taxodiaceae family (now part of the Cupressaceae) including the redwoods, the Japanese red cedar (*Cryptomeria japonica*) and Chinese firs (*Cunninghamia* spp.).

So why be dioecious? There are several possible answers. Guaranteed out-crossed offspring ensures the highest genetic quality, and dioecy is the only way to absolutely ensure this. As we'll see in Chapter 9, high genetic quality is important in a tree defending itself against attack by fungi and insects, and there is evidence that dioecious plants are prey to a smaller number of pathogens (see Williams *et al.* 2011). A more convincing answer, favoured by Darwin, comes from the well-established link between dioecy and the production of large fleshy fruits containing large seeds. Such fruits are costly to produce (see below) and it may be that trees that do not produce pollen can invest more heavily in bigger seeds and nutritious fruits, and are likely to be more successful (see Chapter 8). However, some studies in rainforest (e.g. Queenborough *et al.* 2009) have found that dioecious trees do not produce bigger seeds than near relatives that are monoecious; in this case it may be superior seed number that is the main advantage. There is also evidence that seeds from dioecious plants are more fertile (e.g. Spigler *et al.* 2012). Finally, dioecy may reduce seed predation because if only half of the trees of a species are producing seeds, they will be harder to find.

Whatever the reason, the dioecious habit can be useful to us. Male poplars are planted in urban areas to avoid the troublesome fluffy wind-blown fruits, and female sumacs (*Rhus typhina*) are planted to ensure showy fruit clusters instead of the dull green male blossoms. Female crack willows (*Salix fragilis*) are usually planted along British rivers because they are supposed to pollard better than males.

## Trees that swap sex

When you think that single-sex flowers are formed by the loss of one sex or the other, it is perhaps not surprising that there are trees that slip back into old ways. Monoecious trees, normally with unisexual flowers, can sometimes produce hermaphrodite flowers. Dioecy (having male and female flowers on different trees) is rarely absolute. For example, a male or female tree may have a few hermaphrodite flowers (as has been seen in the shrubby butcher's broom and the N American gambel oak, *Quercus gambelii*), or flowers of the opposite sex: yew (*Taxus baccata*) may occasionally produce the odd flower or whole branch of the opposite sex. Others go further and change sex completely, once or repeatedly. Male trees of the normally female Irish yew (a clone originally from one female tree: *Taxus baccata* cv 'Fastigata') and female trees of the normally male Italian poplar clones (*Populus* x *euroamericana* 'Serotina') are known. Cycads in cultivation have also been reliably reported to change sex, usually following a period of stress such as transplanting, damage, drought and frost (Figure 5.15).

Others are yet more complicated and seem to write the rules as they go along. The persimmons (*Diospyros* spp.) from N America and Asia can have a few branches of the opposite sex on an otherwise unisex tree; some are consistently monoecious (producing flowers of both sexes) while others produce both sexes some years but not others; and (rarer) a few hermaphrodite flowers may be produced on otherwise male or female trees. The European ash (*Fraxinus excelsior*) and some maples take a bit of beating in their apparent total sexual confusion. As Mitchell 1974 describes, in the European ash some trees are all male, some all female, some male with one or more female branches, some the other way round, some branches male one year, female the next, some with perfect flowers, and the variations may differ on a single tree from year to year.

## The cost of sex

All this gender swapping may seem strange to those reared on human genetics where gender is determined by X and Y chromosomes and cannot be readily changed in an individual. In trees and other plants gender is not as genetically fixed as in higher animals and is frequently affected by environmental

Figure 5.15 A female cone of the Zululand cycad (*Encephalartos ferox*) in the process of breaking up to release its bright red glossy seeds. Male and female cones are borne on separate plants but individual plants can swop gender. Brisbane Botanic Gardens, Australia.

conditions (perceived perhaps through the quantity of food stored in the tree). Indeed the proportion of female flowers on monoecious trees generally increases with better growing conditions and increasing age (younger and smaller plants tend to be mostly or wholly male). This reflects the larger cost involved in maternal reproduction; larger, older and better-growing trees have more resources available to invest in the expensive production of fruits and seeds. Changing from male to female is also a way of going out with a bang: the moosewood (*Acer pensylvanicum*) grows around the Great Lakes of N America, invading gaps in forests; as the young trees are gradually shaded by other more vigorous trees they change from male to female putting all their remaining resources into seeds; it is now or never!

The cost of femaleness is also seen in dioecious species. Trees such as yew, ginkgo and poplars show 'male vigour': the males tend to be taller, flower at an earlier age and live longer because of lower reproductive effort.

In the perverse way of nature, there are examples of trees that are predominantly *female* early on in life. This is true of the New Caledonian pine (*Araucaria columnaris*), a relative of the monkey puzzle, a few pines and young individuals of the N American bigtooth maple (*Acer grandidentatum*). These trees are wind-pollinated and since removal of pollen needs high winds and deposition lower speeds, perhaps short trees tend to be better at accumulating pollen and do better as females.

## Seeds without pollen: apomixis but not parthenocarpy

It is possible for seeds to form without pollen (apomixis). This results in the offspring being identical to the mother. The rose family (Rosaceae) is rich in trees predisposed to apomixis such as species of *Amelanchier* (mespils, Saskatoon berry), *Crataegus* (hawthorns), *Sorbus* (whitebeams, rowans) and *Malus* (apples). Strictly speaking, all but the last need pollen for the seed to develop properly but all the genetic material comes from the mother. This ability seems to arise from hybridisation (producing triploids and tetraploids – 3 and 4 times, respectively, the normal number of chromosomes in each cell) which leaves the new trees infertile (as in many hybrids such as the mule) but they can produce seed anyway by cutting out the need for fertilisation. The result is a stand of trees in which all individuals look the same, but most are likely slightly different from the next group down the valley. Since these 'clones' have fewer differences than is normal between species they are usually referred to as microspecies, although some botanists would claim that they are still all the same species, and others that they are valuable and discrete species!

Apomixis is common in hardwoods but does not appear to be normal in conifers. However, a strange case of apomixis, called paternal apomixis, has been found in a rare Mediterranean conifer, *Cupressus dupreziana*. Pollen and ovules normally contain a single DNA strand of each chromosome (described as haploid) and when the pollen fertilises the ovule, the strands zip up to produce the normal double helix of DNA (referred to as diploid). In the case of this conifer, the pollen remains diploid and when it lands on a female flower of a *Cupressus* species it grows into an embryo inside the seed without fertilising the host embryo and so with no genes involved from the maternal plant. You could perhaps argue that the male is parasitic on the female; the tree equivalent of the cuckoo. The new seedlings look identical to the male parent regardless of what the maternal species might be (see Pichot *et al.* 2001).

Parthenocarpy is similar to apomixis (there is no fertilisation from pollen, and the fruit still grows, as in apomixis) but here no seeds develop ('parthenos' meaning virgin). Many common temperate trees (including maples, birches, ashes, elms, hollies, firs and junipers but not beeches or oaks) are known to produce fruits without seeds. It is possible that cuttings taken from wild parthenocarpic figs led to the cultivation of seedless figs over 11 000 years ago in the Jordan Valley. Parthenocarpic fruit are usually smaller than normal but this has been exploited to produce seedless varieties of clementine, navel oranges, bananas, apples and pears. The black flecks along the middle of a banana are all that is left of the ovules; wild bananas can be 90% seed with dozens of large seeds and very little flesh (although the ones I've tried in

Honduras have been intensely flavoured). Species that are not parthenocarpic can be encouraged to be so if they are given hormones to replace those normally produced by the seed; hence we have seedless varieties of such things as cherries and mangos.

## From flower to fruit

In most hardwood trees, the ovules (young seeds) are fertilised within a matter of hours or days of the arrival of the pollen on the stigma. During this time the pollen grows a thread-like tube through the style and down to the ovule, a journey of a matter of centimetres. In conifers the pollen does the same but in the more open environs between the cone scales. Incidentally, pollen grains of cycads and the ginkgo/maidenhair tree (*Ginkgo biloba*) release motile sperm cells which must swim to the ovule. This is an evolutionary remnant typically associated with more primitive plants such as ferns and horsetails.

The time taken from the formation of flower buds (normally the autumn before they open; see Chapter 6) to the release of ripe seeds varies tremendously. Elms (*Ulmus* spp.) are very quick, shedding their seeds in May to July, just 8–10 weeks after flowering in early spring. The majority of temperate trees, however, disperse their seeds in the autumn. Thus, the whole cycle from formation of the flower buds to release of seed takes around 1 year. Conifers such as western red cedar (*Thuja plicata*), true firs, larches, spruces, incense-cedar (*Libocedrus decurrens*) and the coastal redwood (*Sequoia sempervirens*) produce seeds in the same time span.

Other conifers, including, most pines (Figure 5.16), the true cedars, false cypresses (*Chamaecyparis* spp.), junipers, and the giant sequoia (*Sequoiadendron giganteum*) as well as a few hardwoods (notably a number of N American oaks, including the scarlet and red oaks, *Quercus coccinea* and *Q. rubra*), take 2 years from flower bud to seed fall. In Europe, the cork oak (*Quercus suber*) produces both annual (maturing the same year) and biennial acorns (maturing the following year); Turkey oak (*Q. cerris*) is consistently biennial and holm oak (*Q. ilex*) produces just annual acorns. In pines, fertilisation takes place usually in the spring, 1 year after pollination. A number of pines also delay seed fall but this is usually linked with 'serotiny' where seeds are held in the cone for sometimes decades until released by the heat of a fire (see Chapter 9).

## Types of tree fruit

As mentioned in Chapter 1, one of the most fundamental differences between conifers (really the gymnosperms, which literally means 'naked seeds') and hardwoods (really the angiosperms) is the type of fruit.

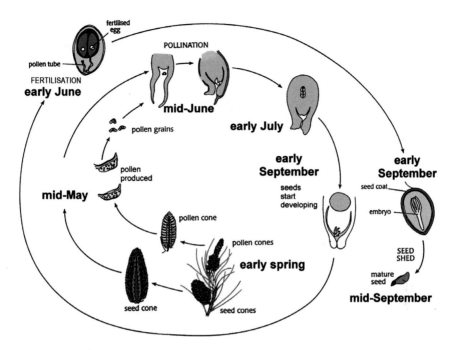

**Figure 5.16** The timetable of reproduction in a pine, starting with flower formation in early spring one year and ending almost 2 years later with the seeds being shed. Modified from: Ledig, F.T. (1998) Genetic variation in *Pinus*. In: D.M. Richardson. *Ecology and Biogeography of Pinus*. Cambridge University Press, Cambridge.

The scales of conifer cones can be bent apart to reveal the seeds without physically breaking anything apart: the seeds are naked. This is true even in the fleshy or fibrous cones of the junipers. The yew berry is really a bare seed with a fleshy red outgrowth from the base of the seed (an aril; see later) and the fleshy ginkgo 'fruit' is made by the seed coat becoming fleshy, so it is still a naked seed.

In hardwoods, the seed is completely enclosed by the fruit and cannot be seen without breaking into the fruit whether this is a dry nut or a fleshy plum. This is because after fertilisation, the ovules become the seeds (Figure 5.1) and the ovary becomes the fruit, completely surrounding the seeds. Box 5.3 shows how hardwood tree fruits are classified and the character-istics of each. You will see that everyday language tends to blur the botanical distinctions between the different fruits. For example, the stone in a date is really just the seed whereas the stone of a plum is the seed plus a hard inner part of the fruit (which can be cracked open to reveal the seed or kernel). Similarly, many things are sold as 'nuts': some are, such as hazel nuts, but walnuts and almonds are strictly speaking drupe stones where the fleshy part of the fruit has been removed (it's still there in pickled walnuts) leaving the hard inner part of the fruit containing the seed. Brazil 'nuts' on the other

---

**Box 5.3** Fruit types found in hardwood trees

The fruit (botanically called the pericarp) is made up of three layers. In dry fruits they appear as one but in fleshy fruits the outer skin (epicarp) encloses the fleshy layer (mesocarp), and the inner layer (endocarp) can take a variety of forms being hard and woody in drupes to juicy in a berry.

**Dehiscent dry fruits (open as they mature)**

*Legume* The usual fruit of the pea family; it splits along both sides into two halves revealing the seeds, e.g. laburnum, gorse and acacia. Some tropical species (e.g. sea-bean; Figure 5.22) break crossways between seeds into small segments.

*Follicle* Like a legume but splitting along just one side, e.g. grevillea.

*Capsule* Fruit splits along a number of weak lines to let the seeds out which are side by side in the fruit, e.g. willow, horse chestnut, spindle, eucalypts and kapok tree (Figure 5.19).

**Indehiscent dry fruits (the fruit stays around the seeds without opening)**

*Achene* Single seed with a dry fruit but not as hard and woody as a nut, e.g. cashew. In the cashew, the achene grows beneath a fleshy 'apple' (Figure 5.20).

*Nut* A common fruit of trees composed of a single seed and a dry fruit that may be hard as in hazel or comparatively soft and easily peeled off as in the acorn. A nut may have a bract or bracts (modified leaves) only loosely attached as a wing (e.g. lime and hornbeam), or more highly modified and enclosing as hazel, the cup of an acorn, and the hard spiny covering of a beech and sweet chestnut (note that these spiny cases do not completely enclose the nut since the stigma and style of the ovary poke out of the end; this separates a nut from the superficially similar capsule of a horse chestnut where the outer spiny case is the fruit itself).

*Samara* The fruit is enlongated into a dry papery wing, as in ash, elm, tulip tree and maple (note the double wing of a maple fruit is really two samaras joined together).

**Succulent fruits**

*Berry* A fleshy fruit enclosing the seeds, e.g. gooseberry, blackcurrant, orange, banana, bilberry and date (the 'stone' is the seed). In a citrus fruit (botanically called a hesperidium), such as a lemon, the outer two layers form the peel (epicarp) and pith (mesocarp) while the inner layer (endocarp) forms fluid-filled hairs or juice-sacs.

---

---

**Box 5.3** (cont)

*Drupe* A fleshy fruit with the innermost layer (endocarp) forming a hard 'stone' around the seed, e.g. plum, cherry, peach, almond, walnut, elder and olive. Sometimes we eat the fleshy part and throw away the stone containing the seed, e.g. plum and peach, or we throw away the fleshy part, crack open the stone and eat the seed, e.g. walnut and almond. Some drupes contain more than one stone, e.g. hawthorn, rowan, holly and medlar. The coconut we buy is just the stone (containing solid and liquid food for the germinating seed), and the coir (removed to make coconut matting and peat-free compost, etc.) is the equivalent of the flesh and skin of a plum.

**False fruits**

Some fruits are supplemented with other structures. In apples and pears (pomes) the base of the flower (the 'receptacle' where all the parts like stamens and petals are joined on) grows up and around the true fruit: the core is the real fruit and the bit we eat is the 'fleshy receptacle'. This is similar in rose hips (containing achenes) and figs (containing drupes) where the receptacle from one (rose) or more flowers (fig) grows up and encases a mass of individual small fruits.

Several flowers can grow together as the fruits develop to produce what appears to be one fruit, e.g. plane, osage orange, mulberry and the cones of alder and banksia: these are multiple fruits. Strictly speaking the false fruits of fig and the double samara of maples could be called multiple fruits.

---

hand are really seeds with a very hard seed coat which grow by the dozen inside a hard woody fruit which is usually thought of as a berry! Another source of confusion is the alder (*Alnus* spp.) which appears to produce cones like a conifer. The cone of an alder is really a woody catkin, and when the 'seeds' fall out they are really complete fruits (small nuts or nutlets) with a dry fruit completely enclosing the seeds.

## What does the fruit do?

The fruit has two roles: protect the seed while it develops and sometimes to help disperse the ripe seeds.

Seeds removed from a tree before they are ripe and able to germinate are a waste for the tree. Moreover, there are many animals angling to eat the nutritious seeds as they develop. In consequence, many fruits (and seeds) are poisonous or astringent when under-ripe, or lacking in nutrients and sugars and so are unattractive to eat (hence the sourness of unripe apples and the high tannin content of unripe persimmons). As the seeds mature and the fruit ripens, the flesh softens, storage materials such as starches and oils convert to sugars, astringent compounds decrease, and to advertise these changes, the skin colour changes.[2] Some fruits, however, will retain some of their poisonous or unpleasant characteristics which reduces the chance of being eaten by the wrong animal.

Fruit colours that stand out against the foliage are useful for attracting birds since they have good colour vision. These fruits are also usually fairly small so they can be easily carried away. Fruits that attract mammals tend to be less brightly coloured since they don't rely so much on visual cues, and they tend to be larger.

When it comes to seed dispersal, you will see in Box 5.3 that succulent or fleshy fruits have a definite role in aiding seed dispersal by animals. Dry fruits can also aid dispersal by forming, for example, the wing of the maple 'helicopter' (samara). Others, such as nuts, do little except add an extra dry protective layer over the seed. Still others, especially those dry fruits containing more than one seed, play no part; they stay on the tree and release the seed to make their own way (technically the fruits that shed seeds are called dehiscent and those that keep the seed inside are indehiscent). Thus the pea-like pods of laburnum split open on the tree, showering the seeds to the ground.

Sometimes where the fruit is not very appetising or nutritious, the seed itself grows a colourful protein-rich covering called an aril. In the yew this forms the bright red 'berry' around the naked seed. In hardwoods the brightly coloured aril can be seen when the fruit opens as in the mace around a nutmeg and the orange aril in some *Euonymus* species (such as the European spindle tree, *E. europaeus*, where the pink aril aids dispersal by birds).

Conifer cones are a special case. Here specialised mechanical tissue at the base of each scale controls closing when humidity is high and opening when it is dry. The cones open wider each time, so seed not lost during the first seed fall is released later: 70–90% of seed usually falls in the first 2 months, the rest falling over winter. To aid the shedding of seeds, most cones gradually bend over from the upright position of pollination to hanging downwards at maturity. Those of the true firs (*Abies* spp.), true cedars (*Cedrus* spp.) and the monkey

---

[2] Colour in a fruit may be throughout, just in the juice (as in a blood orange) or just in the skin (apples). Two pigment groups give most colours, anthocyanins (red, purple, blue) and anthoxanthins including carotenes (pale ivory to deep yellow). Pigment concentrations are often responsible for different colours: 'black' cherries are actually purple and 'white' actually pale red, while yellow cherry cultivars lack the red anthocyanin in the flesh and have a yellow carotene in the skin.

Figure 5.17 Cones of different sizes. (a) Sugar pine (*Pinus lambertiana*), (b) Coulter or big-cone pine (*P. coulteri*), (c) grey or digger pine (*P. sabiniana*), (d) Monterey pine (*P. radiata*), all from California, (e) dwarf mountain pine (*P. mugo*) from Europe, and (f) jack pine (*P. banksiana*) from Canada. The number of needles found in a bundle of leaves varies from 5 (a), 3 (b-d) and 2 (e, f). From: Thomas, P.A. & Packham, J.R. (2007) *Ecology of Woodlands and Forests*. Cambridge University Press, Cambridge.

puzzle relatives (*Araucaria* spp.) are exceptions and stay pointing up. Since their scales are pointing upwards, the only way they can effectively lose seed is for the whole cone to disintegrate on the tree, leaving the central axis on the branch pointing up like a candle. This is why Victorian gentlemen would spend happy hours shooting cones out of the tops of these trees: it was the easiest way to get a more or less intact specimen!

Cone size in pines is very variable (Figure 5.17) but patterns can be seen in their size. Typically the pines with five needles in a bundle (often called the soft pines) have soft cylinder-shaped cones up to 15–50 cm long and even 60 cm in the sugar pine (Figure 5.17a). The 'hard' pines with needles in twos or threes produce harder, rounder cones. Two-needle pine cones (Figure 5.17e, f) tend to be fairly small, golf to tennis ball size, while three-needle pine cones (Figure 5.17b-d) are larger, up to bowling ball size. The largest cone of the three-needle pines is the Coulter or big-cone pine, 25–30 cm in length and weighing up to 2.3 kg each. Why so big? Pines with the biggest cones tend to live in the driest places (the big-cone pine grows in the very arid San Gabriel

Mountains of California) where food for herbivores is in short supply. With lots of hungry animals around, it makes a worthwhile investment to produce large, expensive cones to protect the seeds. Certainly, this appears to have been the driving force behind big cones from when pine first evolved. Early in their evolution, male and female conifer cones were of a similar size but the female cone became bigger (by enlargement of the scales) and woodier in the Jurassic period (180–135 million years ago). At this time the long-necked sauropods *Diplodocus* and *Barapasaurus* appeared and, although there's no evidence that they browsed trees, this may have been the cause. But early mammals and birds could equally well have been the culprits. However it began, this adaptation of cones to resist herbivores eating seeds is plain to see today.

## Mast years

Tree such as birches, aspens, willows and elms produce a large and fairly constant number of seeds each year (around 250 000 in alder, *Alnus glutinosa*, and 20 million in the foxglove tree, *Paulownia tomentosa*, for example). Others, especially conifers and those hardwoods with relatively large seeds (such as oaks, beeches and ashes), show greater variation from year to year, often rhythmically. Acorn production in one British oak tree can vary from almost none to over 50 000 and occasionally 90 000 per year. The good years, in which flower and seed production are exceptionally high, are referred to as mast years. For beech (Figure 5.18) and oak in Britain mast years are usually every 2–3 and 3–4 years, respectively, with exceptionally high seed production occurring every 5–12 and 6–7 years, respectively. The N American aspen (*Populus tremuloides*) has been recorded as producing a maximum of 1.6 million seeds every 4–5 years. Even apples tend to produce biennially good crops, although climate change now appears to be altering this.

Masting can only happen when there are sufficient food reserves in the tree. Large-seeded trees invest an enormous amount of energy into growing seeds and in good mast years a tree puts everything into producing a large quantity of seed, so much so that growth of the tree can be reduced that year. In beech, ring growth is smaller in a mast year (Drobyshev *et al.* 2010) and in the dipterocarps of Southeast Asia there may be no growth at all in that year. With such effort it is perhaps not surprising that there are rarely two mast years in a row. The tree must rebuild its stored food reserves ready for the next big year, and in those intermediate years it may produce just a modest amount of seed or none. However, a few trees, such as holm oak (*Quercus ilex*), do not seem to accumulate stored carbon before a mast year but instead produce a larger number of leaves in a mast year, in effect funding

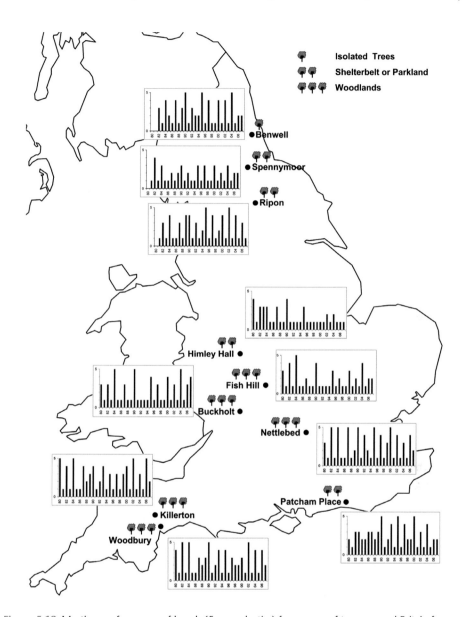

Figure 5.18 Masting performance of beech (*Fagus sylvatica*) for groups of trees around Britain for the years 1980 to 2007. The amount of mast is expressed on a five-point scale using the average number of nuts collected in a 7-minute period (1, fewer than 10 nuts collected; 2, 10–50; 3, 51–100; 4, 101–150; 5, more than 150). From Packham, J.R., Thomas, P.A., Lageard, J.G.A. & Hilton, G.M. (2008) The English beech masting survey 1980–2007: variation in the fruiting of the common beech (*Fagus sylvatica* L.) and its effects on woodland ecosystems. *Arboricultural Journal*, 31, 189–214.

the extra seeds by growing extra leaves. Maximum foliage density is reached in mast years and, being evergreen, is followed the next year by increased leaf shedding, fewer small branches, narrower tree rings and fewer acorns. Many factors affect how much food can be accumulated in a tree including soil, aspect and amount of shading (acorn yields are higher from trees with spreading crowns and conifer seed-yields are increased by thinning the stand). But most variable from year to year is the weather. Growing conditions affect not just how much food can be spared for storage in any one year but also when these stored reserves can be used. Bad weather, either preventing buds from forming or killing them later, will prevent masting even when the tree has ample reserves. Thus in oaks, variation in seed numbers can be linked to the weather (especially temperature) of the 2 years before the masting event.

Since weather is the main governing factor, mast years should be synchronous across large areas. Generally this is true although local variations in weather will inevitably play a large role in how well even geographically close trees are synchronised, and it is not uncommon for even different parts of southern Britain to be out of phase (e.g. 1986, 7th bar from the left in Figure 5.18). Moreover, an attack of herbivores in one area, for example, may prevent trees masting in synchrony with others around it.

This explains how mast years arise but not why. It is worth noting that seed production in non-masting species is also affected by weather but without the huge variation in number seen in masting species, so masting appears to be a deliberate strategy. Why doesn't an oak produce a smaller, more constant and regular number of acorns every year, more like birches and elms? The answer appears to lie in the fact that the large seeds of masting species such as oak are eagerly sought after by seed-eating herbivores like squirrels and wood pigeons (the latter can take 100 acorns per day). Masting swamps the seed eaters with more seeds than they can consume (technically called 'predator satiation') leading, hopefully, to a few surviving to germinate. Studies of beech in Britain have shown that up to 100% of beech seed is eaten by mice and birds in years when there is a poor crop but over 50% of seed may be left at the end of a winter following a mast year. This number is left despite the large flocks of bramblings and great tits attracted from mainland Europe in good beech mast years. Oak and beech often only produce seedlings following mast years.

There can be other factors involved. For example, pinyon pine (*Pinus edulis*) and a number of other pines around the world have evolved large seeds without wings. They are spread primarily by corvid birds in much the same way that jays and squirrels move acorns in Britain (see Animal dispersal below). The birds carry seeds away and bury them to eat during the winter, but in mast years some will be forgotten and left to germinate. In the case of the pinyon pine (at least) the birds do not feed on trees with low seed numbers, so masting acts to *attract* these seed-eating birds as well as ensure that some are left uneaten to germinate.

Janzen (1971) suggested how masting could have evolved. If the weather either destroys a flower crop or fails to provide a flowering cue one year this leads to a larger crop next year with more of the seed surviving predators and going on to form the next generation. This leads to selection of those plants most sensitive to the disruptive weather event, leading to future masting.

So why don't all trees have mast years? Masting occurs in those trees with largish seeds that are wind-dispersed or where dispersal is via mammals or birds (like beech and oak) that carry off the seed to eat but store some in caches in the ground which are subsequently forgotten. Conversely, masting is not prominent in trees which spread their seeds using animal-eaten fruits: it would be self-defeating if fruits were left uneaten. Nor is it common in trees with small seeds such as birch and willow, which presumably do not attract such herbivory. The tropics have an extra problem: in areas where the animals are many and diverse, predator satiation by mast seeding does not occur presumably because the predators can move and change diets in enough numbers to soak up the increased quantities of seed. But in the animal-poor dipterocarp forests of SE Asia there *is* masting. Although the flowering of different species may be staggered (Figure 5.12), the seed fall is remarkably coordinated to drop large numbers of seeds at the same time. Even with an influx of migrant animals not all the seeds are eaten.

From this discussion it can be seen that a good fruit crop is unlikely to be a sign of Mother Nature providing for a bad winter ahead; fruit production is a facet of how good it has been not how bad it is going to be, and generally it is the summer before the current one that is responsible for how good the seed crop is.

## Dispersing the seeds

On the whole, trees use the same methods of moving seeds as they do pollen. Trees that invade open areas (such as birches, ashes and many pines), which obviously tend to grow in open and relatively windy places, have come to rely on wind to disperse fruits and seeds. Trees of woodland are faced with relatively calm conditions and a need for large, heavy seeds (see Chapter 8), and consequently they tend to be dependent upon the specialised collecting habits of mammals and birds. For similar reasons, many woodland shrubs bear fleshy fruits. But there are always exceptions! Seed dispersal in animal-rich tropical forests is primarily by animals (typically 50–75% of trees) and even amongst trees that inhabit the windy areas at the top of the canopy, only 18–30% of species are dispersed by wind.

Working out how far fruits and seeds travel can be very hard to do, and there are basically four ways. You can chase seeds (watch them, or track where seeds end up from a lone tree), use seed traps placed at different distances from a tree, and now you can look at the genetics of fruit found on the ground. Since the fruit is made by the mother-tree (unlike the seeds which are a genetic combination of both parents), it is relatively easy to trace a fruit back to the mother. Finally, for long-distance transport, it is possible to radio-track seed dispersers such as birds and mammals, or, if large enough, even the seeds themselves.

## Wind dispersal

Wind dispersal is most effective if the seeds are liberated as high as possible and have ways of slowing their fall to allow the wind to blow them a long way. Some trees, such as eucalypts and rhododendron rely on fine dust-like seed. Slowing the speed of fall with larger seeds is accomplished by extending the dry fruit into the samara wing of maples, ashes and tulip trees (*Liriodendron* spp.). If the fruit itself is not much help in flying, a bract can be added as a wing (e.g. hornbeams or limes), or the seed itself can be given a wing (conifers), or a plume of long hairs (willows, poplars, balsa, the Indian bean tree and its relatives, *Catalpa* spp., and the kapok tree, *Ceiba pentandra*; Figure 5.19).

Hairs act simply as parachutes but the wings have a more complicated job: they act to spin the seed round, like a helicopter, reducing the vertical speed of fall. The size of the wing fits closely to the size of seed. If the seed is too big or small for the size of the wing the whole thing just plummets like a brick (try nibbling bits off a maple wing and throwing it in the air).

Figure 5.19 Fruits of the kapok tree (*Ceiba pentandra*) photographed in Singapore Botanic Gardens. The fruit is a follicle (see Box 5.3) which splits along several lines of weaknesses to let out the fluffy seeds.

Many trees let go of the seed when leafless to allow them to go further: from maples shedding fruits in the autumn to the kapok tree which flowers and sheds its plumed seeds when leafless (each tree may only flower every 5–10 years, but can produce 800 000 seeds when it does; is it any wonder that enough kapok could be collected to stuff mattresses and life-jackets before artificial replacements were mass-produced). Trees do not just drop their winged seeds at random; it is much more calculated. As the fruits hang on the tree, the separation layer (similar to that developed in leaves in the autumn; see Chapter 2) develops rapidly with low humidity, which, in temperate areas, is typically in the early afternoon when wind speeds are highest. The residual amount of force required to remove the fruit (which can be twice the average wind speed) ensures that it is plucked off by gusts of high wind, helping it to go at least twice the distance that one would predict from average wind speeds. Strong gusts of wind also tend to propel the seed above the canopy where wind speeds are higher, and they travel even further. Along coastlines seeds tend to be released during the day when there are on-shore winds and less often at night when there are off-shore winds.

Just how far wind-blown fruits and seeds travel is difficult to calculate, as mentioned above, and even harder to interpret (the odd seed that travels great distances may be very important in coping with climate change; see Chapter 9). Wind-dispersed seeds are most common in pioneer species like birches, ashes and many pines (although the two N American redwoods are forest species with small winged seeds, that do best on open burnt areas) which you would expect to benefit from spreading evenly over a large area to increase the chance of finding a new bare spot to invade. And this is what happens. Variation in air turbulence is effective at dropping seeds at varying distances from the parent. Most winged fruits and seeds, being relatively heavy, don't go very far and typically drop within 20–60 m of the parent although there are records of sycamore (*Acer pseudoplatanus*) travelling 4 km and poplars 30 km. Jeffrey pine seeds (*Pinus jeffreyi*) in Nevada have been seen to be carried up to 25 m further once they hit the ground by seed-caching small mammals. Small light seed like birch or heather seeds still in their capsules can be blown many kilometres over hard-packed snow to cover an area more than three times that by wind dispersion alone.

## Animal dispersal

Although animals can sometimes be inadvertent spreaders of fruits and seeds, as for example, when they are caught on fur and feathers and are preened out later, most movement by animals is based on supplying them with food. Seeds are bigger than pollen and therefore insects play a fairly small role: larger

(b)

(a)

Figure 5.20 Cashew fruits (*Anacardium occidentale*), originally from Brazil but now planted around the tropics. The cashew 'nut' which we eat (really the seed) is inside a dry fruit (an achene; see Box 5.3) borne below the 'apple'. In (a) the cashew seed (cut in two halves) has been extracted from the achene. The fleshy apple above, which is edible, is what attracts animals to carry the cashew seed away. The achene itself is highly toxic to protect the seed (it is dropped as the apple is eaten) and in humans causes swelling and blistering. The apple grows from the swollen stem of the fruit; in (b) several developing fruits can be seen to the left before the apple has completely grown. Photographed in Honduras.

animals are needed. And these large animals have allowed the evolution of large fruits which in turn has allowed seeds to become larger, helping the new trees to become established. Fruits, especially those in the tropics, differ enormously in size, abundance and nutritional properties, consequently utilising a vast range of different animals, but particularly birds. Fruit attractive to birds tends to be presented in the morning, with little smell but highly coloured; red fruit in the summer and on evergreen shrubs (wild cherry, holly, yew) and black fruit in autumn (brambles, elder, bird cherry, ivy) to ensure they stand out. Similar examples could be used from the tropics. Mammals are attracted primarily by smell; think of the fragrance of apples and pears. Some fruit is aimed at general feeders and tends to offer a general diet of carbohydrate, which needs supplementing elsewhere. Others, such as the avocado (*Persea americana*), offer a much more complete package of nutrition and utilise the undivided services of a more restricted number of seed distributors.

Seeds in fleshy fruits are usually hard and often toxic (especially in the tropics). So, once the animal has eaten the fruit, the seeds are either spat out, regurgitated (as the splendid quetzal bird of Central America does to wild avocado stones), dropped (as in the cashew nut; Figure 5.20) or they are swallowed and have to survive crushing by teeth or gizzards to pass through the gut and appear in the droppings. In all these cases the seeds hit the ground hopefully some way from where they started. In the UK, elder saplings (*Sambucus nigra*) are often seen below starling roosts. The journey through the animal can be eased by speed or by extra protection. 'It is no surprise that syrup of figs eases constipation: [strangler] fig trees planned it that way. From the fig tree's point of view, the effect of fig fruit on the bowels encourages its seed to be dispersed widely in all directions' (Mitchell 1987). Protection can come from hard layers such as with the hard inner shell of a drupe. The passage through the animal can be distinctly beneficial; abrasion of the seed coat of holly (*Ilex aquifolium*) by passage through the gizzard of birds improves the germination.

Seeds with dry fruits, such as nuts, are eaten for the seed itself. This may seem somewhat self-defeating for the tree since a digested seed cannot grow. But the system works by depending upon the clumsiness or forgetfulness of animals. Seeds may be dropped as they are being carried away as happens with small birds feeding on European beech nuts and with parrots feeding on the pods of New World *Parkia* trees. Other animals hoard caches of the seeds for later (especially during mast years), some of which are forgotten, providing the seed with a convenient system of being planted ready to germinate. This is seen with various pines around the world which have developed large wingless seeds moved by corvid birds; the tough woody cannonball-like fruits of the brazil nut tree which only the agouti can chew into and who then caches the seeds in the ground; acorns moved by jays, pigeons and squirrels in the UK; and smaller nuts moved by small mammals. Incidentally, acorns rapidly lose moisture and viability when left on the soil surface, so burying by jays is especially important. A pair of jays may store several thousand acorns in an autumn (Chapter 8 tells the story of how jays influence the size of acorns). The grey squirrel introduced to Britain, on the other hand, frequently bites the end off acorns which prevents them germinating.

Elsewhere seed-caching birds can move huge numbers of seeds. In western USA, the Clark's nutcracker (a corvid) collects seeds of various pines to cache in the ground over winter (Figure 5.21). With the pinyon pine, *Pinus edulis*, each bird carries away around 55 seeds (up to a maximum of 95 seeds) each trip, taking them up to 22 km away. A flock of 150 birds investigated in 1969 stored an estimated 3.3 to 5 million seeds, weighing between 650 and 1000 kg.

(a)

(b)

Figure 5.21 (a) Single-leaf pinyon pine (*Pinus monophylla*) mixed in with juniper (*Juniperus osteosperma*) in the White Mountains, California. (b) The small cones carry relatively large seeds (over a cm long) which have no wing and a thin seed coat, ideally suited to be carried away by the Clark's nutcracker. The seeds sit within deep hollows on the cone scales so they do not easily fall out and are readily available to the birds.

Seed movement is not always straightforward. For example, the Australian quinine bush (*Petalostigma pubescens*) has a three-stage dispersal process. The round fruits (2–2.5 cm diameter) are swallowed by emus which strip the flesh from the drupe stone. Once passed through the gut, there may be over a thousand seeds in a single emu scat, which could lead to intense seedling competition. But all is not lost: the stone of the drupe exposed to the sun explodes 2–3 days later and fires the seeds 1.5–2.5 m away. The seeds have a conspicuous fatty appendage (eliasome) at one end and are carried off by ants into nests where the ants eat the appendage and discard the seed, leaving it underground where it gets protection from mice and fires, and a favourable germination site. Although spectacular, it is not so different from the broom and gorse of Europe where the pea-like pods hanging on the bush explosively scatter the seed as they dry in the sun, and the seeds are similarly carried off, and conveniently buried, for a similar fatty appendage. Other examples could also be included: many pines are first moved by wind but then cached by small mammals.

How far are seeds taken by animals? Most information stems from studies done in Europe where mammals and birds take acorns, hazelnuts and berries in the order of 10–30 m from the tree. Radio-tracking of acorns moved by jays in Spain found that acorns of cork and holm oaks are normally taken 3–550 m from the tree with an average distance of just 69 m. Although, as noted above, some birds take pine seeds a number of kilometres away. Big seeds tend to get taken less far. The important thing to remember here is that seeds do not necessarily have to go long distances since the likelihood of a gap in the canopy (by, for example, a tree falling over) is as great near as far. And despite the heavy losses expected by seeds being eaten this can work very well: in 1968 Mellanby found up to 5000 oak seedlings per hectare in a bare field from acorns carried by birds.

## Dispersal by water

A wide range of trees have seeds or fruits that float. This includes the common (and very invasive) tree of heaven (*Ailanthus altissima*) now found on every continent except Antarctica. Its seeds are spread mainly by wind but those that are blown into water can float for up to 20 days and still be viable, and may even germinate better and quicker afterwards. Many trees that have floating seeds are not necessarily adapted to being moved by water, they just happen to have fruits that will float. For example, the cottony hairs of willow seeds help them to float for a week or so.

Trees that habitually grow beside water, however, have often evolved fruits that float very well and this may be their main means of movement. These often have small winged fruits with an oily outer coat that are capable of floating in still water often for more than a year and can be moved by running water and wind drift over still water considerably further than they can be blown through the air. A number of tropical legumes which grow beside rivers drop their pods into the water where they break into small one-seeded segments and float away.

Figure 5.22 A sea-bean found on the coast of Cornwall. This is a seed from the Caribbean liana *Entada gigas*. The seeds grow within pods that botanically are legumes (see Box 5.3).

Many wash up close by but some go further, even out to sea where they may float for more than a year and still be viable; indeed many of these seeds are found worldwide. It is not unusual to find 'sea-beans' (seeds from the pods of *Entada gigas*) washed up on British beaches looking like fat chocolate coins up to 4–6 cm in diameter and 1.5 cm thick, having travelled from the Caribbean on the Gulf Stream (Figure 5.22). Water probably holds the long-distance seed movement record: coconuts, wrapped in their fibrous coir-float with an impermeable skin, can sail the oceans between continents on journeys of over 3000 kilometres.

Over $100\,000\,km^2$ of the Amazon jungle are flooded each year during the wet season just when the fruits of many trees and vines ripen. The fruits of rubber trees (*Hevea brasiliensis*) explode on the tree, hurling the seeds 10–30 m out over the water. The seeds have a shell nearly as hard as that of a brazil nut but fish with molar-like teeth, resembling those of a horse, can crack them open. Many seeds are lost in this way but inevitably some survive floating around for the next 2–4 months. Seeds of other species, encased in fleshy fruits (built-in buoyancy aids), are either swallowed by fish or drop to the bottom as the flesh decays; either way the seed is moved. Folklore has it that electric eels wrap themselves around the base of submerged palms and give them an electric shock – up to 500 V – to dislodge fruit, although it is doubtful that the palms would be conductive enough for this to work.

## The cost of reproduction

Producing flowers and fruits is expensive. Young trees can't afford to do it and older trees pay a cost in doing it. Reproduction slows height and girth growth, and ring-widths may be halved in good seed years. Perhaps it is not surprising that a few trees, notably some palms (for example *Tahina spectabilis* of Madagascar), will save up their resources to flower just once before they die. In Scots pine, cone production has been estimated to reduce the amount of wood grown by 10–15% during an average year. Trees start at a disadvantage because the flowers take the place of some leaves: male cones in lodgepole pine (*Pinus contorta*) result in 27–50% reduction in the number of needles on that branch (female cones, hanging on branch ends, replace few needles). But leaves immediately adjacent to fruits compensate by working harder, increasing their photosynthetic rate by up to 100%. The value of this is seen in trees growing in marginal climates for flowering. A classic example is the small-leaved lime (*Tilia cordata*) in the Białowieża Forest in Poland (pronounced Bee-ow-a-vey-sha); Donald Pigott (1975) found that the south sides of trees (warmer and sunnier) produce 80 seeds per 100 leaves, while on the north side it was eight seeds per 100 leaves. Nor are flowers and fruits helpless. The green sepals of apple flowers have been measured to contribute 15–33% of the carbohydrate needed during flowering and fruit set. Green fruits also contribute to their own growth. This has been measured at just 2.3% in burr oak (*Quercus macrocarpa*) of eastern N America rising to around 17% in Norway spruce (*Picea abies*), 30% in Scots pine (*Pinus sylvestris*) – conifer cones are green when young – and 65% in Norway maple (*Acer platanoides*) with its large green samara wings.

## Why grow more flowers than are needed?

It is common for trees to produce mature fruits from only a small proportion of their female flowers. Pines may produce mature fruit from over three quarters of their flowers going down to less than 1 in 20 in mango (*Mangifera indica*) and teak (*Tectona grandis*) and only 1 in 1000 in the kapok (*Ceiba pentandra*) in tropical Africa. Part of this failure can be due to a lack of pollination. Normally one or more fertilised seeds are needed to stimulate fruit development via hormone release from the seed, and it is not uncommon for fruits and cones to be dropped if they contain too few seeds (the threshold number can vary annually in any one tree). However, even where pollination fails it is sometimes seen that fruits develop anyway (see parthenocarpy, earlier in the chapter).

On the whole, lack of pollination is a small part of the answer. The overriding reason that flowers fail to produce fruit is because the young fruit abort.

This may be due to frost or drought damage, or attack by fungi or animals. But there is a lot of evidence that plants actively discard excess young fruit. Aborting unwanted fruits early is good husbandry since fruits are costly to produce. But this begs the question: why produce unwanted flowers and fruits in the first place? One answer is to do with 'plant architecture'. A flower head may be composed of many flowers (possibly to present a good display to attract pollinators), each capable of producing a fruit but for which there is physically not enough room for all to develop.

A more general answer is that, firstly, it is a way of producing as many fruits as possible in a good year. This is a similar process to that seen in birds, which produce an optimistic number of eggs and the smallest chick(s) dies if there is not enough food and survives if there is. In the same way a tree can take advantage of good years when there is plenty of spare food to grow extra fruit. Young fruits are 'competing' with the rest of the tree for resources, and so in times of shortage, fruits may be 'starved to death' (see Chapter 6). Flowers at the base of an inflorescence have first claim on food and water and therefore tend to be the ones that survive, so as with birds it is not a case of all starving, but the weakest going to the wall while the others prosper. Secondly, aborting fruit allows the tree to keep the best fruit. Trees appear able to detect, for example, fruits heavily infested with seed-eating grubs and selectively abort them. There is some evidence that cross-pollinated flowers begin fruit growth quicker than self-pollinated flowers, which therefore have first call on resources and are the ones to survive. In this way the tree will grow the preferred cross-pollinated seeds in preference but will fill up with self-pollinated flowers if there are enough resources available.

In a handful of pines, the opposite problem sometimes occurs where far more flowers are produced and survive than is normal. In these cases of 'aggregated cones' there may be more than a hundred cones in a whorl compared to the normal 1–5. The cones are typically smaller than normal but they may still use up so much food that the branch above the cluster dies.

## The next generation

What happens once the seeds have been spread around? See Chapter 8!

## 🍃 *Further Reading*

Ashton, P.S., Givnish, T.J. & Appanah, S. (1988) Staggered flowering in the Dipterocarpaceae: new insights into floral induction and the evolution of mast flowering in the aseasonal tropics. *American Naturalist*, 132, 44–66.

Bazzaz, F.A., Carlson, R.W. & Harper, J.L. (1979) Contribution to reproductive effort by photosynthesis of flowers and fruits. *Nature*, 279, 554–555.

Borchert, R., Renner, S.S., Calle, Z. *et al.* (2005) Photoperiodic induction of synchronous flowering near the equator. *Nature*, 433, 627–629.

Camarero, J.J., Albuixech, J., López-Lozano, R., Casterad, M.A. & Montserrat-Martı, G. (2010) An increase in canopy cover leads to masting in *Quercus ilex*. *Trees*, 24, 909–918.

Dick, J. McP., Leakey, R.R.B. & Jarvis, P.G. (1990) Influence of female cones on the vegetative growth of *Pinus contorta* trees. *Tree Physiology*, 6, 151–163.

Drobyshev, I., Övergaard, R., Saygin, I., *et al.* (2010) Masting behaviour and dendrochronology of European beech (*Fagus sylvatica* L.) in southern Sweden. *Forest Ecology and Management*, 259, 2160–2171.

Greene, D.F., Quesada, M. & Calogeropoulos, C. (2008) Dispersal of seeds by the tropical sea breeze. *Ecology*, 89, 118–125.

Horsley, T.N. & Johnson, S.D. (2007) Is *Eucalyptus* cryptically self-incompatible? *Annals of Botany*, 100, 1373–1378.

Janzen, D.H. (1971) Seed predation by animals. *Annual Review of Ecology and Systematics*, 2, 465–492.

Kay, Q.O.N. (1985) Nectar from willow catkins as a food source for Blue Tits. *Bird Study*, 32, 40–45.

Kikuchi, S., Shibata, M., Tanaka, H., Yoshimaru, H. & Niiyama, K. (2009) Analysis of the disassortative mating pattern in a heterodichogamous plant, *Acer mono* Maxim. using microsatellite markers. *Plant Ecology*, 204, 43–54.

Kislev, M.E., Hartmann, A. & Bar-Yosef, O. (2006) Early domesticated fig in the Jordan Valley. *Science*, 312, 1372–1374.

Kowarik, I. & Säumel, I. (2008) Water dispersal as an additional pathway to invasions by the primarily wind-dispersed tree *Ailanthus altissima*. *Plant Ecology*, 198, 241–252.

Lanner, R.M. (1996) *Made for Each Other: a Symbiosis of Birds and Pines*. Oxford University Press, Oxford.

Lee, W.Y., Lee, J.S., Lee, J.-H., Noh, E.W. & Park, E.-J. (2011) Enhanced seed production and metabolic alterations in *Larix leptoepis* by girdling. *Forest Ecology and Management*, 261, 1957–1961.

Mellanby, K. (1968) The effects of some mammals and birds on regeneration of oak. *Journal of Applied Ecology*, 5, 359–366.

Mitchell, A.M. (1974) *A Field Guide to the Trees of Britain and Northern Ireland*. Collins, London.

Mitchell, A.W. (1987) *The Enchanted Canopy*. Fontana, London.

Obeso, J.R. (1997) Costs of reproduction in *Ilex aquifolium*: effects at tree, branch and leaf levels. *Journal of Ecology*, 85, 159–166.

Ohashi, K. & Thomson, J.D. (2007) Trapline foraging by pollinators: its ontogeny, economics and possible consequences for plants. *Annals of Botany*, 103, 1365–1378.

Pichot, C., El Maâtaoui, M., Raddi, S. & Raddi, P. (2001) Surrogate mother for endangered *Cupressus*. *Nature*, 412, 39.

Piggot, C.D. (1975) Natural regeneration of *Tilia cordata* in relation to forest structure in the forest of Białowieża. *Philosophical Transactions of the Royal Society, Series B*, 270, 151–179.

Pons, J. & Pausas, J.G. (2007) Acorn dispersal estimated by radio-tracking. *Oecologia*, 153, 903–911.

Queenborough, S.A., Mazer, S.J., Vamosi, S.M., *et al.* (2009) Seed mass, abundance and breeding system among tropical forest species: do dioecious species exhibit compensatory reproduction or abundances? *Journal of Ecology*, 97, 555–566.

Silvertown, J.W. (1980) The evolutionary ecology of mast seeding in trees. *Biological Journal of the Linnean Society*, 14, 235–250.

Silvertown, J. (2010) *An Orchard Invisible, a Natural History of Seeds*. University of Chicago Press, Chicago.

Soons, M.B. & Bullock, J.M. (2008) Non-random seed abscission, long-distance wind dispersal and plant migration rates. *Journal of Ecology*, 96, 581–590.

Spigler, R.B. & Ashman, T.L. (2012) Gynodioecy to dioecy: are we there yet? *Annals of Botany*, 109, 531–543.

Stephenson, A.G. (1981) Flower and fruit abortion: proximate causes and ultimate functions. *Annual Review of Ecology and Systematics*, 12, 253–279.

Terry, I., Walter, G.H., Moore, C., Roemer, R. & Hull, C. (2007) Odor-mediated push-pull pollination in cycads. *Science*, 318, 70.

Thomas, P.A., El-Barghathi, M. & Polwart, A. (2007) Biological Flora of the British Isles, *Juniperus communis* L. *Journal of Ecology*, 95, 1404–1440.

Thomas, P.A., El-Barghathi, M. & Polwart, A. (2011) Biological Flora of the British Isles, *Euonymus europaeus* L. *Journal of Ecology*, 99, 345–365.

Vander Wall, S.B. (1994) Removal of wind-dispersed pine seeds by ground-foraging vertebrates. *Oikos*, 69, 125–132.

Vander Wall, S.B. & Balda, R.P. (1977) Coadaptations of the Clark's nutcracker and the piñyon pine for efficient seed harvest and dispersal. *Ecological Monographs*, 47, 89–111.

Williams, A., Antonovics, J. & Rolff, J. (2011) Dioecy, hermaphrodites and pathogen load in plants. *Oikos*, 120, 657–660.

Yeang, H.-Y. (2007) Synchronous flowering of the rubber tree (*Hevea brasiliensis*) induced by solar radiation intensity. *New Phytologist*, 175, 283–289.

# Chapter 6: The growing tree

The growth of trees leads to greater changes in size than in any other organism: for example, the mass of a giant sequoia (*Sequoiadendron giganteum*) can increase by 12 orders of magnitude from seed to mature tree (that's a billion, billion times bigger, equivalent to an average seed of 0.005 grams becoming a tree of 5000 tonnes, which is the right order of magnitude; see below). Not surprisingly, common questions to ask are 'how quickly will my tree grow?' and 'how big will my tree eventually get?'. In this chapter we will look at these and related questions, and the reasons behind the answers.

## Speed of growth

### Height

You have probably seen films where the hero in the Orient is strapped over a bed of growing bamboo as a means of torture and eventual death, speared by the hard growing shoots. The reason this works is the extraordinarily fast growth of over half a metre per day. While tropical vines and lianas can grow almost as fast, trees proper can't equal this rate but can nevertheless be impressively quick, especially when young. A number of tropical species can add 8–9 m to their height in a year. A New World relative of the elm, *Trema micrantha*, has been seen to grow 30 m in 8 years (an average of 3.75 m per year) and a eucalypt (*Eucalyptus deglypta*) in New Guinea reached 10.6 m in just 15 months. *The Guinness Book of Records* has in the past quoted the air-speed record for a tree as a specimen of *Albizia falcata* (now called *Paraserianthes falcataria*) planted in Malaysia which grew 10.74 m (35 ft 3 in) in 13 months! As you would expect, there is a lot of variation among species. Trees that invade gaps in tropical forests, and need to grow quickly to win the race to the top, grow faster (an average of 1.5–4.0 m in height per year) than later species that can afford to slowly plod upwards through the shade at 0.5–1.2 m per year.

Temperate trees tend to grow more slowly. Vigorous young trees in Britain grow something between 15 and 50 cm per year (a tortoise-like 1–2 mm per day) although 1.5 m is not impossible, and coppice shoots (with their established roots) can grow 3 m in the first year. As in the tropics, pioneer trees tend to grow faster than those which invade ready-grown forests. There's an old Lancashire saying that a willow will be worth a horse before an oak will be worth a saddle: the pioneer willow grows faster than the oak. Growth in any tree is also affected by age. As trees get taller, the height increase inevitably slows down before finally stopping once it reaches its maximum height.

Measuring minimum growth in trees is much like the slow cycling record which became pointless when someone learnt to stay stationary! Where the environment is so inhospitable that new growth is negligible or constantly killed back, trees may not gain height at all for many years. Trees at northern tree-lines can be hundreds of years old but no more than a few tens of centimetres high.

## Width

As we saw in Chapter 3, a tree trunk must normally keep getting wider by forming new rings to ensure water transport up the tree. In temperate trees, the average width of new rings is 0.1 to around 5–6 mm, although some trees such as young giant sequoias (see below) can grow such wide rings that the tree diameter increases by several centimetres each year. The smallest rings found so far, just 0.05 mm wide, have been measured in three different species spread around the world: white cedar (*Thuja occidentalis*) growing on cliffs of the Niagara escarpment, bristlecone pines (*Pinus longaeva*) in California and yew (*Taxus baccata*) in Britain. For these trees it can take 10 years to add 1 mm to the trunk diameter. In the white cedar, an individual little more than 5 cm in diameter was 422 years old, an average ring width of just over 0.05 mm. All these examples are growing on very inhospitable dry rocky areas.

Size of the trunk can be useful for ageing a tree, not just by counting the rings, but by measuring the circumference. Alan Mitchell, a past doyen of British trees, proposed a general and useful rule that temperate trees increase in circumference (girth) at 2.5 cm (1 in) per year when growing in the open, reducing to half an inch in woodland. Remarkably, as a very rough rule, it works (corresponding to an average ring width of 4 mm per year). Generally, young trees grow faster than this, and mature trees slower, averaging out to the inch rule except in very young and old trees (more later). There are, of course, exceptions: a number grow faster, increasing at 5–7.5 cm (2–3 in) in girth per year, including the giant sequoia (*Sequoiadendron giganteum*), coastal redwood (*Sequoia sempervirens*), cedar of Lebanon (*Cedrus libani*), Sitka spruce (*Picea sitchensis*), Douglas fir (*Pseudotsuga menziesii*), southern beeches (*Nothofagus* spp.), Turkey oak (*Quercus cerris*), tulip tree (*Liriodendron tulipifera*) and

London plane (*Platanus* x *hispanica*). Others grow slower than expected; trees such as Scots pine (*Pinus sylvestris*), horse chestnut (*Aesculus hippocastanum*) and common lime (*Tilia* x *europaea*).

## Champion trees in size

Trees are the biggest living things on earth, no matter how you measure. There are 46 different species that are capable of growing over 70 m tall in the wild (see Tng *et al.* 2012), concentrated along the west coast of N America, the coast of Australia and in SE Asia, particularly Borneo. The tallest living trees are the coastal redwoods (*Sequoia sempervirens*) on the Californian coast. In the first edition of this book, the record was 112.2 m (368 ft), measured in October 1996. Since then, a taller tree, called Hyperion, has been measured at 115.7 m (379 ft 7 in) by Steve Sillett and Michael Taylor in August 2006 (Figure 6.1). There are now four of these trees known to be taller than 113 m (370 ft). Outside the conifers, the tallest tree is a mountain ash (*Eucalyptus regnans*), called Centurion, in Tasmania which is currently 99.6 m (326 ft 8 in) tall, although there are records of another that was 114.3 m (374 ft 11 in). The tallest tree ever is reputed to have been another mountain ash in Victoria, Australia measured at 132.6 m (435 ft) in 1872 and thought to have been over 150 m (500 ft) at its peak. Uncorroborated records of Douglas firs in British Columbia over 140 m also exist. Strictly speaking, the *longest* woody plant is a rattan (a climbing palm; note that rattan furniture comes from a number of

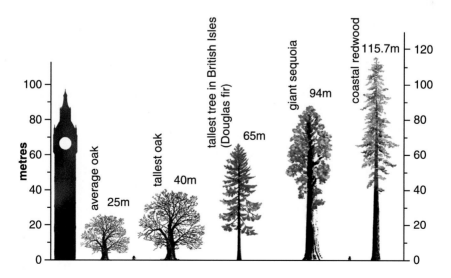

**Figure 6.1** Tall trees. An average and the tallest oak in the UK give an impression of just how tall the world's tallest trees are compared to Big Ben in London (98 m).

*Calamus* species) once measured as 171 m (560 ft) long, but being a vine it has no responsibility for its own support and so doesn't really count.

California also boasts the world's largest living things in bulk: the giant sequoias (*Sequoiadendron giganteum*) growing in the Sierra Nevada Mountains. The largest individual is General Sherman in Sequoia National Park (Figure 6.2) which has an impressive height of 83.8 m (275 ft) – although they can be up to 94.8 m (311 ft) – which, combined with a trunk-width of 11 m (36.5 ft) at the base, results in an estimated volume of 1487 cubic metres ($m^3$) or 52 500 cubic feet. This is the estimated volume for just the trunk. Since giant sequoia wood has a greenwood density of around 0.44 tonnes per $m^3$, the trunk alone probably weighs around 650 tonnes. Adding the weight of the branches is largely guesswork, so estimates of the total weight of General Sherman have ranged wildly from 1250 to 5500 tonnes with the true figure probably being nearer to the lower end at around 1250–1400 tonnes. As an idea of how big the tree is, in 1978 the General dropped a branch 45 m (150 ft) long and nearly 2 m (7 ft) in diameter: a branch larger than any tree east of the Mississippi River. For comparison, blue whales, the largest animals, weigh in at around a mere 100 tonnes apiece! Several fungal colonies of *Armillaria* spp. found in America and China cover huge areas (typically more than 800 ha) and may weigh more than General Sherman but although each colony is all genetically identical, are all the bits still connected together as one organism or is it a clone of now separate individuals? In the same way, aspen clones (*Populus tremuloides*) – suckers from the same original tree – can cover vast areas. The biggest reported is a male clump in Utah called Pando (Latin for *I spread*) which covers 43 ha (106 acres) with more than 47 000 stems and an estimated weight (above and below ground combined) of more than 6000 tonnes. It is all the same genetically (so has grown from one original plant) but again, is it still all one big tree, or a collection of genetically identical but physically separate clones?

For comparison with General Sherman, the tallest coastal redwood, Hyperion, has been estimated to contain 527 $m^3$ (18 600 cu ft) of wood in the trunk. At the other end of the world, New Zealand is noted for the huge kauri trees (*Agathis australis*). The largest one still living, Tāne Mahuta (Lord of the Forest), at 51.2 m (168 ft) tall and a diameter of 4.91 m (16 ft 1 in), has an estimated trunk volume of 245 $m^3$ (8650 cu ft) and 517 $m^3$ (18 250 cu ft) if the branches are included (Figure 6.3).

To put these giants into perspective, European and N American hardwood trees commonly reach 30 m and under favourable conditions may reach 45 m or so (see Figure 6.1). A 100-year-old oak in the UK will normally have 1–2 $m^3$ of wood in the trunk and 2–4 $m^3$ in the whole tree. When fresh out of the tree and full of water a cubic metre of oak weighs almost 1 tonne

Figure 6.2 General Sherman, the world's biggest tree by volume. The General is a giant sequoia (*Sequoiadendron giganteum*) in Sequoia National Park, Sierra Nevada Mountains, California. Photograph by Alan Crowden.

Figure 6.3 Tāne Mahuta, the largest living kauri tree (*Agathis australis*) in the Waipoua Forest, on the NE side of North Island, New Zealand.

(c. 980 kg); when seasoned (air-dried) this drops to around 720 kg per m$^3$ so the trunk of our 100-year-old tree would contain around 0.75–1.5 tonnes of dry wood. The tallest trees in the British Isles are Douglas firs at Stronardron in Argyll, Scotland and Lake Vyrnwy Estate (pronounced vurn-wee) in Powys, Wales, both 63.79 m (209 ft) high. The tallest oak in Britain is a mere 43 m (141 ft). But we do have some particularly wide trees. The fattest British oak is the Fredsville Oak in Kent which is only 24 m (79 ft) tall but a huge 3.8 m (12 ft 6 in) wide and estimated at 93.4 m$^3$ (3300 cu ft) in volume, putting it amongst the biggest trees in the country. Better still, the Fortingall Yew on Tayside, Scotland, which, although now only represented by two living fragments (look forward to Figure 9.3b), had an original diameter that can be traced in the ground of 5.4 m (17 ft 8 in), and a lime was measured by Kannegiesser in Germany in 1906 with a similar diameter. You should try measuring these out to appreciate just how big they are. This is beaten by the African baobab (Figure 6.4) which often achieves a diameter of 10 m. Livingstone spoke of a tree in which 20–30 men could lie down with ease! And don't forget the banyan trees mentioned in Chapter 4 which can cover several hectares. The kauri called Te Matua Ngahere (Father of the Forest) in NW New Zealand at 16.41 m (54 ft) diameter is probably the widest tree in the southern hemisphere.

You might think that tropical rainforest trees would be up there amongst the giants. A few are: the tallest tropical tree ever found appears to be a relative of the monkey puzzle (*Araucaria hunsteinii*) in New Guinea at 89 m (292 ft). Surprisingly, though, the tallest tropical trees usually average no more than 46–55 m. Like the coastal redwoods, rainforest trees are often remarkably slender for their height. The fattest can be as large as the Fortingall yew above but diameters greater than 1 m are uncommon.

Figure 6.4 Baobab tree (*Adansonia digitata*) in Zimbabwe. Photograph by Rosemary Coleshaw.

# What limits the size of a tree?

Why does a tree stop growing upwards? There has long been a debate over whether the slowing of a tree's vertical growth is due to age or size, and most research suggests that it is size that matters (Niklas 2007 gives a good account). This leads to the next question: why should size be a problem, surely trees are designed to be as tall as possible? Three linked answers can be given, to do with site quality, water and investment in mechanical strength.

The maximum height for a particular species is predictable from the quality of the site it is growing on; the more nutrients and moisture available (within limits), the taller the eventual height of the trees. Indeed, the height a tree species will reach at a set age is used by foresters to classify the quality of an area (look ahead to Figure 6.16 as an example).

Within this site quality, tree height usually correlates particularly well with the availability of water. Water shortage is felt more frequently at the top of a tree because the water has to be pulled up further. As described in Chapter 3, the tree can control water loss by closing the stomata, and foliage at the tops of trees is capable of working at lower water content but eventually water stress will slow and stop further growth. The drier the environment, the harder it is to get water to the top of the tree, and so the sooner upward growth will be stopped. As an example, coastal redwoods grow taller than 100 m on the deep moist alluvial soils where they are watered by the frequent coastal fogs, but they reach only 30 m on the drier inner edge of the fog belt. However, water stress is not the entire answer to tree height because no matter how much water a tree is given it still has a more or less fixed upper limit: an oak will never grow as tall as a redwood.

This leads to the third constraint to tree height which involves investment in mechanical strength. As a tree gets taller, the wind bending the tree has more leverage on the base in the same way that a longer spanner puts more turning force on a stiff bolt. So, the tree is more likely to snap the taller it is unless, of course, the trunk is made fatter to withstand the extra force. For most trees, a doubling in height requires the diameter at the base to increase by between 3 and 8 times for the tree to remain reasonably safe from breaking when bent by the wind. Obviously, a taller tree has therefore to invest a disproportionately higher amount of wood in the trunk and eventually it becomes uneconomical to carry on growing taller. There is variation between types of tree; some such as birches tend to be fairly narrow for their height while others such as oak err more on the side of caution and are fatter for the same height. In effect they have different compromises between cost of growth and safety (Box 6.1). And there is huge variation between individuals of the same species as each responds to the conditions around it. A tree

## Box 6.1 Just how safely designed is my tree?

How fast a tree grows in height and width differs among species; a birch will grow tall more rapidly than an oak. In effect, these represent two ends of a wide spectrum of compromise between speed and longevity. A birch, which invades open areas, works with a lower safety margin, putting more emphasis on quick height growth with a tall thin trunk of less durable wood and rapid reproduction, growing tall and reproducing before it is outcompeted by slower growing, shade producing trees. To help this, the wood is lower in density and has fewer defences, the trunk is comparatively thin for its height, and so is cheaper to grow. It also makes it less durable but that's okay because the birch is only likely to be able to grow for a comparatively short time and is normally dead within a few hundred years at most. With that goes the greater risk of catastrophe and early death: it's a boom and bust strategy.

At the other end there is the oak which has greater investment in dense, durable, highly-defended wood, with a comparatively wide trunk when old that will be more likely to survive anything that the physical environment and living organisms can throw at it, including rare catastrophic events. So growth is slower and the oak settles into a long game, going for longevity over a thousand years or so. The negative side is that it will be less competitive and is likely to suffer more extreme competition, especially when young.

The net result of these two strategies is in many ways the same; a back-of-an-envelope calculation suggests that both the short-lived birch and the long-lived oak produce somewhere around 25 million seeds over their life (birch: 250 000 seeds per year over 100 years of reproductive life; oak 50 000 seeds per year over 500 years).

It is important to bear in mind that this compromise between speed of growth and longevity *is* a compromise. This means that trees are never engineered to be able to withstand all extremes of weather. If they were built that solidly, they would be hopelessly outcompeted by others. So when we see a tree that is broken or uprooted by wind it does not necessarily mean that the tree was diseased or otherwise damaged; the wind may well just have been too strong (this is discussed further in Chapter 9). As Mattheck and Breloer (1994) point out when considering the legal position of, say, a car being squashed by a broken tree:

> ... it is perfectly normal for trees occasionally to break without there having to be anyone to blame. The breakage of a tree is the natural price that the species must pay for achieving an energy-saving, lightweight structure.

sheltered in the middle of a plantation will be thinner at the base compared to the same height tree growing on an exposed hill. The tallest coastal redwoods grow in forests which maintain moist conditions and buffer trees against the wind and so can afford to be narrower for their height, investing a lot of their energy in growing upwards (see Ennos 2012 for more detailed information). Each individual tree is constantly monitoring the stress it is under and will be fine-tuning its compromise between growth and durability at every stage of its life. Trees are dynamic things and while they may look unchanging they are in reality constantly monitoring and adjusting their woody skeleton to match their surroundings.

These answers are linked because height can often be limited by a combination of water and wind. At high altitude and towards the poles, trees are shorter. This is partly due to higher wind speeds but also, certainly in Norway spruce, at lower temperatures the xylem grows less well at the top of the tree and so has a less efficient transport system which reduces the ability to grow (Petit *et al.* 2010).

Ultimately, a tree grows tall to out-compete its neighbours, and big to hold as many leaves as it can, but there comes a point where the return is less than the investment. Evolution has honed the best compromise for each species in each area, producing a genetically-determined normal maximum height. You can see that this is genetic because seed from high altitude and high latitude trees planted in favourable areas will still produce shorter specimens than seed of the same species from less extreme areas.

As the young tree grows, its width is delicately balanced to be thick enough to withstand the rigours of the environment without being wasteful and overly thick (with the caveat, as noted above, that different species have different cost/safety compromises). Although trees reach a maximum height, they do carry on getting fatter each year, almost without limit. This produces the magnificent squat hulks so characteristic of the British landscape where many old trees have been preserved (Figure 6.5). From what we have said before this would seem mechanically to be a waste. As explained in Chapter 3, however, a tree has no choice but to produce new wood to maintain the water-filled link between the roots and the leaves. And it has the added advantage of maintaining a shell of sound wood in the face of rot at the centre of the tree. But this need to produce new wood each year is also partly responsible for the death of the tree (see Chapter 9).

## What controls tree growth?

Trees have evolved to grow tall in competition for light. When a tree is young and in its formative stage, the amount of wood grown per year increases as the

**Figure 6.5** An old oak (*Quercus robur*) in Cheshire, England.

canopy grows larger and produces more sugars. In this juvenile stage of its life, growth rings tend to be fairly even in width each year (leaving aside the variation caused by weather; see below) so the volume of each annual coat of wood over the tree gets larger every year. This is useful since the extra wood each year can carry extra water to the developing canopy, and the trunk cross-sectional area does indeed tend to increase in proportion to the total leaf area of the canopy. As the tree reaches maturity (40–100 years for many temperate species), upward growth slows down and will eventually stop. At this stage the canopy is probably as large as it will get and so the amount of sugars produced each year becomes fairly constant. As a consequence, the volume of wood grown each year becomes more or less constant, and so inevitably, since that wood has to cover a larger tree each year, the width of the rings tends to decline progressively with age, as we've noted above. Senescence is reached when the canopy starts dying back and rings become even smaller, and normally when they are less than 0.5 mm wide the tree is in trouble.

But what controls how the tree progresses through these stages? Büsgen and Münch, two German foresters in the early part of the twentieth century noted that growth in trees depends on 'internal disposition and external influences'. The external influences are obvious enough: growth requires sufficient supplies of light, water, carbon dioxide and nutrients (discussed below), and an amenable climate (see Chapter 9). But the internal disposition is also important. Within a tree there are conflicts on where to use limited resources. Many parts of the tree produce food (carbohydrates) by photosynthesis: mainly the leaves but also the bark, flowers, fruits and buds. Different parts of the tree (described as sinks in the language of physiology) compete strongly for these resources, and a balance has to be kept between the growth of different parts to ensure long-term survival. Too rapid height growth might leave a tree with no stored food for next year's spring leaves. Too many fruits on a young tree might disastrously stunt growth. Fortunately, the tree has a highly organised internal control system of allocating resources to growth, maintenance and reproduction by switching on and off different sinks at different times. The signals for this switching come from the external environment.

## Internal control

Without nerves to co-ordinate activity, a tree relies on hormones (growth regulators and inhibitors) which circulate through the tree to co-ordinate activities. Three types of growth regulator act to promote various aspects of growth: auxin (indoleacetic acid) produced primarily by shoot tips and leaves (and the basis of commercial rooting compounds); gibberelins, produced in the same place as auxins plus the root tips; and cytokinins produced particularly by the root tips and young fruit. The gas ethylene, should also be included since it is involved in regulating wood formation (it is in high concentrations in sapwood and may be involved in responses to bending pressures and other mechanical disturbance, and in heartwood formation) and is involved in fruit ripening (one rotten apple – producing ethylene – spoils the barrelful). There are many minor inhibitors in plants but the main one is abscisic acid (ABA), produced in leaves, seeds and other organs. ABA generally slows things down or stops them, causing bud and seed dormancy, and stimulating the shedding of leaves. These hormones are, however, just a part of the story. They are part of a complex set of feedback loops that control growth, which involve an array of genes and enzymes that influence, and in turn are influenced by, the carbohydrate and nutrient supply within the tree (see Halford & Paul 2003 for more details). This allows the tree to fine-tune growth much more quickly and precisely than would be possible just with hormones.

## Balancing the roots and shoots

As touched on in Chapter 4, this is an important element in the internal disposition of a tree. The shoots provide food for the roots and in turn the roots provide water and minerals to the shoots, and so there needs to be a balance between the two parts (usually referred to as the root/shoot ratio). Too many roots become a burden on the limited sugars produced by the canopy; too few roots means water stress for the canopy. Trees have the ability to fine-tune the root/shoot ratio to prevailing conditions of light, water and nutrient availability: this is orchestrated by the hormones and feedback loops described above. For example, if part of the canopy is broken off, some of the roots will die. If the roots have problems growing because of shallow soil or competition with other plants, the canopy remains small. The balance of the ratio is at least partly under genetic control: trees from arid areas tend to have higher root/shoot ratios than specimens from moister areas even when grown under similar conditions, and tree species that invade open areas tend to have proportionately more roots. Also, trees that have genetically determined higher tannin levels in their leaves (which has the side effect of slowing down their decomposition and thus nutrient recycling) have more fine roots (less than 2 mm in width) to help gather more of the scarce nutrients. But the precise balance struck will also vary depending upon conditions at any one time. For example, drought leads to a higher proportion of roots because

Figure 6.6 An uprooted seedling of oak (*Quercus robur*) showing the balance between the shoot and major roots. Despite care, the majority of the fine roots (less than 2 mm in diameter) have been lost. Keele University, England.

water must be found even if these roots are a drain on the sugar-producing capabilities of the canopy.

We see the importance of the balance between roots and shoots most acutely when it comes to planting trees. A tree dug from a nursery may lose up to 98% of its roots (see Chapter 4 and Figure 6.6). This is why we normally plant trees in the dormant season to allow some growth of new roots before the canopy comes into leaf and demands water, but is it any wonder that newly planted trees can easily die of drought? Likewise, young trees planted in containers may be fine for a few years but as the canopy gets bigger the restricted roots cannot supply enough water to support a large canopy, resulting in small leaves and a tree that remains stunted or dies. If it dies, people often can't understand why: 'It survived that really dry spell a few years ago, why has it died now?'. Similarly, although the interactions of dwarfing root stocks with the shoot grafted on top can be quite complicated, they basically operate via this ratio. An apple tree, for example, grafted onto a dwarfing root stock, which produces few roots, will keep the grafted shoot (scion) to a small size making it easier to pick fruit. A corollary of all this is that if the roots of a tree are damaged by, for example, cable laying along urban streets (see Figure 4.9), the solution to improving the chance of the tree surviving is to reduce the size of the canopy by pruning.

## *External factors*

As noted above, a tree needs optimum supplies of light, water, carbon dioxide and nutrients for optimum growth. A lack of one or more will inevitably slow growth. Other constraints such as drought, wind, fire, insects and disease are considered in Chapter 9.

Light is often the most crucial factor that comes to mind. Fortunately, trees have a number of ways of countering too little light. A tree growing in shade will be 'drawn up' to the light and will be taller and thinner with shorter branches than a tree in open sunshine (in a paradoxical way, light tends to stunt growth). Trees also vary greatly in their response to shade. Foresters usually express this in terms of shade tolerance. The shade-intolerant (light-demanding) species, such as aspen, false acacia (*Robinia pseudoacacia*), larch and many pines, are affected by shade much more than shade-tolerant species such as beeches, sugar maple (*Acer saccharum*), hemlocks (*Tsuga* spp.), firs (*Abies* spp.) and western red cedar (*Thuja plicata*). Photosynthesis works in the same way in both shade-tolerant and shade-intolerant trees but the shade-tolerant species have more effective mechanisms for coping, such as sun and shade leaves (Chapter 2) and a different canopy 'architecture' (Chapter 7).

Trees, in common with other plants, need nitrogen, phosphorus, potassium (the NPK mixture), calcium, magnesium and sulphur in largish quantities (the 'macronutrients') and smaller quantities of iron, manganese, boron, zinc, copper, molybdenum and chlorine (the 'micronutrients'). The most paradoxical nutrient is nitrogen: 78% of the atmosphere is composed of nitrogen and yet it is often the nutrient most in short supply. The problem is that plants cannot use elemental nitrogen until it has been converted into a form such as nitrate or ammonia, either in the atmosphere by lightning or by bacteria (see Chapter 4).

Lack of any of these nutrients leads to characteristic symptoms and reduced growth. It is not just the amount but also the relative proportions of nutrients that are important. Too much of one nutrient can cause as many problems as too little. For example, as explored in Chapter 9, too much nitrogen can kill a tree. Additions of nutrients will generally speed growth but different species may respond better than others. Studies of the Arctic dwarf birch (*Betula nana*) have shown that artificially fertilising areas of tundra results in the birch producing proportionately more shoots than other shrubs so that the birch comes to physically dominate by being denser.

Plants will go a long way to scavenge back nutrients from being lost; before leaves are shed, nutrients are salvaged back into the tree (see Chapter 2), and it has been observed in teak (*Tectona grandis*), which normally grows on nutrient-poor soils, that nutrients (especially nitrogen) are removed from the body and stalk of the wind-dispersed fruits before being shed (Karmacharya & Singh 1992). In most trees the concentration of nutrients such as nitrogen and phosphorus is greatest in the trunk and gets less in progressively smaller branches and roots. Is this a defence against nutrient loss by keeping less of the valuable supply in parts of the trees that are more likely to be lost?

Too little (or too much) water also limits growth. In Chapters 2 and 4 we saw that drought and flooding play equally important roles in reducing the growth of most trees. Indeed, repeated water stress tends to produce a typical shrubby appearance. Although carbon dioxide is not usually limiting to growth (air always contains a small – 0.039% – but adequate supply) drought can affect its uptake. A water-stressed plant will by necessity limit the opening of the stomata to reduce water loss, with the consequence of also limiting carbon dioxide uptake.

The effect of climatic change is discussed in Chapter 9, but it is worth mentioning here something about temperature. Foresters, and now those interested in climate change, have shown a strong interest in the effect of temperature on growth. Photosynthesis and other physiological processes will occur at temperatures from not far above freezing to around 35°C. Not surprisingly, the optimum temperature for growth increases as you move from the poles towards the tropics. Whatever the optimal temperature for growth, for many species around the world, hardwoods and conifers, growth is highest under fluctuating temperatures with the nights cooler than the day, probably because of reduced respiration and conservation of food at night. But this is not always the case. Plenty of tropical plants have been found which do better with warmer night-time temperatures; many plants do most of their growing at night and so warm nights perhaps promote leaf production and therefore increase the plant's subsequent capacity for photosynthesis.

In passing, it is worth noting that some apparently insignificant features of the external environment can affect tree growth. Sigrid Dengel and colleagues (2009) found a link between tree growth in Scotland and galactic cosmic radiation, of all things. The four peaks in radiation since the early 1960s have coincided with wider growth rings. The authors suggest that high cosmic radiation produces more condensation nuclei in the atmosphere leading to more clouds, which in turn scatter the sun's light so that a tree gets it from all sides, increasing photosynthesis. A link has also been found between increased diameter growth of various trees at high tides. This is attributed to the beneficial effect of lunar gravity on growth, perhaps coupled in some way with lows in geomagnetic activity that accompany the tide cycles (Barlow *et al.* 2010).

## Growth in the real world

Trying to work out what is causing a tree to grow slowly can be difficult for a number of reasons.

1.    A number of small 'problems', fairly minor by themselves, can amalgamate leading to a large effect. For example, bonsai trees are kept small by a combination of restricted growth of roots, genetic selection of stock, low nutrients and sometimes limited water supply. Also, a significant but not disastrous lack of, say, nutrients when coupled with, say, pollution can lead to a very unhealthy tree. And stress of one sort may predispose trees to disease and insect attack because defences are weaker (for example, drought often leads to fungal outbreaks), masking the original cause of stress.

2.    The problem may be sporadic and so go unnoticed: harsh temperatures, high winds, drought, etc.

3.    Being so big, different parts of the tree can be exposed to different conditions. Air temperatures vary more than in the soil so while roots are comfortably cool, the temperature of the cambium in the trunk can be more than 30°C on the south side of a tree and half that on the north side. In the American Midwest, where water is limiting, the north side of a tree grows larger (longer branches and roots) because the south side experiences more water stress. In damp England you expect the opposite. These variations in conditions may not be obvious to a human on the ground.

4.    Symptoms may not point directly to the real cause. If a young tree is not growing very well, we automatically reach for the nitrogen-rich fertiliser. And yet in studies done on tree planting in urban areas it has been shown that when nitrogen is applied around trees less than 20 years old the grass benefits far more than the trees! Weed control around the tree is far more effective in stimulating growth than applying fertiliser. (Older and weed-free trees do, however, respond readily to nitrogen.) Competition for nutrients in the soil is a pitched battle that should not be underestimated. We know that certain trees tend to grow naturally on certain soils: in Britain ash is found on nutrient-rich soil (moist river valley bottoms to dry limestone), birch and pine on nutrient-poor sands and peats, yew on dry chalk and limestone. But if we protect them from competition, we can grow all these trees on a wide range of soils. It is also important to realise that the factors limiting seed germination are often different from those limiting subsequent growth. Alders (*Alnus* spp.) can be planted as saplings on remarkably dry soil but wouldn't grow there from seed; the seed and young seedling need waterlogged soil to do well. On a larger scale, the southern limit of the coniferous forest across Canada is dictated by moisture,

the northern limit by summer temperature but these factors act primarily on young trees; established trees can persist and grow beyond these limits.

## Growth rings and dendrochronology

We saw above that in a young tree the annual growth rings tend to be a constant width and, once the tree reaches maturity, they then become progressively narrower over the years. Having said this, in any one tree there will be some year to year variation in ring width. Some of this can be due to shading; a tree overtopped by others may have very narrow rings which, if its neighbours are removed, will become much wider in subsequent years. Also, as we saw in Chapter 5, heavy seed crops can reduce ring width as food supplies are used to grow seed rather than wood. In fact, this can occur in trees like yew where male trees will have wider rings than female trees due to the burden of producing seeds. However, other things being equal, the interaction of climate with tree growth is well illustrated in the size of rings. In favourable years trees grow wider rings than in unfavourable years.[1] But what precisely is it that the trees are responding to?

In cold environments, ring width often correlates closely with summer temperature whereas in arid environments it is, not surprisingly, precipitation that affects rings most, with a negative correlation to temperature (i.e. the warmer it is, the narrower the rings) since this increases evaporation. In less extreme areas different tree species each have their own sensitivity to a mix of temperature and precipitation. For example, in Northern Ireland ring-widths in beech (*Fagus sylvatica*) and ash (*Fraxinus excelsior*) are strongly influenced by summer rainfall and soil moisture in early summer and very little by temperature. Conversely, Scots pine is sensitive to maximum summer temperature and soil temperature (García-Suárez *et al.* 2009). The responsiveness of trees to climate also varies depending upon where they are. While beech in the mild climate of Northern Ireland does not respond to temperature, in the more continental climate of Central Europe ring width is negatively affected by minimum March and maximum August temperatures (i.e. the coldest and hottest temperatures) and favoured by May and July precipitation.

The relationship between climate and ring width has led to the discipline of dendrochronology. Trees of the same species growing in the same area tend to produce similar sequences of wide and narrow rings. On the core taken from

---

[1] Wider and narrower rings are measured in relation to each other. The absolute width of rings varies by species, age of the tree and growing conditions. In any one tree the rings can be bigger on a side that gets more light or due to the presence of reaction wood; rings also tend to be wider at the base of the crown compared to the base of the trunk.

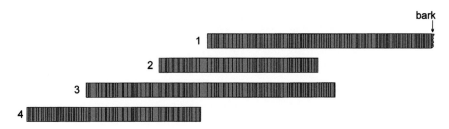

**Figure 6.7** Dendrochronology: matching patterns of tree rings from different trees to enable dating of unknown pieces of wood. In a living tree, such as tree 1, the year a particular ring grew can be calculated by counting back from the bark. If the pattern of wide and narrow rings can be matched to pieces of dead wood (2–4), they can also be dated. In this way long 'chronologies' can be built up and used to date pieces of wood of unknown history; an example of this is shown in Figure 9.12. (Note that the 'crossmatching' between cores is never quite as perfect as shown here; there is always some variation from tree to tree requiring the use of computers to find the best match.)

tree 1 in Figure 6.7 we can find exactly what year any one ring was growing by counting back from the bark, and if we are dealing with a long-lived species we might get ring sequences many hundreds of years long (what Fritz Schwein-gruber, a Swiss dendrochronologist, has described as a 'tree's private diary'). Now say we find a piece of wood that is dead (core 2 in Figure 6.7) and we have no idea when the wood was growing. If we find ring patterns that match with those somewhere along our first core then we can accurately date the rings in the previously unknown piece of wood. By adding new pieces of wood of the same species to the sequence we can build up 'chronologies' that extend back sometimes thousands of years. In Britain the longest chronology for oak now stretches back from the present day to over 7000 years ago. This is possible because oak can be found in modern woodlands, historic buildings and older archaeological sites, and is also naturally preserved for thousands of years in peat bogs (see Figure 9.12). Other chronologies go back even further. The bristle-cone pine chronology in western USA is almost 9000 years long and the combined oak and pine chronology of Central Europe now goes back to 10 461 BC.

These chronologies have provided a powerful device for ageing pieces of wood from old buildings and archaeological sites, and for checking the accuracy of carbon dating. They are also being used to reconstruct climate for times before records were kept, a useful tool in looking at the effect of past climatic changes and making predictions for the future. There are some limitations: a piece of wood from another part of the country, growing under different weather conditions, may be hard to fit against a chronology, and it can be difficult to match ring patterns between different species since each reacts in its own way to the same weather. But with more sophisticated computer matching, this has become more realistic because they can pick out very small underlying patterns from the local 'noise'.

223

As well as acting as precise dating tools, tree-ring chronologies can also tell us a lot about past climates. In Scandinavia a chronology for Scots pine has been constructed stretching back 1400 years. The most recent part of this chronology mirrors summer temperature records that have been kept during the historical period, and it has been possible to infer yearly mean summer temperatures for the full 1400 year chronology, giving climate records for a period when none were kept by humans. Thus it is possible to pick out the Little Ice Age (AD 1550–1850) and also the colder and warmer climatic episodes from prehistory. Ulf Büntgen, a Swiss palaeoclimatologist, and his colleagues (2011) have suggested that the rise of the Roman Empire in Europe was associated with a string of mild summers and that its demise was associated with a much more variable climate with prolonged droughts and cold periods. Their information comes from the tree rings of oaks in France and Germany (which reflect rainfall patterns) and of stone pine (*Pinus cembra*) and larch (*Larix decidua*) in the Austrian Alps, which provide a record of temperature.

## Buds and tree growth

As we saw in Chapters 1 and 3, trees grow in two ways: elongation of the branches (primary growth) and a later fattening of the same branches and trunk (secondary growth). Secondary growth is dealt with in Chapters 3 and 4. Here we will look at the mechanics of primary growth and the important role played by buds.

### Buds

In climates with no distinct seasons, growth may be continuous all year round. This is seen, for example, in some eucalypts and mangrove species. But the norm is rhythmic growth where a period of elongation is followed by a resting stage coinciding with a less favourable time of year. During the resting stage the vulnerable growing tip (the 'meristem') is usually protected from cold, desiccation and insect attack by being enclosed in a bud. As the end of the growing season approaches, the last few leaves (in most plants) are modified into thickened bud scales which remain on the plant when other leaves fall, enclosing the delicate tip. The defences of the tough, waterproof, overlapping bud scales (officially called cataphylls) are often supplemented by resins, gums and waxes (hence the sticky buds of horse chestnuts). That bud scales are really modified leaves is more obvious when the bud starts growing in spring. Figure 6.8 shows a horse chestnut bud opening and you can see that some of the upper scales bear leaf blades at their ends which may stay on the plant

(a)                                                    (b)

Figure 6.8 (a) A bursting bud of horse chestnut (*Aesculus hippocastanum*) showing the transition from bud scales through bud scales with part of a leaf at the tip to proper leaves. This demonstrates that the bud scales are really modified leaves. (b) A few weeks later and the bud scales will soon be dropping off but those with a partial leaf will remain for much longer and act like a small leaf. Keele University, England.

longer than the rapidly shed scales. The whole leaf does not have to be used: in the tulip tree (*Liriodendron* spp.) and limes (*Tilia* spp.) the highly toughened stipules (two appendages at the base of a leaf; see Chapter 2) of the upper leaves serve as bud scales.

A number of trees, even temperate ones, don't go to all the fuss of using bud scales. At the end of the growing season the newest semi-grown leaves simply stop growing and form a 'naked bud' (Figure 6.9). In the spring these young leaves will resume growth as if nothing had happened. Presumably the risk of damage or death during the dormant period is counterbalanced by not wasting energy growing scales which are simply shed in spring. Still, eucalypts that have naked buds have an insurance policy: the main buds are accompanied by smaller 'concealed' buds covered by the leaf base. It must also be said that some conifers, especially in the cypress family, including the false cypresses (*Cha-maecyparis* spp.), junipers (*Juniperus* spp.) and thujas (*Thuja* spp.), do not have distinct buds at all; instead they produce new growth from meristematic tissue

Figure 6.9 Naked buds – with no bud scales – of the wayfaring tree (*Viburnum lantana*). At the end of the growing season the newest leaves simply stop growing and form a 'naked bud'. The leaves resume growth in the spring from where they left off. From: Oliver, F.W. (1902) *The Natural History of Plants*. Vol. 1: *Biology and Reproduction of Plants*. Gresham, London.

hidden under the skin of the twig. Other conifers that do have buds don't have them in every leaf axil (see Powell 2008): there would be many times more than there would be room to grow. Some hardwoods may also appear not to have lateral buds but these are hidden. For example, in the London plane (*Platanus* x *hispanica*) and the Manitoba maple (*Acer negundo*) the buds are hidden inside the base of the leaf stalk which in effect forms a cone over the bud. In black locust/false acacia (*Robinia pseudoacacia*) the buds are not only hidden in the leaf stalk but are buried under the bark so even when the leaf falls they still appear to be missing.

## The new shoot

Dissect a bud and you find next year's shoot preformed in miniature. The twig and leaves (and maybe flowers or just flowers) are there waiting. So when spring comes and the bud bursts open, the new shoot can grow out quickly because it is just a case of expanding what is already there, in the same way that inflating a balloon is quicker than starting with a bowl of rubber solution and having to make the balloon first.

Buds of different trees vary in just how much of the next year's growth is preformed. In fact buds can be divided into three types of growth. In trees such as ashes, beech, hornbeams, oaks, hickories, walnuts, horse chestnuts/ buckeyes and many maples and conifers, every part of next year's shoot is preformed. Since everything is preformed, spring growth occurs in a single, rapid flush and is all over in just 10 days to a few weeks and then the terminal bud immediately takes on its winter appearance (Figure 6.10). These species are described as showing fixed or determinate growth. If dissecting buds to prove this to yourself, choose buds at the ends of branches since long-shoot buds tend to be more preformed than short-shoot buds.

In many others, however, only some of the leaves are preformed. Once these have expanded, given suitable weather, the shoot will continue to produce other leaves from scratch. As an example, the tulip tree (*Liriodendron tulipifera*) usually produces 14–20 leaves on a shoot over a summer of which eight are normally preformed. The 'early' preformed leaves are sometimes a different shape from the 'late' leaves (called heterophylly) so you can see just how many were preformed; early leaves may be less deeply lobed or divided as in sweet

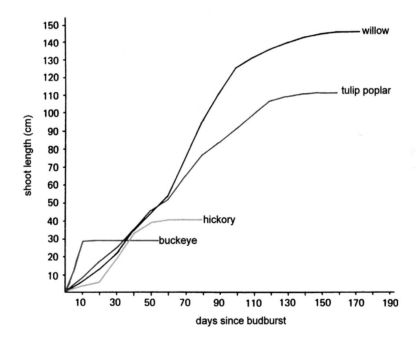

**Figure 6.10** The growth of shoots of a number of trees in Georgia, USA. Measurements of shoot length were made every 2 weeks on trees ranging from 8 to 15 years old. Buckeye (*Aesculus georgiana*) and hickory (*Carya tomentosa*) stop growing in a few weeks (fixed growth) whereas the tulip tree (*Liriodendron tulipifera*) and willow (*Salix nigra*) continue growth for a greater part of the summer (free growth). Modified from: Zimmermann, M.H. & Brown, C.L. (1971) *Trees: Structure and Function*. Springer, Berlin. With kind permission from Springer Science and Business Media.

gum (*Liquidambar styraciflua*) and ginkgo (*Ginkgo biloba*), or proportionately wider as in silver birch (*Betula pendula*). Trees showing this free or indeterminate growth (sometimes also called continuous growth) include elms, limes (lindens), cherries, birches, poplars, willows, sweet gum, alders, apples and conifers such as larches, junipers, western red cedar (*Thuja plicata*), the coastal redwood (*Sequoia sempervirens*) and ginkgo. Many of these you will notice are early invaders into open areas. Growth continues for longer than in determinate species but still normally stops well before the end of the growing season, giving time for next year's early leaves to be preformed in the new buds.

Both fixed and free type trees will sometimes show a second burst of 'lammas' growth from the terminal buds (called lammas because it usually occurs around Lammas Day, the 'bread-feast' harvest festival, traditionally 1 August or the seventh Sunday after Trinity). This is typical of young oaks but is also seen in elms, hickories, beech, alder and a number of conifers including Scots pine (*Pinus sylvestris*), Douglas fir (*Pseudotsuga menziesii*) and Sitka spruce (*Picea sitchensis*). In oaks lammas growth can be identified by the more deeply incised leaves; in pines there may be more needles per bunch than normal but shorter, giving a tufted appearance.

The third strategy is intermediate, growing by recurrent flushes (also called rhythmic growth) such that there are several cycles of growth and bud formation in a year. This is commonly found in the fast-growing southern pines from the warm south and east USA – e.g. loblolly (*Pinus taeda*), shortleaf (*Pinus echinata*) and Monterey pines (*Pinus radiata*) – and the Caribbean (Caribbean pine, *Pinus caribaea*), and more tropical species such as cocoa (*Theobroma cacao*), rubber tree (*Hevea brasiliensis*), avocado (*Persea americana*), mango (*Mangifera indica*), tea (*Camellia sinensis*), lychee (*Litchi chinensis*) and citrus species, plus the European olive (*Olea europaea*). Tea, for example (Figure 6.11), generally has four flushes of leaves per year (just as well for the tea producer since it is the new shoot tips that are used). Certainly in tea and rubber, these flushes do not seem to be induced in response to the external environment and so must be internally controlled. Sometimes the recurrent flushing can merge towards continuous free growth. When this happens in pines it results in a long branchless leader with a tuft of needles at the end, aptly called 'foxtail' growth.

## Value of these strategies

Trees with fixed growth, producing all their leaves in one go in spring, would seem to be under a distinct disadvantage. Firstly, it makes them less responsive to good growing conditions in the current year; the amount of growth this year is correlated to the weather of the previous summer when the shoots and

(a)

(b)

Figure 6.11 (a) Tea plantation near Jinja, Uganda with (b) the tea plants (*Camellia sinensis*) showing one of the several flushes of leaves they have per year.

leaves were preformed in the buds. This could mean that the benefits of a good summer will be diminished if the tree is handicapped by having small buds with few preformed leaves as a result of a previous bad summer. The tree might compensate by growing longer shoots and making its few leaves larger but it is unlikely to completely make up the loss.

So where does the advantage of preforming all next year's growth lie? The answer takes us back to the differences between ring-porous and diffuse-porous wood discussed in Chapter 3. Ring-porous trees (such as oak and ash) have to grow a new ring of wood before they can produce leaves which means they generally leaf-out later than diffuse-porous trees. Lo and behold, most ring-porous trees make up for this by having fixed growth which means they produce a full set of leaves without any further delay. The advantages of not losing any more time must outweigh the disadvantages. But this is probably not the whole answer because elms, ring-porous trees, have free growth, and diffuse-porous trees like limes, planes, tupelos (*Nyssa* spp.), magnolias and the sweet gum (*Liquidambar styraciflua*) leaf-out even later, apparently wasting some of the growing season. This late-leafing habit could be because these trees were part of the tropical and semitropical flora in the Tertiary (spread over the last 65 million years) compared to the more Arctic connections of the earlier-leafing diffuse-porous trees.

This still begs the question of why show fixed growth? It makes sense to produce a full set of leaves rapidly but why not tack on a few more later in the summer if the conditions are good? Why stop growth of the branches so early in the summer when growing conditions are just reaching their best and extra valuable height could be gained by continuing a little longer? Even in trees with free growth, branches normally stop growing well before the end of the growing season. Why have this period of 'summer dormancy'?

The answer, unfortunately, is still not clear but may be to do with maintaining sufficient food reserves within the tree. The spring growth in deciduous trees may sufficiently deplete the food reserves that further growth is prevented until they are replenished. This doesn't mean that spring growth completely drains *all* the reserves of a tree: some reserves are kept for emergencies so that, for example, a tree defoliated in spring can grow a second flush of leaves. But it is probably sufficiently fine a balance (like aiming to keep a bank balance just in the black despite emergencies) that once the new branches (and roots) are grown, and energy is put into maintaining the tree, flowering, fruiting and replenishing the food reserves, there is little left over for further branch growth that year except under exceptionally good conditions or severe defoliation (such as lammas growth, described earlier). Bear in mind that in large old trees (where more than 90% of the mass may be in the woody skeleton), respiration of the living tissue takes one to two thirds of the carbohydrate produced (tropical trees tend to be at the higher end). This may explain why young trees (with a lower respiratory burden) and coppice shoots (with abundant food stored in the roots) show growth for longer in a season than a mature tree.

I am sure this is still not the whole story. For one thing, studies have shown that reserves are rapidly replaced and are normally complete by around late July in temperate areas, and may not be appreciably depleted by spring growth in the first place, especially in those trees that flower later in the year. Also, evergreen species, which finance new growth from the photosynthesis of old needles rather than reserves, still show a halt in growth early in the season. Alternative arguments can be built around the avoidance of waste. If new outer leaves on vigorous, never-ending growth start shading those produced earlier in the year then the tree is wasting resources on inferior, shaded leaves. Moreover, there will come a point in the summer where new leaves will cost more to grow than they can hope to recover in photosynthesis in the short time left in the season. Indeed, those trees with repeated flushes of leaves are found in areas of long or continuous growing seasons and many are either typical of forest gaps (and so less likely to suffer from self-shading) or regularly shed their leaves.

Last of all, reduced root growth in summer (and therefore reduced shoot growth to keep roots and shoots in balance) and mid-summer water shortage at the top of mature trees has also been implicated in the argument, which helps add to the complexity! Like many complex questions, the truth undoubtedly lies somewhere in a mixture of these possible explanations.

## Phenology: timing and pattern of annual growth

Phenology is the study of timing of recurring natural events such as flower and leaf production. As we saw in Chapter 5, trees and other plants are capable of measuring day (or rather, night) length. This daily measuring can impose a yearly cycle on the tree, dictating when, for example, in the year it will flower (Lüttge & Hertel (2009) give a good review of growth rhythms in trees). The precise timing of events is usually modified to a greater or lesser extent by the weather patterns of each year, depending upon the species and which event is being looked at.

Shoots and root growth are both affected by temperature, but in temperate areas, shoots need air temperatures above 10°C to begin growth while roots need soil temperatures usually above 5–6°C. So which starts first depends upon which warms first above those thresholds. In temperate areas root growth normally starts up to a week or so before shoot growth although deeper, colder roots may not start until the same time or later than the shoots. In warmer environments or mild winters the roots might never have stopped growing and in Arctic tundra, which starts spring with frozen soil, shoots come first followed by the roots. Roots stirring in spring before the shoots makes good sense because a high priority in spring is water to expand the new shoots. Roots don't do anything special to survive winter, they merely stop growing. Even the fine roots normally overwinter although there are exceptions: the common walnut (*Juglans regia*) loses more than 90% of its fine roots which may explain the unusually slow leaf development in spring. Once the buds break, root growth slows down (Figure 6.12). This may partly be due to limited resources being commandeered by the active above-ground growth or due to soils getting a little too warm or dry for root growth. Either way, root growth speeds up again in the autumn.

Activity above ground starts with the breaking of the buds. For trees like beeches and rhododendrons, the increasing day length of spring triggers bud burst. For the majority, however, bud burst is triggered primarily by temperature. Just how warm it needs to be depends not just on species but also upon geographical location (trees get going at lower temperatures in cold areas) and how cold the winter has been (which affects winter dormancy; see below). Like many other processes in plants, bud burst is a compromise between opening early enough to make the most of the growing season but not so early as to overly increase the risk of frost damage. Each species of tree has arrived at its own compromise and so, in any one area, different species tend to leaf-out in the same order each year, spread over 2–3 months, but the whole sequence may be shifted by several weeks depending upon the warmth of the spring. Having said that, there is often some small variation in sequence since different species respond to different combinations of conditions; hence the old British rain predictor, 'oak before ash in for a splash, ash before oak in for a soak'. The cause of this is not water but the fact that oak is more sensitive to spring temperatures. In Norfolk, eastern England, oak came out first 60% of the years between

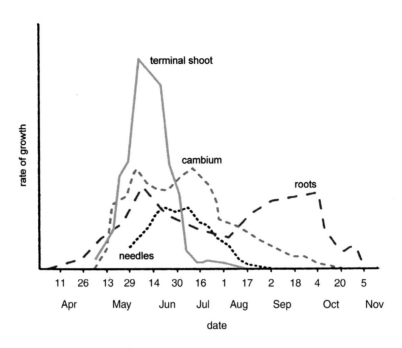

**Figure 6.12** The timing of growth of different parts of a 10-year-old eastern white pine or Weymouth pine (*Pinus strobus*) in New Hampshire, USA. The roots begin elongating before the shoots and show a mid-summer dip in growth. The needle length and trunk diameter (cambium growth) keep increasing over much of the summer but around 90% of all shoot growth (including height) occurs in just 6–9 weeks early in the season. Modified from: Kienholz, R. (1934) Leader, needle, cambial and root growth of certain conifers and their relationships. *Botanical Gazette*, 96, 73–92. © 1934 by the University of Chicago. All rights reserved.

1750 and 1958, but in the last 40 (warmer) years this has risen to more than 90% of the years (see http://www.naturescalendar.org.uk). Even within a species, different individuals can have their thermostats set slightly differently; for example there are two sycamore trees (*Acer pseudoplatanus*) of similar size that I pass on the way to work and every year the same one is regularly in full leaf in spring while the other is just putting out a few tentative leaves. For the same reason, the middle tree of three birches outside my house always starts later and finishes earlier each year (Figure 6.13). Age also makes a difference; saplings in deciduous forests generally start growing 1–3 weeks earlier in spring than the mature trees above them. This makes sense since these saplings will receive 33–98% of their year's sunlight during this early spring period before the developing canopy above robs them of sunlight.

Many tropical trees put out new leaves regularly over the whole year but those in more seasonal climates react to environmental signals. In dry tropical

**Figure 6.13** Three silver birches (*Betula pendula*) at Keele University, photographed in the autumn. The middle one has already lost most of its leaves while the two either side are still in full leaf. The same thing happens in spring in reverse: the outer two are in full leaf while the middle tree is just opening its buds. The middle tree has its thermostat set slightly higher and starts to respond to its surroundings at higher temperatures.

forests new leaves appear on the bare trees at the start of the rainy season. But the buds break just *before* the rains, and are really induced by slight changes in temperature.

Once the buds begin to open, a hormonal signal is passed to the cambium to start producing new wood (xylem) and inner bark (phloem). In conifers and diffuse-porous trees (like beech and birch; see Chapter 3), phloem is produced right away, ensuring the mobilisation of stored food to the growing points. New wood is not such a high priority in these trees because the wood from previous years is quite capable of conducting enough water before the new leaves are in full swing. So new wood growth starts at the buds and progresses down to the base of the tree over several weeks and reaches the end of the roots in another 4–6 weeks, depending upon soil temperature. The same is not true for ring-porous trees such as oak and elm. As we saw in Chapter 3, these trees move the bulk of their water through the ring formed the same spring. These trees begin new wood formation rapidly through the whole tree, up to 2–3 weeks before bud break. Cambium growth is stimulated by hormones (auxins) from developing buds, so how can the trees start so early? The answer is that the cambium is sensitive to the very low levels of auxins produced by the buds just beginning their development. Xylem growth in most trees may go on till mid-summer, tapering off into autumn (Figure 6.12). When the cambium is most active it is a physically weak tissue and the bark is most easily peeled from the tree; the season for peeling oak bark for tanning is reckoned to be from May to June. (Incidentally, if you want the bark to stay on a piece of wood for ornamental purposes, don't cut the tree during this period; if you do it won't be long before the bark simply falls off.) Growth of new wood, in conifers growing in cold areas at least, generally reaches a maximum at the time of longest day length (not necessarily the warmest part of the year) which probably helps to ensure that all xylem cells are complete before winter. Phloem production continues for a shorter time than xylem because there is less of it produced.

As the season warms, the buds on a tree progressively open in a more or less regular pattern starting at the terminal bud and working back along the branch. Traditionally it is said that bud expansion starts at the top of the crown but it is my experience that, in woodlands at least, leaves often appear first on the lower branches. Generally, leaves and shoots grow more at night when there is less water stress. There are, of course, plenty of exceptions to this in both temperate and tropical trees for a variety of reasons. For example, in Sweden the leaves of the common osier used for basket weaving (*Salix viminalis*) have been found to grow more during the late afternoon/early evening because of cold nights.

Trees are masters of forward planning. As the current set of buds open and the new shoots grow out, next year's buds are already forming. If the tree is old enough, some of these buds may contain flowers. In temperate hardwoods and conifers, flower buds are usually formed in late summer or autumn, opening the following spring or summer. There are (of course) exceptions: some warm temperate trees (e.g. buddleia, fuschia, hibiscus and *Catalpa* spp.) form their buds in spring just prior to flowering and, at the other end of the scale, eucalypts start their flowers 2 or more years before opening. In the more equitable climate of the tropics, flower buds are often initiated more evenly over the year, taking just a few months to fully form, and these may open straight away (as in fig species) or be accumulated until environmental conditions are right for them to open, which may be sometimes three times a year (as in strangler figs) or only once every 5–6 years as in the rainforests of Indonesia. Generally the better the growing season when the buds are formed, the greater the number of flowers. Paradoxically, imposing a stress on trees, such as drought, can also stimulate a 'stress crop', almost as if the tree is making a valiant effort to produce a crop of seeds in case it is its last! Ring-barking of the trunk has also been used to induce flowering in fruit trees and conifers. In this case the idea is to keep more of the carbohydrates produced by the leaves in the canopy, preventing them being 'wasted' on the roots. This is a dangerous operation which can kill the tree if taken too far.

The trigger for flower buds to open may be different to that for leaf buds, explaining why flowers can open either before or after the leaves. As explained in Chapter 5, in trees the most general trigger of flower opening is temperature. The required temperature changes can be quite subtle; the mass flowering of many trees together in the tropical forests of Southeast Asia (Figure 5.12) appears to be triggered by a decrease in minimum night-time temperatures of just 2°C for three or more nights. In addition to temperature, evolution has equipped plants with a range of specialist triggers to ensure they flower and fruit at the best time. Triggers may be rainfall in arid areas, fire for some eucalypts, or even, as in the case of one particular Central Australian shrub, complex interactions between time of year, rainfall, insects and birds (see Friedel *et al.* 1993).

This brings us to the end of the growing season, the shedding of leaves and the start of dormancy for the buds. A few deciduous trees like the tulip trees (*Liriodendrom* spp.) and honey-locust (*Gleditsia triacanthos*) will keep growing until stopped by the cold days of autumn. Most trees, however, detect the shortening days and, as described in Chapter 2, this allows time for nutrients to be removed from leaves and safely stored before the leaves are shed, and the orderly formation of the abscission zone, before the leaves are killed by wintery conditions. Temperature may play a role but this is usually restricted to low temperatures speeding up the shedding process once it is triggered by day length. For example, in the eastern cottonwood (*Populus deltoides*) short days act as the signal for dormancy and the development of cold hardiness. A study by Park *et al.* (2008) showed that genes for cold hardiness were switched on under the short days of autumn well before low temperature exposure.

Proximity to the poles does, however, alter the triggers. Although most plants monitor the length of the night, nearer to the poles (where night length can be very short in summer) a number of trees, including downy birch (*Betula pubescens*), bay willow (*Salix pentandra*) and Norway spruce (*Picea abies*), measure daytime length, particularly the amount of far-red light they are receiving. (You'll remember from Chapter 5 that day length is monitored by the phytochrome system, which reacts to the amount of red and far-red light.) Moreover, cold hardiness can be triggered by temperature alone, allowing trees such as willows to respond to sudden unseasonal cold weather that would otherwise catch them out, especially near the poles.

During summer, the buds formed during the year have not been truly dormant; rather they have been suppressed by the leaves around them.

Trees: Their Natural History

If the leaves are removed or damaged, or growing conditions are bolstered by high rain or fertiliser, the buds will grow at once. In autumn, however, there comes a period of internally imposed dormancy where even if favourable conditions are given the buds will still not grow. Dormancy is most commonly broken by exposure to low temperatures (below 5°C) for something around 300 hours in Britain. There is, of course, a good deal of variation in how much chilling is needed by different trees. For example, beech (*Fagus sylvatica*) needs a lot of winter chilling and currently only just receives enough in British winters. They will still open their buds in spring but require a large amount of spring warmth (almost as if the trees need extra convincing that winter really has finished). Other species that grow in colder areas, such as European larch (*Larix decidua*) and wild cherry (*Prunus avium*), will open their buds after less than half the winter chilling needed by beech. However long it is, this dormant period ensures that the trees don't start growth during a warm autumn or a mild mid-winter spell. Once the dormancy requirements have been met, the tree is once again able to respond to favourable conditions. Wilson (1984) quotes the German custom of collecting cherry twigs on St Barbara's Day (4 December) so that they will flower at Christmas; if they are collected in November they won't have been chilled enough and so won't flower when given warm conditions. For the same reason potted Christmas trees need care. By the time we bring them indoors they may have completed their dormancy requirements and start growing, which won't do them any good when taken back out in the New Year (it is best to keep them cool but frost-free for a month or so after Christmas to convince them that winter is coming back again).

## Tinkering with phenology

Phenology can cause us problems. Northern trees moved into southern gardens will often drop their leaves annoyingly early, and southern trees taken north will continue to grow later in the season than is often good for them. Trees growing near street lights will perceive long days late into the autumn and often keep their leaves long after others around have shed theirs. Fortunately, most trees, even when given long days and moderate temperatures, will eventually go into autumn of their own accord. If due to global warming British winters keep getting shorter and warmer, as seems to be happening, we can expect additional changes in spring phenology. The UK Phenology Network, organised in 1998 with the support of the Woodland Trust is using records posted online by amateur recorders and school students to look at these changes. And indeed, in recent decades,

236

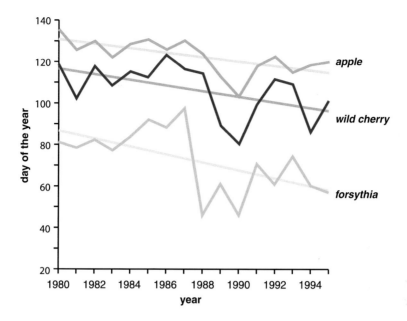

**Figure 6.14** The start of flowering of forsythia (*Forsythia* sp.) in Cologne, Germany, wild cherry (*Prunus avium*) in Zurich, Switzerland and apple (*Malus domestica*) in Berlin, Germany between 1980 and 1995. From: Roetzer, T., Wittenzeller, M., Haeckel, H. & Nekovar, J. (2000) Phenology in Central Europe – differences and trends of spring phenophases in urban and rural areas. *International Journal of Biometeorology*, 44, 60–66. With kind permission from Springer Science and Business Media.

trees like oak, hawthorn (*Crataegus monogyna*) and blackthorn/sloe (*Prunus spinosa*), whose chilling requirements are easily met, have been opening their buds earlier with higher spring temperatures. Apples, cherries and forsythia flower 1–3 weeks earlier in spring than they did 15 years ago (Figure 6.14). Oak in southern England that leafed-out around 2 May in 1950, was by 2000 doing so in the first week of April. An article in *Arnoldia* in 2007 tells the story of Mr. Fujiro Shinagawa a well-known child psychologist in Japan who was born on 15 April 1916 when the cherry blossom was at its peak, and he remembers the cherries always being in flower on his childhood birthdays. When his daughters were at school in 1955–65 the cherries were in flower on 8 April just when the school opening ceremonies were happening. But by the time his grandchild went to school in the 1990s in some years there were often no flowers left on the trees at this time. There are fewer data for autumn, but what there are suggests that such things as leaf fall are getting later.

## Lifetime changes in growth

As I used to sit exhausted and the children came and asked me to play games *again*, I wondered where they got their energy from. If trees were capable of thought, they might well think the same of their offspring. Trees go through a juvenile period in which reproduction, while not impossible, is rare, the rooting of cuttings is easy and, most importantly here, height growth is rapid (Figure 6.15). Young conifers, and probably other trees, are top-heavy with a concentration of leaves at the top of the canopy, intercepting as much light as possible. But inevitably, as the trees mature they slow the pace to reach quiet middle age. In deciduous trees like oak, the juveniles show free growth and grow quickly but this gradually reduces to the few short weeks of 'fixed growth' seen in adults. Juvenile tropical trees may shoot up vertically for several years before slowing down, branching and changing from evergreen to deciduous. Trees like tea, which flush several times in a year, slow from the youthful exuberance of 8–10 flushes each year to the sedate normal of four. Leaf shape, shade tolerance and the whole shape of the canopy can change with maturation (see Chapters 2 and 7, respectively).

If you plot the height of a tree over its life against its age, it will form an S-shaped curve (Figure 6.16). That is, growth is initially slow in the seedling, increasing to a maximum at the end of the juvenile phase when it has a good-sized canopy but only a small amount of woody and other

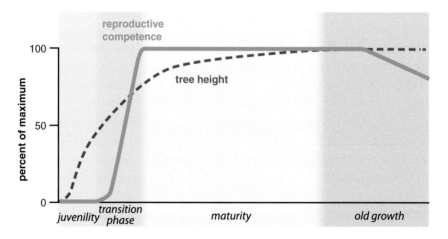

**Figure 6.15** Growth and reproduction of a tree through its life. Tree height is shown by the broken line and reproductive effort by the solid line. It should be noted that the exact relationship between height growth and reproduction varies greatly between species and between individuals of the same species, depending upon growing conditions. Redrawn from: Bond, B.J. (2000) Age-related changes in photosynthesis of woody plants. *Trends in Plant Science*, 5, 349–353, with permission from Elsevier.

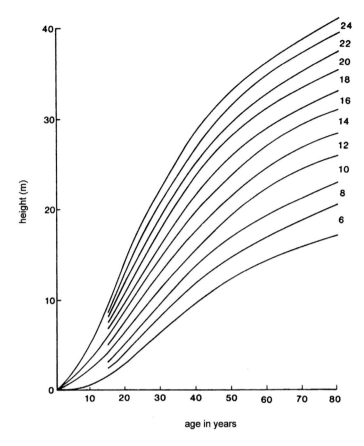

**Figure 6.16** Height growth with age for Sitka spruce (*Picea sitchensis*) under different growing conditions. The numbers at the end of the lines are the number of cubic metres of wood that are added to a hectare of trees per year (referred to as 'yield classes'). The higher the number, the better the growing conditions. Note that however fast the trees grow, the curve is still the S-shape. From: Hibberd, B.G. (1991) *Forestry Practice*. Forestry Commission, Handbook 6, HMSO, London. Reproduced under the Open Government Licence.

tissue to keep alive. As the tree gets larger with a bigger woody skeleton (Figure 6.17) – which takes more energy to maintain (see Chapter 9) – and as it reaches its maximum height determined by genetics and water supply (see the beginning of the chapter), the height and mass gains level off.

As the tree becomes still older, it develops characteristics of 'old growth'. It may still be flowering almost as well as when younger but it shows other changes. Branches with leaves become thicker, leaves become smaller and thicker and, in evergreens, they live a little longer. And in many species, the canopy changes shape (see the next chapter for further discussion on this).

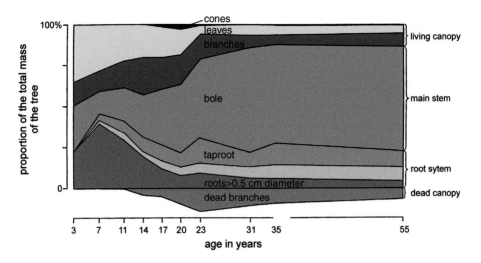

Figure 6.17 Changes in the proportion of mass of various parts of a Scots pine (*Pinus sylvestris*) as the tree ages. From: Ovington, J.D. (1957) Dry matter production by *Pinus sylvestris* L. *Annals of Botany*, 21, 287–314.

## The start of reproduction

The change from juvenile to mature adult (Figure 6.15) is normally defined by the progressive build-up of flowering. However, some trees will delay first flowering for many years after they are 'vegetatively mature' and others will flower while growing rapidly and thus still classified as juveniles. Pioneer trees that invade open areas and so need to produce seeds quickly before they are squeezed out will flower early, usually within their first decade (Box 6.2) although as an extreme, teak has been seen to flower at just 3 months under optimum nursery conditions. By contrast, long-lived trees that invade at a later stage put their resources into dominating before looking towards the next generation and flowering. Thus beech and oak in Britain are normally around 60 years old before they flower. Before you start e-mailing me about precocious trees you know of, I should point out that many factors affect age of first flowering. The better the growing conditions the younger a tree will flower. Shade is a primary factor. Oaks in the open can begin flowering at less than 40 years old but in heavy shade they may be approaching a century. Male flowers are often produced first because they are cheaper to produce when the tree is still putting most of its resources into growing (see Chapter 5).

| Box 6.2 Approximate age of first fruiting and optimum seed-bearing age | | | |
|---|---|---|---|
| Common name | Scientific name | Age of first seed production (years) | Optimum seed-bearing age (years) |
| **Hardwoods** | | | |
| Alder | *Alnus* spp. | 12 | – |
| Ash | *Fraxinus excelsior* | 20–40 | 40–100 |
| Beech | *Fagus* spp. | 40–60 | 80–200 |
| Birch | *Betula* spp. | 7–40 | 20–70 |
| Coastal redwood | *Sequoia sempervirens* | 5–15 | 250+ |
| Elm | *Ulmus* spp. | 15–40 | 40+ |
| False acacia | *Robinia pseudoacacia* | 6 | 15–40 |
| Field maple | *Acer campestre* | 25 | 50–150 |
| Hazel | *Corylus avellana* | 10 | – |
| Hickory | *Carya ovata, C. laciniosa* | 40 | 60–200 |
| Honey-locust | *Gleditsia triacanthos* | 10 | 25–75 |
| Hornbeam | *Carpinus betulus* | 20–30 | – |
| Horse chestnut | *Aesculus hippocastanum* | 20 | 30 |
| Giant sequoia | *Sequoiadendron giganteum* | 10 | 150–200 |
| Lime | *Tilia* spp. | 25 | 50–150 |
| Poplar, willow | *Populus, Salix* spp. | 5–10 | 25–75 |
| Sweet gum | *Liquidambar styraciflua* | 20–30 | up to 150 |
| Sycamore | *Acer pseudoplatanus* | 20–40 | 40–100 |

## Box 6.2 (cont)

| Common name | Scientific name | Age of first seed production (years) | Optimum seed-bearing age (years) |
|---|---|---|---|
| Tulip tree | *Liriodendron tulipifera* | 15–20 | up to 200 |
| Oak, English | *Quercus robur* | 40–60[a] | 80–120+ |
| Oak, red | *Quercus rubra* | 25 | 50+ |
| Sweet chestnut | *Castanea sativa* | 30–40 | 50+ |
| Walnut | *Juglans regia* | 4–6 | 30–130 |
| **Conifers** | | | |
| Douglas fir | *Pseudotsuga menziesii* | 30–35 | 50–60 |
| Fir | *Abies* spp. | 20–45 | 40–200 |
| Larch | *Larix* spp. | 15–25 | 40–400 |
| Pine | *Pinus* spp. | 5–60 | 15–160+ |
| Spruce | *Picea* spp. | 10–40 | 60–250 |
| Western red cedar | *Thuja plicata* | 20–25 | 40–60 |
| **Tropical trees** | | | |
| Pioneers | | 2–5 | – |
| Teak | *Tectona grandis* | 5–15 | – |
| Most species | | 10–30 | – |
| Dipterocarps of Malaysia | *Dipterocarpus* spp. | 60 | – |

Actual age in any one tree may vary either way of the figures given here depending upon growing conditions.

[a] coppice shoots of oak may begin fruiting at 20–25 years of age.

Based on Skene, M. (1927) *Trees*. Thornton Butterworth, London, UK; Hibberd, B.G. (1986) *Forestry Practice*. Forestry Commission Bulletin 14. HMSO, London; and Burns, R.M. & Honkala, B.H. (1990) *Silvics of North America*. Vol I, *Conifers*; Vol 2, *Hardwoods*. USDA, Forest Service, Agricultural Handbook 654.

##  *Further Reading*

Augspurger, C.K. (2008) Early spring leaf out enhances growth and survival of saplings in a temperate deciduous forest. *Oecologia*, 156, 281–286.

Barlow, P.W., Mikulecký Sr, M. & Střeštík, J. (2010) Tree-stem diameter fluctuates with the lunar tides and perhaps with geomagnetic activity. *Protoplasma*, 247, 25–43.

Bond, B.J. & Franklin, J.F. (2002) Aging in Pacific Northwest forests: a selection of recent research. *Tree Physiology*, 22, 73–76.

Bret-Harte, M.S., Shaver, G.R., Zoerner, J.P., *et al.* (2001) Developmental plasticity allows *Betula nana* to dominate tundra subjected to an altered environment. *Ecology*, 82, 18–32.

Büntgen, U., Tegel, W., Nicolussi, K., *et al.* (2011) 2500 years of European climate variability and human susceptibility. *Science*, 331, 578–582.

Büsgen, M. & Münch, T. (1929) *The Structure and Life of Forest Trees* (3rd edn). Chapman & Hall, London.

Cedro, A. & Iszkuło, G. (2011) Do females differ from males of European yew (*Taxus baccata* L.) in dendrochronological analysis? *Tree-Ring Research*, 67, 3–11.

Costes, E. & García-Villanueva, E. (2007) Clarifying the effects of dwarfing rootstock on vegetative and reproductive growth during tree development: a study on apple trees. *Annals of Botany*, 100, 347–357.

Čufar, K., Prislan, P., de Luis, M. & Gričar, J. (2008) Tree-ring variation, wood formation and phenology of beech (*Fagus sylvatica*) from a representative site in Slovenia, SE Central Europe. *Trees*, 22, 749–758.

Dengel, S., Aeby, D. & Grace, J. (2009) A relationship between galactic cosmic radiation and tree rings. *New Phytologist*, 184, 545–551.

DeWoody, J., Rowe, C.A., Hipkins, V.D. & Mock, K.E. (2008) "Pando" lives: molecular genetic evidence of a giant aspen clone in Central Utah. *Western North American Naturalist*, 68, 493–497.

Ennos, R. (2012) *Solid Biomechanics*. Princeton University Press, Princeton.

Fischer, D.G., Hart, S.C., Rehill, B.J., *et al.* (2006) Do high-tannin leaves require more roots? *Oecologia*, 149, 668–675.

Friedel, M.H., Nelson, D.J., Sparrow, A.D., Kinloch, J.E. & Maconochie, J.R. (1993) What induces central Australian arid zone trees and shrubs to flower and fruit. *Australian Journal of Botany*, 41, 307–319.

Friedrich, M., Remmele, S., Kromer, B., *et al.* (2004) The 12,460-year Hohenheim oak and pine tree-ring chronology from central Europe – a unique annual record for radiocarbon calibration and paleoenvironment reconstructions. *Radiocarbon*, 46, 1111–1122.

García-Suárez, A.M., Butlera, C.J. & Baillie, M.G.L. (2009) Climate signal in tree-ring chronologies in a temperate climate: a multi-species approach. *Dendrochronologia*, 27, 183–198.

Halford, N.G. & Paul, M.J. (2003) Carbon metabolite sensing and signalling. *Plant Biotechnology Journal*, 1, 381–398.

Hodge, S.J. (1991) Improving the growth of established amenity trees: fertilizer and weed control. *Arboriculture Research Note* 103–191.

Karmacharya, S.B. & Singh, K.P. (1992) Production and nutrient dynamics of reproductive components of teak trees in the dry tropics. *Tree Physiology*, 11, 357–368.

Kelly, P.E., Cook, E.R. & Larson, D.W. (1992) Constrained growth, cambial mortality, and dendrochronology of ancient *Thuja occidentalis* on cliffs of the Niagara escarpment: an eastern version of bristlecone pine? *International Journal of Plant Science*, 153, 117–127.

King, D.A., Davies, S.J., Tan, S. & Nur Supardi, M.N. (2009) Trees approach gravitational limits to height in tall lowland forests of Malaysia. *Functional Ecology*, 23, 284–291.

Kozlowski, T.T. (1992) Carbohydrate sources and sinks in woody plants. *Botanical Review*, 58, 107–222.

Larjavaara, M. & Muller-Landau, H.C. (2010) Rethinking the value of high wood density. *Functional Ecology*, 24, 701–705.

Li, A., Guo, D., Wang, Z. & Liu, H. (2010) Nitrogen and phosphorus allocation in leaves, twigs, and fine roots across 49 temperate, subtropical and tropical tree species: a hierarchical pattern. *Functional Ecology*, 24, 224–232.

Lüttge, U. & Hertel, B. (2009) Diurnal and annual rhythms in trees. *Trees*, 23, 683–700.

Mattheck, C. & Breloer, H. (1994) *The Body Language of Trees: a Handbook for Failure Analysis. Research for Amenity Trees, No 4.* HMSO, London.

McDonald, A.J.S., Stadenberg, I. & Sands, R. (1992) Diurnal variation in extension growth of leaves of *Salix viminalis. Tree Physiology*, 11, 123–132.

Mølmann, J.A., Junttila, O., Johnsen, Ø. & Olsen, J.E. (2005) Effects of red, far-red and blue light in maintaining growth in latitudinal populations of Norway spruce (*Picea abies*). *Plant, Cell and Environment*, 29, 166–172.

Niklas, K.J. (2007) Maximum plant height and the biophysical factors that limit it. *Tree Physiology*, 27, 433–440.

Park, S., Keathley, D.E. & Han, K.-H. (2008) Transcriptional profiles of the annual growth cycle in *Populus deltoides. Tree Physiology*, 28, 321–329.

Petit, G., Anfodillo, T., Carraro, V., Grani, F. & Carrer, M. (2010) Hydraulic constraints limit height growth in trees at high altitude. *New Phytologist*, 189, 241–252.

Powell, G.R. (2008) On buds man. *Forestry Chronicle*, 84, 590–594.

Primack, R. & Higuchi, H. (2007) Climate change and cherry tree blossom festivals in Japan. *Arnoldia*, 65, 14–22.

Roetzer, T., Wittenzeller, M., Haeckel, H. & Nekovar, J. (2000) Phenology in central Europe – differences and trends of spring phenophases in urban and rural areas. *International Journal of Biometeorology*, 44, 60–66.

Rossi, S., Deslauriers, A., Anfodillo, T., *et al.* (2006) Conifers in cold environments synchronize maximum growth rate of tree-ring formation with day length. *New Phytologist*, 170, 301–310.

Sparks, T.H. & Menzel, A. (2002) Observed changes in seasons: an overview. *International Journal of Climatology*, 22, 1715–1725.

Thomas, P.A. & Polwart, A. (2003) Biological Flora of the British Isles *Taxus baccata* L. *Journal of Ecology*, 91, 489–524.

Tng, D.Y.P., Williamson, G.J., Jordan, G.J. & Bowman, D.M.J.S. (2012) Giant eucalypts – globally unique fire-adapted rain-forest trees? *New Phytologist*, 196, 1001–1014.

Van Pelt, R. (2001) *Forest Giants of the Pacific Coast.* University of Washington Press, Seattle.

Wilson, B.F. (1984) *The Growing Tree.* University of Massachusetts Press, Amherst.

# Chapter 7: The shape of trees

The whole point of a woody skeleton is ultimately to get the leaves above competitors to ensure a lion's share of the light. And from this simple goal comes an enormous range of tree shapes, from the unbranched stems of palms and tree ferns to the tall spires of conifers, the broad spreading crown of oaks and the multiple stems of an old yew. What governs the shape of trees? How are trees organised to display what often looks like an impossibly large number of leaves?

## Trees of distinctive shape

It is usually possible (but not always!) to identify a conifer from a distance by its conical outline. Within the cone there are usually horizontal plates of foliage showing where the branches are produced in whorls around the main stem, usually one whorl per year (Figure 7.1). This contrasts with the wide dome of a hardwood where the leading shoot of the young tree gives way to a number of strong branches giving the whole canopy a rounded shape.

Within these two main shapes it is possible (with a little practice) to distinguish different species simply by their shape. This book is not the place to list the distinctive features of common species but one example will illustrate the point. In common lime (*Tilia* x *europaea*) the main branches develop in great arching curves which in time lose the terminal buds. New growth comes from near the top of the branch end resulting in another arch, creating the effect of multiple rainbows joined together at their ends (Figure 7.2). Also, epicormic buds (Chapter 3) characteristically produce a mass of sprouts around the base of the trunk and frequently a congested growth of small twigs in the centre of the canopy. Some of these young growths escape to produce vertical branches through the crown, parallel to the main stem.

Since different tree species have a distinctive inherent shape, it is perhaps obvious that shape is controlled genetically. This implies that the shape has been honed over evolutionary time to fit aspects of the tree's environment and, as discussed below, this appears to be broadly true. But two trees of the same species will never be exactly identical. This can be partly due to genetic differences (in the same way that two brothers may share family features but

(a)

(b)

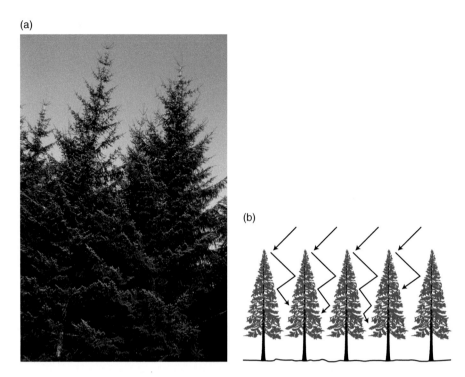

**Figure 7.1** (a) Sitka spruce (*Picea sitchensis*) showing the typical spire of conifers with whorls of branches. Perthshire, Scotland. (b) This conical shape is superbly adapted for absorbing light coming from a low angle, as in the northern conifer forests of Canada and Siberia. Based on Walker, D.J. & Kenkel, N.C. (2000) The adaptive geometry of boreal conifers. *Community Ecology*, 1, 13–23.

still be distinct enough to be easily told apart). But the differences will also be due to the interaction of the tree with its environment, which is also discussed below. In biological parlance, we can see the effect of the tree's genotype (its genetic make-up), and how this interacts with the environment to produce the phenotype, the actual tree we see.

## Why have these distinctive shapes?

Shape is a compromise between displaying leaves without undue self-shading, the needs of pollination and seed dispersal, optimum investment in the woody structure ('biomechanics') and coping with the surrounding environment (high winds, poor soil, etc.). In the tropics, where you would expect most variation in shape because of the huge number of species,[1] trees have been

---

[1] A hectare (2.47 acres) of tropical rainforest usually contains 60–150 different species and occasionally up to 300. By contrast temperate forest averages around 25–30 species in a hectare. The temperate forests of N America contain fewer than 400 species of tree whereas Madagascar alone has 2000!

(a)

**Figure 7.2** The distinctive shape of common lime (*Tilia* x *europaea*). (a) The typical arching curves of the main branches; because the end buds tend to die, new growth comes from near the branch end resulting in another arch, creating the effect of multiple rainbows joined together at their ends. The tree in (b) has the characteristic congested growth of small twigs in the centre of the canopy. What you can't see are the mass of shoots that normally spring from the base of the tree. Staffordshire, England.

(b)

Figure 7.2 (cont)

categorised into just 25 or so shapes (see Hallé *et al.* 1978). This suggests that there are a limited number of shapes that can produce a workable compromise.

The nature of the compromise is often hard to pinpoint but some gross generalisations are possible. Conifers of high latitude and altitude are typically steeply pyramidal with short downward sloping branches to help shed snow (Figure 7.1a); this shape also helps intercept the maximum amount of light from the sun low on the horizon. More than this, dense stands of even-sized trees (as occur after a forest fire) are very efficient at capturing light since the light is reflected further and further down into the stand (Figure 7.1b) so that it is almost fully absorbed. Dave Walker and Norm Kenkel at the University of Manitoba point out that this explains why conifer stands look so dark in satellite images; they reflect very little light. This works in the same way that paint on a stealth aircraft absorbs rather than reflects radar! For the tree, not only does this supply the maximum amount of light for photosynthesis but since this captured energy is eventually released as heat, the stands can be 5–10 °C warmer than non-treed areas, helping to extend the growing season. Conifers on dry sites have a similar shape but in this case it appears to be an adaptation to intercepting the *least* light (and therefore heat) at noon when it is hottest, saving on the need for cooling by water loss. By contrast, the broad crowns of most hardwoods and some spruces and firs are associated with moist sites, deep shade (a wide canopy intercepts maximum light) or harsh tree-line conditions (to keep low and out of the wind). It is also a good shape for intercepting most light in cloudy climates where the light is diffuse and in effect comes from over the whole sky. Because in Britain we have an almost uniformly moist (and mostly cloudy) environment, most of our trees take on the same, roughly spherical, shape (Figure 7.3). Pines further south in

Figure 7.3 The spherical shape of most trees from moist, cloudy environments. In this case they are in the grounds of the Anglican Cathedral, Liverpool, England.

**Figure 7.4** Stone pine (*Pinus pinea*) growing in Tuscany, Italy showing the aerofoil shape of the canopy which helps in conserving water.

savannah-type climates, such as the Mediterranean stone pine (*Pinus pinea*), develop a flat-topped, umbrella shape (Figure 7.4) which helps resist drying winds (the aerofoil shape of the canopy allows leaves to hide behind each other out of the wind) and maximises convective heat loss by allowing free passage of air up through the canopy.

## The dynamic tree: reacting to the changing world

Having said that a species has a characteristic shape adapted to the type of environment, it is obvious that the shape of an individual tree will vary with conditions. Light, wind, snow, herbivory, fire, root health and many other factors can all influence shape. Trees grown in the open, for example, have a bushy crown with wide spreading branches while those in shade are taller and narrower with fewer side branches, in an effort to reach light. But it is not just light. As a tree grows, the canopy increases in width (usually in proportion to the trunk diameter) unless it meets other trees. The canopy of each tree tends to remain discrete and does not intergrow with neighbours (referred to as 'canopy shyness'; Figure 7.5). (The reasons for this shyness are described below.) So the first trees in an area grow wide branches, claiming a large volume of air. Later trees have to worm their way into the spaces left between the older trees and so often end up very narrow and tall.

Figure 7.5 The canopy of a red pine (*Pinus resinosa*) forest showing the 'canopy shyness' between individual trees. New Brunswick, Canada.

Mountains provide ample evidence of the effect of climate on tree shape. Puffing up a mountain you notice that, as a result of lower year-round temperature and increased wind, conifers become shorter and more squat. Quite literally, shaking (or even rubbing) a tree stunts its growth: a glasshouse study by Neel and Harris in 1971 showed that shaking sweet gum trees (*Liquidambar styraciflua*) for just 30 seconds a day reduced the height growth to less than a third that of the unshaken controls, partly because the shaken trees stopped growth and produced terminal buds. Experiments have shown that contact from just spraying water onto a plant is enough to induce reduction in height, a change in leaf shape and an increase in root growth. This makes good mechanical sense: the stronger the wind (or shaking) the more leverage is exerted on the tree so the optimum design will be more squat (see the discussion on tree height and engineering in Chapter 6). This also explains why staking a planted tree can cause problems. If the base of the tree is held firmly, it doesn't detect movement and so stays thinner than in an unstaked tree (see Watt *et al.* 2009 for an example), and is more likely to snap when the stake is removed.

Higher up the mountain, trees only survive in tight, isolated clumps usually called Krummholz (German for 'crooked wood') although purists will sometime use the word Kruppelholz. Harsh winds carrying ice particles in the

(a)

(b)

Figure 7.6 Subalpine fir (*Abies lasiocarpa*) in the Canadian Rocky Mountains demonstrating the effect of hostile conditions at high altitude. (a) A 'flagged' specimen, bare on the side facing prevailing winds; (b) a larger clump of Krummholz with the prevailing winds moving into the picture. Note the skirt of healthy foliage around the bottom showing where snow protects against winter winds.

winter wear away the waxy coating on needles like sandpaper, leaving them open to death by dehydration. This often produces 'banner' or 'flagged' trees looking like a flag blowing in the wind with branches surviving only in the lee of the prevailing wind (Figure 7.6). Similar flagging can be seen in other windy places such as on cliffs by the sea where the wind is aided in its damage not by ice but by salt spray. New trees can only establish in the shelter of others (often by layering – rooting – of the branches), leading to clumps of stems huddled together. As the stems on the windward side are slowly killed and new ones grow on the lee side, the whole clump moves downwind. In N America this has been estimated at 2–7 m per century. At the bottom of these clumps, healthy branches survive in a thick prickly skirt: these are protected from the ravages of the wind by snow lying on the ground. Higher still up the mountain and the skirt is the only part to survive and the noble spire of lower altitudes has been reduced to a prostrate shrub by environmental conditions.

But trees are not passive in the face of changes to their shape made by the environment: they react. If branches are lost, new ones can be grown from stored buds or new adventitious ones (Chapter 3). Alternatively, an existing branch can be used to fill the gap, by reorientation through bending (see reaction wood in Chapter 3). Experiments have shown that removing one branch from a whorl results in the branches on either side bending to take up the space (Westing 1965). Branches can also take on a more major role. This is seen to perfection when a tree falls over yet remains at least partially rooted. What were minor side branches find they are physically the new leaders and get a new lease of life and start growing as miniature trees, eventually producing a line or thicket of what appear to be individual trees. Incidentally, new or 'reassigned' branches tend to grow in the same shape as a seedling would; this programmed growth is referred to as 'reiteration', repeating the same shape over and over again.

## Biomechanics and gravity

Gravity plays a big part in determining the shape of individual trees. This can be understood using the principles of 'biomechanics' fostered by a German physicist, Claus Mattheck. One of the simplest is the principle of minimum lever arms. This is less complicated than it sounds. As mentioned in the previous chapter, the longer a branch, the more likely it is to break (unless serious investment in wood takes place) because it is acting like a longer lever (in the same way that a longer spanner puts more turning force on a stiff nut). This acts to limit the ultimate length of a branch: a compromise between displaying leaves and keeping the lever to a minimum. There are, of course, ways of cheating: Chapter 4 describes how banyans use pillar roots to support tremendously long branches, and if branches graft together (as in Figure 7.7) this support can allow them to become abnormally long.

**Figure 7.7** Two oak (*Quercus robur*) branches grafted together. Just as roots pressing against each other will weld together, so will branches. This is often claimed to be impossible because the constant movement by wind prevents the tissues joining, but with enough pressure the union is possible, as demonstrated by the picture! Cheshire, England.

The weight of these branches also acts to bend the trunk. Trees on the whole grow to keep the net effect of these levers to a minimum: in a normal tree the weight of the branches is balanced either side of the tree so that there is no net force acting to bend the tree. Figure 7.8 shows a tree that has lost its top leaving one branch sticking sideways. The power of the lever is calculated simply as the weight of the canopy multiplied by the effective length of the lever. Very soon, the branch will begin to bend upright even when the tree is in the open and would get no more light by doing so. The advantage is obvious: it is to reduce the lever arm that is otherwise constantly putting a strain on the tree, pulling it sideways, in the same way we would when carrying a heavy bucket. The tree can detect the strain imposed by a lever arm and, other things being equal, will act to minimise the lever arm by bending branches using the 'reaction wood' described in Chapter 3.

The same principle applies when whole trees lean. In Figure 7.9 the leaning tree will gradually bend to bring the centre of gravity above the base, to reduce the lever arm to a minimum. As the diagram shows, we do the same thing when carrying a heavy load on our backs and we bend forward to bring the centre of gravity of the load above our feet. Otherwise we are constantly straining to stay upright and a little shove will have us

Figure 7.8 After the loss of the top part of the tree, a single branch acts like a lever bending the stem where the twisting force exerted (*M*) is calculated simply as the mass of the branch (*F*) multiplied by the length of the lever (*l*, really the distance from the branch base to the branch's centre of gravity). The tree reacts by bending the branch upwards to reduce the length of the lever (the twisting force becomes zero), thus bringing the centre of gravity of the branch over the base of the trunk. This is analogous to the ease with which we can carry a heavy bucket by keeping it closer to our body (although not necessarily on our head!). From: Mattheck, C. & Kubler, H. (1995) *Wood – The Internal Optimization of Trees*. Springer, Berlin. Fig. 18, page 23. With kind permission from Springer Science and Business Media.

over. In a tree this results in the characteristic 'J' or 'S' shape. At this point you may be thinking that you have seen plenty of trees that have been far from upright with the centre of gravity way off to one side. And you would be right. Sometimes, in the compromise determining tree shape, other things are more important. For a tree leaning out into the centre of a gap in the forest the extra light outweighs the strain of the lever. (Also, thick stems are less able to bend and a tree may be unable to bend to completely remove the lever arm.) You may be wondering how a tree with no brain works out these complicated solutions; all is revealed below in *How does the tree control shape?*.

## Buds, branches and tree shape

Standing on a cliff looking at trees shaped by strong winds leaves us in no doubt that growing conditions can alter the shape of a tree. But while

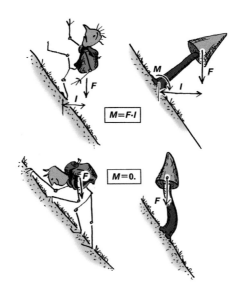

Figure 7.9 In a similar way to Figure 7.8, a leaning tree will straighten to reduce the twisting force exerted by gravity, in the same way that we bend forward when carrying a heavy pack to bring the centre of gravity over our feet. From: Mattheck, C. & Kubler, H. (1995) *Wood – The Internal Optimization of Trees*. Springer, Berlin, Fig. 19, page 23. With kind permission from Springer Science and Business Media.

individual trees may vary a lot depending upon damage and growing conditions, the characteristic shape will shine through given half a chance showing that it is genetically determined. How do trees manage to grow into such characteristic shapes? The answer lies in looking at the building blocks used. Animals, on the whole, have a fixed shape that simply gets bigger over the years. A baby and an adult human have all the same bits; they are just bigger in a grown-up. Plants on the other hand work by repeatedly adding together small modules (called reiteration), much like making a daisy-chain longer by adding more daisies rather than making each daisy bigger. The basic module of a tree is most easily thought of as the leafy twig grown in one year.

## A twig and its buds

Starting at the end of a branch and following back, it is usually possible to see the bud scale scars forming a circle around the twig (and therefore sometimes called the girdle scar) which mark where last year's terminal bud was (Figure 7.10). Thus, this end portion of the branch is what grew in the latest season and is the basic module of the tree. The young twig bears a number of leaves, and as these fall off a leaf scar remains as shown in Figure 7.10. Nestling between the leaf and the twig is a bud (these are in the axils of the leaf and so

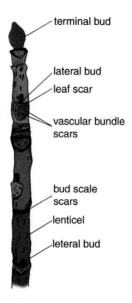

terminal bud

lateral bud

leaf scar

vascular bundle
scars

bud scale
scars

lenticel

leteral bud

Figure 7.10 A branch in winter. The successive sets of bud scale scars mark how much of the branch grew each year. The leaf scars mark the positions of the previous year's leaves and the bud associated with each leaf is clearly to be seen. From: Nadakavukaren, M. & McCracken, D. (1985) *Botany: An Introduction to Plant Biology.* West Publishing, Saint Paul.

are usually called axillary or lateral buds).[2] These buds contain next year's growth, i.e. a new twig complete with leaves and buds[3]. It is worth pointing out here that new leaves can only be grown on new twigs; a bud never gives rise to just leaves although sometimes the twig may be inconspicuous; see the discussion on short shoots below. It is now possible to see how the tree gets bigger each year by each bud growing out into a new module, which bears buds that grow out into further modules and so on. It is usually possible to trace your way back along a branch and see perhaps the last 5–10 years' worth of growth, each year marked by the bud scale scars. This adding of new modules happens at the top of the tree and along each branch so the whole tree gets bigger. Bear in mind that the new twigs (modules) are the finger-like projections of the new coat of wood added over the whole tree so last year's twig will now be fatter and have two rings, the twig from the year before will

---

[2] As always there are exceptions! In dawn redwood the bud is either beside or underneath the compound leaf. But since the 'leaf' is really a small twig with many leaves, perhaps it doesn't count (but is worth remembering for natural history quizzes!).

[3] Occasionally in fast-growing individuals of temperate trees such as birch, alder, cherry and the tulip tree (and more commonly in tropical trees) buds will grow out the same year as they are formed (referred to as sylleptic shoots as opposed to a shoot from a dormant bud, a proleptic shoot).

be fatter still and have three rings, and so on, the branch getting thicker as you trace back to the trunk (see Figure 3.2).

Since the buds are in the axils of leaves, the arrangement of leaves (or the 'phyllotaxy') can play a large role in determining the basic branching pattern. Leaves and buds are basically arranged around a twig in a spiral but often this spiral is modified to produce leaves that are effectively alternate along a twig or, by reducing the length of twig between two or more leaves to a minimum, opposite or even whorled. Thus trees with alternate leaves (like oaks and beeches) tend to have alternate branches, and those with opposite buds such as ashes and maples expand as a series of pairs of opposite branches which can be traced back through the whole tree. Thus small-scale differences at the branch level influence the larger-scale shape of the crown.

The length of new branches and their angle from the parent branch will also help determine the overall shape of the tree, although this does vary with branch size. A large branch will tend to have a shallow angle to the branch it arises from whereas a smaller branch will diverge at a wider angle until a small stick springing from a large branch is often at right angles to the large branch. But why? Tree branching patterns are often considered to be analogous to the branching of blood vessels since both involve plumbing for moving fluid around. Most theories suggest that the angles of branches are to ensure energy efficient flow, so major junctions are at a shallow angle while smaller junctions involving less fluid can be at a sharper angle. Imagine the traffic jams that would happen if there was a right angle bend in a major highway, but this is fine on farm tracks. The analogy of tree branches with blood vessels and highways is not perfect because inside the tree there are no large wide tubes carrying fluid, rather every branch carries bundles of numerous small-diameter tubes, but the principle of encouraging maximum flow for the least energy expenditure still applies in the main trunk and branches. Certainly, Leonardo da Vinci noticed that the cross-sectional area of the branches stays the same above and below a division. There may also be other rules at work. For example, a main driving force behind shape may be to minimise the length of branches needed since they are expensive things to grow (more on this can be found in Philip Ball's book on *Branches*).

## A hopeless tangle

We have seen above that the tree is built up by continually adding new modules, just like a wall is built by adding new bricks, small in themselves but capable of adding up to a large imposing structure. But if every bud on a tree grew into a branch the canopy would soon become a hopeless tangle of dense branches. A 100-year-old oak should have 99 orders of branching rather than the 5–6 that actually exist. Temperate trees rarely show more than 5–8 orders, and tropical trees 2–3 or at most 4 orders of branching (tropical trees generally have bigger leaves requiring a less-fine network of branches to hold them). The potential tangle is prevented in three main ways.

## Dealing with too many buds

Sometimes the tree is helped by outside conditions. Spring frosts will, to our annoyance, readily kill buds when they are losing their cold hardiness. Abrasion of buds, leaves and twigs is also implicated as a cause of canopy shyness (Figure 7.5) where the crowns of neighbouring trees (and branches on the same tree) rub together in the wind and do not intermesh. White spruce (*Picea glauca*) growing up under aspen in Canada can suffer in the same way from being whipped by swaying aspen branches. But trees are not at the mercy of outside influences to solve the problem of too many buds. It has been shown in silver birch (*Betula pendula*), and is undoubtedly true of other trees, that fewer buds develop in parts of the crown that are already dense or where the crowns of different trees start meeting, presumably because of low light intensities leading to shorter branches (and so fewer buds) and more bud death. This helps explain why trees grown close together show a strong tendency to grow more on those sides which face away from neighbours, and another reason why trees show crown shyness.

Buds can also fail to develop as a normal part of the tree's growth. The spines of hawthorns (*Crataegus* spp.) and honey-locust (*Gleditsia triacanthos*), and the bundles of needles on pines are all modified branches which stop growth after a while and lose the ability to produce new buds. This simple solution of not producing buds is taken to the extreme in conifers. Conifers like spruces (*Picea* spp.) and firs (*Abies* spp.) with many small needle-like leaves do not produce a bud in the axil of every leaf[4]: if they did they would end up with an impossible number of buds and potential branches. Rather, they concentrate resources into relatively few buds at the end of each year's branches. As mentioned in the last chapter, some conifers, especially in the cypress family do not produce any distinct buds at all.

Buds that *are* produced but are excess to requirement can be deliberately aborted. In oaks there is a steady rain of aborted buds from the canopy throughout the growing season as up to 45–70% of buds are lost. An alternative to getting rid of buds is to save them for later, keeping them dormant for

---

[4] Juvenile growth of pines, larches and cedars bears single needles which *do* have a bud in almost every axil. These buds produce the bunches of needles characteristic of mature growth. These needles in turn show the lack of buds in their axils as expected in conifers.

many years as an insurance policy in case of need (epicormic buds; see Chapter 3). In oak the small buds at the base of a twig are usually kept in this way. Moreover, since bud scales are really modified leaves they themselves have axillary buds which usually remain small and largely unseen but quite capable of growth. The value of these stored buds is seen in pruning trials. The kermes oak of Mediterranean areas (*Quercus coccifera*) has been shown to produce new shoots from stored buds and survive for 4 years despite all new shoots being clipped off every 15 days. The lack of buds on many conifers shows in their difficulty in breaking from old wood if heavily pruned; this is especially noticeable in the cypress family, particularly Lawson's cypress (*Chamaecyparis lawsoniana*) which consequently makes a poor choice for hedging. Fortunately for landscapers, some conifers can form new buds in the apparently 'empty' leaf axils. Yew (*Taxus baccata*; Figure 7.11) and coastal redwood (*Sequoia sempervirens*) grow buds and store them ready-made, while others, including the giant sequoia (*Sequoiadendron giganteum*) and white cedar (*Thuja occidentalis*) from N America, and the hiba (*Thujopsis dolabrata*) and Japanese red cedar (*Cryptomeria japonica*) from Japan, grow them when needed.

**Figure 7.11** The yew hedge on the right had become grossly overgrown and was cut back to the trunks. Fortunately, yew (*Taxus baccata*), unlike most conifers, has buds in some leaf axils and can usually regrow (although you'll notice some bare areas). This process was helped by drastically cutting only one side of the hedge, reducing the overall stress on the trees. Once this side has recovered the other side will be cut. Crathes Castle, Scotland.

Thus, like many hardwoods, they are capable of forming new branches when heavily pruned or damaged.

Loss of the terminal bud of a branch can have a large effect on the shape of a tree. If the terminal branch lives for many years it will grow strongly in one direction, carrying a number of side branches (called monopodial growth) as seen in trees such as ashes. If it dies, the buds either side usually grow to replace it, giving a fork (referred to as sympodial growth), and if this happens many times a large branch ends up being composed of a series of short sections at angles to each other, as in mature horse chestnuts (look ahead to Figure 7.14).

## Shedding branches

A useful way of preventing too many branches clogging the tree is to get rid of them once they have fulfilled their purpose. This happens as the tree gets bigger and grows new shells of foliage which shade the inner and lower branches. If you look at a large tree the centre of the canopy is made up of large branches (the 'scaffold branches') leading to fine twigs only at the edge of the canopy. In the shaded centre (the 'dysphotic zone' in scientific parlance) the small branches that would once have occupied that space are long gone. Trees like the true cypresses (*Cupressus* spp.), coastal redwood, swamp cypress (*Taxodium distichum*) and the dawn redwood (*Metasequoia glyptostroboides*) regularly shed small twigs complete with leaves towards the end of summer. Other trees shed only those that prove unproductive. If a branch is not producing enough carbohydrate to cover its own running costs, i.e. it needs to be subsidised by other branches around it because, for example, it is being shaded, it will usually be got rid of. In some ways a tree is like a block of apartments: if the rent isn't paid by a tenant, they are thrown out. While this may sound harsh, in a modular system bits can be lost very easily to the overall benefit of the tree. Perhaps a better analogy is social insects where an individual may sacrifice its life for the benefit of the colony. The net effect of shedding branches is that it prevents unproductive branches from being a drain on the tree, it removes potential invasion sites for disease and prevents unneeded branches creating extra wind drag.

Branches are shed for reasons other than lack of light. In dry parts of the world it is common for trees and shrubs to lose their smaller branches to save water. Small branches have the thinnest bark and greatest surface area, and are the source of most water loss once the leaves have been lost. The creosote bush of USA deserts self-prunes in the face of extreme heat or drought, starting from the highest and most exposed twigs and working downwards to bigger and bigger branches; it's a desperate act because if it loses too much wood it dies. Shedding branches can, of course, also be useful for self-propagation.

Most poplars and willows characteristic of waterways will readily drop branches which root when washed up on muddy banks further downstream (see Chapter 8). Finally, buds that have detrimental genetic changes can be got rid of by shedding.

How are branches shed? In the simplest cases, dead branches rot and fall off, or healthy branches are snapped off by wind, snow and animals. Some willows have a brittle zone at the base of small branches that encourages breaking in the wind, seemingly for propagation. Other cases of 'natural pruning' are more startling; elms (*Ulmus* spp.), and to a certain extent others including oaks, deodar (*Cedrus deodara*) and London plane (*Platanus* x *hispanica*), have a reputation for dropping large branches (up to half a metre in diameter) with no warning usually on calm, hot summer afternoons: hence the quote from Kipling 'Ellum [elm] she hateth mankind and waiteth'. Boy scouts should be warned not to pitch tents under elms! Such dramatic shedding appears to be due to a combination of internal water stress coupled with heat expansion affecting cracks and decayed wood.

Branch shedding in many trees is, however, a deliberate act. Branches are shed in the same way as leaves in autumn by the formation of a corky layer which leaves the wound sealed over with cork, which in turn is undergrown with wood the following year. Figure 7.12 shows the typical ball and socket

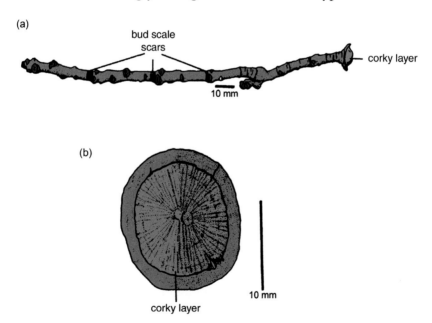

Figure 7.12 A branch shed from an oak (*Quercus petraea*) in the same way that a leaf is shed. Note the typical ball-shaped end (close-up in b) where the branch has been cut off by a corky layer. From: Bell, A.D. (1998) *An Illustrated Guide to Flowering Plant Morphology*. Oxford University Press, Oxford. Reprinted by permission of Oxford University Press.

appearance of branches shed in this way (the process is officially called cladoptosis). In hardwoods, branches up to a metre in length and several centimetres in diameter can be shed, normally after the leaves have fallen in the autumn (maples are unusual in casting branches mainly in spring and early summer). Oaks tend to shed small twigs up to the thickness of a pencil, beeches may shed larger ones and birches dump whole branches of dead twigs. Pines shed their clusters of needles (really short branches) and members of the redwood family (such as the coastal redwood and giant sequoia, and the deciduous dawn redwood and swamp cypress) shed their small branchlets with leaves. Typically in hardwood trees, somewhere around 10% of terminal branches are lost each year through a mixture of deliberate shedding and being broken off. Not surprisingly, shed branches can make up a third of the weight of forest floor litter.

## Length of branches

Another way of reducing potential congestion is to make some branches smaller than others. As already mentioned, branches in the shade grow smaller than those in the sun. But trees can also regulate branch length from within. In conifers, side branches are shorter, so a main branch is longest, side branches are shorter, and side branches on these side branches are shorter still. This makes for a very efficient use of space without self-shading. In many hardwood trees there is a clear distinction between 'long' and 'short' branches or shoots (Figure 7.13). This is true of apple, birch, beech, hornbeam, katsura (*Cercidiphyllum japonicum*) and ginkgo (and to a certain extent in elm, poplar and lime which have a more gradual transition between long and short shoots). The long shoots build the framework of the tree, making it bigger. The job of the short shoots (called spur shoots by horticulturalists) is to produce leaves, and commonly flowers, at more or less the same position every year. The short shoots therefore grow in length each year just enough to produce closely packed leaves and the next set of buds: short shoots can therefore be told by the closely packed sets of bud scale scars from each year (see Figures 7.13, 7.16 and 7.17). The bundles of pine needles and the tufts of needles found in cedars and larches are also borne on short shoots. The cedar and larch short shoots continue to grow new needles for several years but in pines, the short shoot grows only about a quarter of a millimetre and stops, and then falls off intact once the needles die).

To maintain flexibility, any one shoot can swop from long to short or *vice versa* depending upon the internal control, light levels and damage (see below).

(a)

(b)

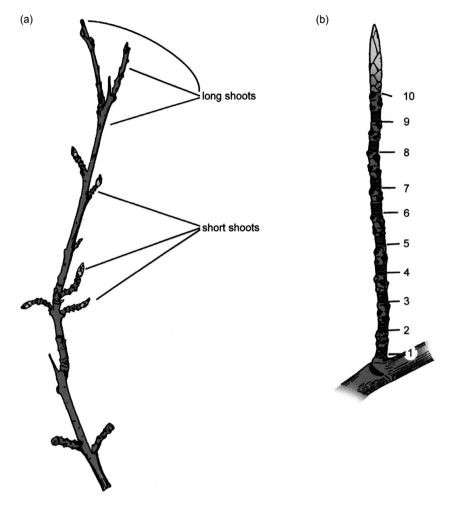

long shoots

short shoots

10
9
8
7
6
5
4
3
2
1

**Figure 7.13** (a) A shoot of purging buckthorn (*Rhamnus catharticus*) made up of long shoots and short shoots; (b) a close-up of a 10-year-old beech (*Fagus sylvatica*) short shoot: annual growth can be identified by the crowded bud scale scars. From: (a) Büsgen, M. & Münch, E. (1929) *The Structure and Life of Forest Trees*. Chapman & Hall, London; (b) Troll, W. (1954) *Allgemeine Botanik*. Ferdinand Enke, Stuttgart.

## Flowering

Flowering can radically alter the shape of a tree because a growing point that ends in a flower dies when flowering/fruiting are over and cannot revert back to growing leaves. This is seen in magnolias, dogwoods, maples and horse chestnuts: the flowers are produced at the end of the shoot which consequently does not produce a new bud. Thus, next year the two side buds (marked a in Figure 7.14) will grow out to resume branch growth, leaving a

**Figure 7.14** Horse chestnut branch (*Aesculus hippocastanum*). The growing point dies once it flowers. Next year's growth will be from the two buds behind (a) leaving forks (b) to mark the site of flowers in previous years.

Figure 7.15 Wych elm (*Ulmus glabra*). Terminal buds, and the next one further back on the end shoot, have produced just leaves (marked a and b, respectively). Buds further back along the twig have produced just flowers (c).

fork to mark where the flowers were (b in the figure), an example of the sympodial growth described above.

The trees above are often described as fruiting on 'new wood', i.e. branches grown in the current year. In many trees, however, the flowers are tucked away further back on the branch, described as fruiting on 'second year' or 'old wood' (but bear in mind that the fruits, like leaves, are really hanging from a small twig extension grown this year). This is seen in, for example, elms and cherries. The terminal buds of each twig (marked a in Figures 7.15 and 7.16), and perhaps one or more further back on strongly growing shoots (b), produce just leaves, while the buds further back along the twig produce just flowers (c). These flowering points will, of course, not be able to produce new buds and so are used just once. After the fruits are shed, that growing point is dead. Scars further down the branch bear testimony to previous flowering. The new leafy shoots grown this year will lay down buds to repeat the process next year.

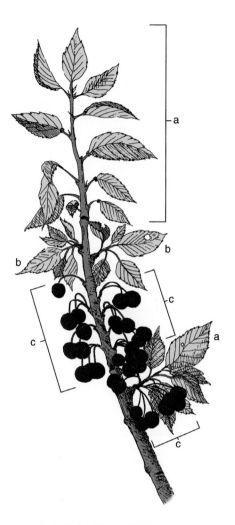

Figure 7.16 Cherry (*Prunus* variety) labelled as Figure 7.15.

Elms and cherries are fairly simple in that their buds produce either leaves or flowers. Some other trees produce 'mixed' buds containing both flowers *and* leaves. This is seen in the plum branch in Figure 7.17. Behind the buds producing just leafy shoots (a, b), the single 'mixed' bud at c has opened to grow both flowers (and hence plums) and leaves. Thus this bud has produced fruit *and* will be able to continue growing next year, unlike the flowering buds in elm and cherry. Further back on wood a year older, shoots show a mix of bud types. The terminal bud of the lowest twig shown is just a leaf bud (l), the other terminals are mixed buds (m), and the side buds further back on these shoots are just flower buds (f). The pear and blackcurrant also produce mixed buds but the flowers are at the end with a ring of leaves at the base so the growing point is still doomed to die. They get round this by using the bud

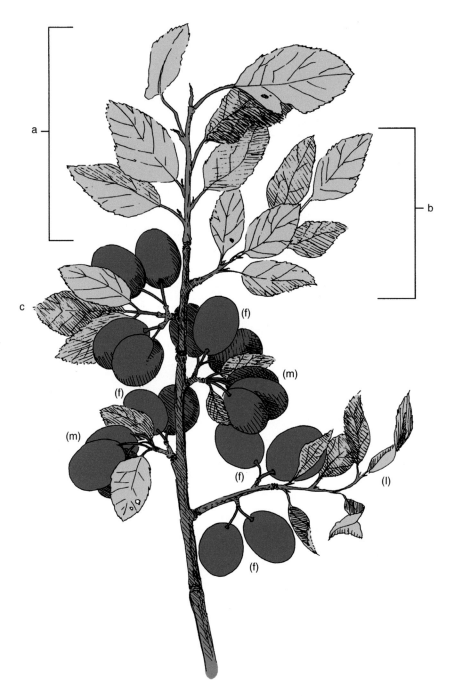

**Figure 7.17** Victoria plum (*Prunus domestica*). Behind the buds producing just leafy shoots (a and b), the single 'mixed' bud at c has opened to grow both flowers (and hence plums) and leaves. Further back on wood a year older, shoots show a mix of bud types: flower buds (f), leaf (l) and mixed buds (m).

associated with the lowest leaf for next year's growth. Many other variations are found, too numerous to mention here. It makes an interesting project to look at different branches and work out the strategy being used. The inevitable conclusion is that different patterns of producing flowers has a profound effect on the pattern of branches and hence the shape of a tree.

## How does the tree control shape?

We have seen that the overall shape of a tree is governed by how buds and branches grow and die. How does the tree control this? The simple answer is that the ends of the branches (apices in botanical parlance) tend to suppress what goes on further back along the branch. Over the whole tree this apical control can become quite complicated and is analogous to the power struggle within a kingdom. The topmost growing tip (the leader) is like the king having great power over the buds and growing tips immediately below. The strongly growing ends of side branches are like the princes of principalities and have great influence over the buds and branches near them but they themselves and their subjects are still under some control from higher up. In effect, every growing tip from the very top apex to the most minor branch end has some control, albeit diminishing in importance just like that from the king to a minor government official. The more minor the bud the more likely it is to be suppressed and the more likely that any resulting branch will be smaller and slower growing (nothing we see in politics is new to nature!). The process is influenced from outside by the amount of light; the more light a branch receives the greater the power of the branch and its buds. Power does change with time. Like an ageing king, the leader(s) becomes less vigorous as the tree approaches it maximum height and loses some of the control, allowing the underling branches (princes) to exert more independence and growth. Complete removal of the seats of power by pruning (or accidental damage) means that one or more lesser branches (or new branches from suppressed buds) will usually bend upwards and compete to be the new leader. Clipping young Christmas trees releases dormant buds making them bushier and more saleable.

Strictly speaking this method of control can be divided into 'apical dominance', which refers to suppression of buds growing out, and 'apical control', which describes the control of subsequent growth from these buds. The reason this is mentioned is not to make your life more complicated but because it helps explain the two basic shapes of trees: the spire characteristic of the conifer and the broad dome of the hardwood. In rounded, domed trees like the oak, apical *dominance* is very high so lateral buds are often suppressed. In future years, however, they are left far enough behind the apex that they are released and grow into branches. Because apical *control* is weak, these new shoots are very good at competing with the original tip and grow as fast as or faster than the leader. This leads to the spire-shaped young tree disappearing into a mass of strong branches giving the rounded shape (officially described as 'decurrent' which my Latin translates as 'running into', describing the way the branches merge). Conversely, in the tall spire of a conifer, apical *dominance* is weak, allowing the majority of buds to grow, producing the regular whorls of branches. But since apical *control* is strong, the leader suppresses the growth of these new branches and keeps them growing out more horizontally rather than upwards, and the spire shape of all young trees is maintained for much of the life (an 'excurrent shape': the leader 'running out beyond'). The canopy is cone shaped since the branches are longer lower down simply because they are older.

The mechanism by which buds regulate growth is complicated but the hormone auxin (produced by the apices) plays a fundamental role in suppressing buds but also in stimulating root growth; the roots in turn produce gibberellins which stimulate the growth of new shoots. This is also tied in with the action of other hormones such as cytokinins (initiators of bud growth in many species) and most likely aided by internal competition between growing points for nutrients and sugars. These hormonal messages from the kings and princes are passed down the tree through the phloem. This also explains how epicormic shoots (see Chapters 3 and 9) are controlled. These grow from dormant epicormic buds when they are 'released', often when the top of the canopy dies or is broken off, removing the hormonal control (Figure 7.18). But these can also be released when the tree is shaded (weakening the apical buds) or suddenly exposed to light or given extra water or nutrients (strengthening the side buds). So thinning a group of trees, which results in many parts of the tree receiving plenty of light, alters the hormonal balance and the buds are released to grow into 'light suckers'. Conversely, when a tree is increasingly shaded it can produce 'agony branches' or 'shade suckers', when the top of the tree is shaded and thus produces less auxin, releasing the branches below.

While auxins are clearly involved in apical dominance (stopping buds from growing out), it is less clear how apical control (the suppression of branch growth) is carried out. Morris Cline and colleagues have shown in a hybrid poplar and to a certain extent in Douglas fir (*Pseudotsuga menziesii*) that auxin appeared to play a small role and that much more important was the ability of the dominant branches to deprive other branches of their fair share of nutrients (nitrogen, phosphorus, etc.), and so grow quicker.

Horticulturalists have cashed in on the role of hormones in growth control by producing chemical growth regulators which interrupt apical dominance

(a)

**Figure 7.18** Epicormic buds released to grow. (a) Death of part of the canopy of this old yew (*Taxus baccata*) in the churchyard at Brailsford, Debyshire, England has allowed a mass of epicormic buds to grow out. The yew is suggested to be 1000 years old. These buds tend to self-thin resulting in a smaller number of larger branches as seen in (b) on an old sweet chestnut (*Castanea sativa*) at Killerton Hall, England.

and control, encouraging bushy growth in ornamental plants (such as azaleas). Incidentally, a number of studies have shown that 'shoot inversion', bending the upper shoot over so that it points downwards, can release apical regulation. This undoubtedly explains the old country custom of snapping over the top branches of hazel (*Corylus avellana*) to improve fruiting: hazelnuts are borne on short shoots which will grow and fruit more prolifically once apical regulation is partly removed. Similarly, espalier-trained plants (where a vertical

(b)

Figure 7.18 (cont)

stem gives rise to several tiers of horizontally trained branches) are productive because tying a branch closer down to horizontal makes it produce more fruit by weakening apical regulation (conversely raising a branch stimulates vegetative growth). This also explains how true weeping trees grow taller; as the leaders get longer and flop over, apical dominance is reduced and lateral buds are released which produce new branches upwards which flop over in turn. Thus weeping trees get taller in a 'stair-step' fashion (the exception are trees such as the deodar, *Cedrus deodara*, which straighten the pendulous tip during new growth; Figure 7.19).

Figure 7.19 The drooping tips of the deodar cedar (*Cedrus deodara*) which straighten during new growth. Staffordshire, England.

As a finale in this section, is there any truth in the following ditty?

*A woman, a dog and a walnut tree,*
*The more you beat them, the better they be*

Taking it as read that the first two are rightly exempt, what about the walnut? In nineteenth century Europe, walnuts were knocked off the tree using long poles. Branch ends were inevitably broken which caused the production of more short shoots (by removal of apical regulation) and thus more flowers and fruit. Hence the custom of beating a barren tree to make it bear fruit. As mentioned in Chapter 3, beating the trunk may also help by damaging the phloem and so reducing the amount of sugars transported to the roots which are then available to be put into flowers and fruit.

## Changes with age

The mature tree is not just a seedling writ large; the shape changes with age, so much so that it may be possible to age tropical rainforest trees purely by their shape. Young trees may be much less branched especially if they have compound leaves which are, in effect, cheap throw-away branches, or if they self-prune (tulip trees, *Liriodendron tulipifera*, are easily identified in American forests because of the long straight unbranched trunk). Tropical rainforest trees can be up to 7 m before branching. We planted a foxglove tree (*Paulownia tomentosa*) on the University campus and I think the kind sponsor was a little perturbed to see just a 3 m pole that had no branches! As the number of branch ends increase, apical regulation is shared between more of them leading to the rounded dome or flat top of maturity (as described above), and an increasing proportion of short shoots (more than 90% of growing shoots). How quickly the change in shape happens depends on species (early in horse chestnuts, late in poplars, never in many conifers) and soil (early on drier, poor soils).

Branch angle also changes with age. As described above, dominant branches tend to have a narrow angle to the trunk but this can change with time. Over the short-term branches have a set angle to which they will try to return using reaction wood if bent by, for example, the weight of snow. But over the long-term, branches tend to sag as they get longer and heavier, helped by the compression of new wood in the branch crotch (in the same way that putting on more and more coats would force your raised arms to be more and more horizontal). Thus younger branches at the top of the tree are most upright and increasingly older branches down the tree are seen to be bent further outwards. The younger outer ends of old branches may object to being dragged down and so old conifer branches which sag below the horizontal often have the young tip turned upwards. You might regard sagging branches in old trees as one of the disadvantages of age like wrinkles and an expanding waistline, but it has been suggested that trees may benefit from their changes. The erect shape of a young tree will encourage rainfall hitting the tree to run down the stem, concentrating it on the developing roots at the base of the trunk. In an older tree, the sagging branches will encourage most water to be shed outwards and drip off the edge of the canopy onto the zone where its absorbing roots are concentrated. This is a gross oversimplification but does illustrate that many changes in shape do have a potential use.

Shape change with age can be even more radical. Perhaps the most classic is in ivy (*Hedera helix*), although it is admittedly a woody climber rather than a tree proper. As a juvenile it is a climber with heart-shaped leaves that clings on with aerial roots. Once the top reaches adequate light it changes to a mature, flowering phase with less deeply lobed leaves. It no longer produces aerial roots but holds its stems erect like any shrub, and it flowers. This is such a definite change that if a cutting of the flowering shoots is rooted it carries on in the same fashion making quite a decent upright bush. Trees also dramatically change shape as a matter of course by abrupt bending. Creeping willow (*Salix repens*), a plant of alpine Europe, starts life by growing vertically but soon bends at the base, thereafter growing prostrate. In some families of trees, and especially amongst conifers, cuttings taken from vertical and horizontal shoots (technically, orthotropic and plagiotropic shoots, respectively) follow the orientation of their original position. Thus a cutting from a horizontal branch will produce a weak tree with a tendency to be prostrate. This 'memory' (or topophysis) of position can obviously greatly affect the shape of new trees and is a wonderful tool for horticulturalists breeding new cultivars. A similar 'memory' has been found in clonal forestry where plantations of new trees are derived from cuttings taken from a particularly good tree (where 'good' may mean superior fruit production, high wood quality or fast growth). There can be an increasing memory of age – a young tree from a cutting acting as if it is still part of an older tree – with consequent changes in shape. Age is important since like Dolly the sheep (born in July 1996 but retained the 'age' of her parent and had lung disease and crippling arthritis by her death in February 2003), cuttings from old plants grafted onto young root stock still 'remember' that they are old, and so this must involve some genetic changes that persist. Thus grafts from older trees grow more slowly, are less branched, and flower and fruit earlier.

Shape can also be dramatically affected over a tree's life by the accumulation of damage, and the addition of new branches on any part of the woody skeleton from dormant epicormic buds. The ability to produce epicormic shoots often also changes with age. Older trees (but not very old trees) are often better able to sprout, probably because of greater food reserves in the roots.

## How are all the leaves exposed to light?

We come back to the problem of how a tree holding perhaps 100 000 leaves can arrange each to catch the optimum amount of light. This would certainly pose a challenge for a solar-power engineer even without the added problems of changes caused by growing new leaves in different places and the constant movement by the wind! The first problem to solve is how leaves on a single twig avoid shading each other. In a vertical branch the leaves are carefully arranged so that they are not directly above each other (their positioning along a spiral can be described by the Fibonacci series, a mathematical progression which predicts the angles between successive leaves). Or, if they are likely to be directly above each other, as happens in a vertical branch of maple, the upper leaves are smaller with shorter stalks (Figure 7.20a). Preventing shading in a similar horizontal branch is not as easy but is solved by the upper leaf of a vertically oriented pair being smaller than its partner beneath and twisted sideways or backwards (Figure 7.20b: this dissimilarity of leaves borne on the two sides is called anisophylly). This is even more pronounced in some tropical trees which on horizontal branches may have two rows of large leaves below

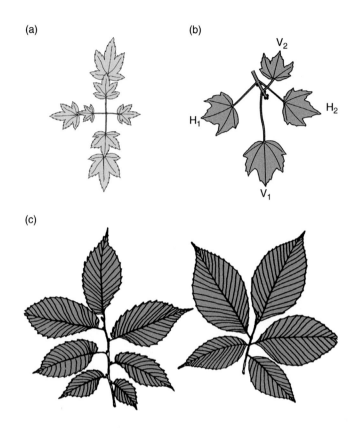

**Figure 7.20** Arrangement of leaves on a branch to reduce self-shading. (a) In a vertical branch with leaves directly above each other (as in maples), the upper leaves are smaller with shorter stalks. (b) In a similar horizontal branch the problem is solved by the upper leaf of a vertically oriented pair ($V_2$) being smaller than its partner beneath ($V_1$). (c) In most trees a flat plane of leaves is produced on a horizontal branch where the leaves fit together like pieces in a jigsaw puzzle as in elm (*Ulmus* spp.) on the left and beech (*Fagus grandifolia*) on the right. From: (b) & (c) Zimmermann, M.H. & Brown, C.L. (1971) *Trees: Structure and Function*. Springer, Berlin, Figs III-27 & III-28a, page 157. With kind permission from Springer Science and Business Media.

and two of very small leaves above. The majority of trees produce a flat plane of leaves on a horizontal branch either by twisting of the leaf stalks or by alignment of the buds and leaves in this flat plane. Here the leaves avoid shading each other by neatly fitting together like pieces in a jigsaw puzzle (Figure 7.20c).

Looking at the whole canopy, some trees simply expand the method used for one twig to produce a single shell of leaves over the entire canopy, fitted together like a domed jigsaw puzzle. These 'monolayer' trees, such as beeches, hemlocks (*Tsuga* spp.) and sugar maple (*Acer saccharum*) are well adapted to shaded environments and are often found growing up beneath other trees in the comparative shade of a forest. On the other hand, trees growing in the open would waste a lot of useful light with just one layer of leaves: most leaves are working flat out in around only 20% of sunlight. To use the otherwise wasted light, layers of leaves are stacked one above the other to give 'multilayer' trees. If you were to drop an arrow from above it would punch through up to 5–6 leaves before hitting the ground (this is often referred to as leaf area index: the number of layers of leaves above any area of ground). It is not always easy to see the multilayers but the impression is one of a fairly open yet deep canopy which you can see into, such as birches and poplars. These trees need a lot of light and so tend to do best in open areas. Fifty-four percent of full sunlight appears to be the watershed; any darker and monolayer trees fare better, any lighter and multilayer trees grow better.

With more than one layer of leaves in a multilayer tree it is important that lower leaves are not sitting in the shadows of the one above otherwise they will not work very well. This is because the green light passing *through* a leaf has had the useful wavelengths of light largely removed and is little use to leaves below. The lower leaves need light that has passed *between* leaves. This could call for an impossibly high level of co-ordination of growth between different parts of the tree. Luckily the solutions are simple.

1.   Individual leaves can change their orientation to point towards where unfiltered light is coming from, possibly from the side. Light foraging (making use of light from gaps) is easier and more common in temperate trees because of their broad shape. Conifers having a dominant trunk have a much more rigid architecture and show only limited flexibility in light foraging through modifications in branching.

2.   As described in Chapter 2 under *Sun and shade leaves*, shaded leaves can use brief sunflecks of light penetrating a swaying canopy to good effect. Poplars and birches may gain from the fast moving mosaic of sunflecks caused by the constant wobbling of their leaves on narrow leaf stalks.

3.   If a leaf or branch is too shaded to be self-sufficient it will, as seen above, be shed.

4.   If the layers are far enough apart the shadows of the upper layer disappear. This is because the sun is a disc rather than a point of light. You can try this with a light and a small coin; hold the coin near a piece of paper and it casts a shadow, move it further away and although the light reaching the paper is dimmer, the shadow disappears. So providing the layers of leaves are far enough apart, the upper layer merely acts as a neutral density filter rather than casting a series of sun and shadow spots. Thus the leaves in the lower layer can be arranged independently of those in the layer above. The minimum distance between layers varies from 50 to 70 leaf diameters in sunny areas to just 1 leaf

diameter in cloudy climates where the light comes from all over the sky (see Horn 1971 for further details).

## Human influence

For centuries we have changed the shape of trees to suit ourselves. Sometimes this is for pleasure as in topiary where a range of supposedly aesthetic shapes are created by clipping such trees as privet (*Ligustrum* spp.), hiba (*Thujopsis dolabrata*) and yew (Figure 7.21a, b), and in others it is to keep the tree small (Figure 7.21c). In other cases it is to ensure that a tree fits with its situation. This might be in terms of crown reduction (reducing the height of the tree by pruning back the outer branches, but never by just 'topping' the tree by cutting off the upper portion of the tree which is ugly, encourages a jungle of new shoots from the top which will need pruning themselves in a few years, and which are easily torn off the tree by wind – see new trees from old in Chapter 8 – and it lets in disease through the horizontal cut). Or it might be by crown thinning to increase the amount of light reaching the ground or to reduce weight on heavy limbs, or even canopy raising where the bottom branches are removed to allow easier access and to create the traditional trunk and canopy shape beloved by park keepers. In yet other cases it is to increase fruit production by training branches (such as the espalier work mentioned above) and pruning to encourage fruit growth and development: apples are bigger and redder if they and the leaves on the same spur (short shoot) are exposed to high light. Whatever the reason, there are a number of general rules that are based on an understanding of how trees grow.

1.   When cutting back a branch, prune back to a side branch which is at least a third larger than the diameter of the branch being removed. This will ensure that the branch produces strong growth the next year with a natural shape.

2.   If you prune the tips off branches this removes the apical dominance and the side buds on branches will grow out to form a dense bushy growth. This is great if that's what you want, such as in a hedge. If, however, you want to thin the canopy you should cut side branches back to the main branch, leaving the apical buds on the long branches intact, so they maintain their dominance.

3.   Never remove more than a quarter of the leaf-bearing crown. If you do, the remaining leaves may be unable to meet the demands of the roots and trunk for sugars and the tree could die or at least be weakened enough to be vulnerable to pests and diseases.

4.   Never remove more than a third of the height of the canopy when raising the canopy, for the same reason as above.

(a)

(b)

Figure 7.21 Yew topiary that varies from (a) informal cloud pruning at Powys Castle, Wales to (b) a much more formal style at Chirk Castle, also in Wales. Other forms of pruning are designed to keep a tree small and compact, including the traditional close pruning of London plane (*Platanus* x *hispanica*) typical of the UK but seen here in Rotorua, New Zealand (c).

(c)

Figure 7.21 (cont)

Time of pruning is also important. Traditionally, trees are pruned in the winter when they are more or less dormant, and certainly never in spring when they have depleted their stored food to produce new growth. Spring pruning can indeed be a problem for those trees that flower before producing leaves (such as cherries and elms) or trees with ring-porous wood (such as oak; see Chapter 3) since they may indeed run short of stored food to replace any lost leaves (see Hirons 2012 for more details). For most other trees this is much less of a problem, and indeed, in plum trees winter pruning lays them open to silver leaf disease (a fungus *Chondrostereum purpureum*) and so it is better to prune from early spring onwards.

Foresters have long controlled crown size and shape by adjusting the spacing of their crops. We have also discovered chemicals which will stimulate or restrict growth: the herbicide maleic hydrazide applied to new epicormic shoots will suppress their development in the following year; gibberellin biosynthesis-inhibiting growth regulators (e.g. paclobutrazol) have been used to reduce twig growth near power lines. We have also been busy producing trees of desirable shape through breeding and selective propagation (see Chapter 9). And we have changed tree shapes inadvertently. Perhaps the biggest influence has been our grazing animals which are capable of keeping

**Figure 7.22** A line of hornbeams (*Carpinus betulus*) planted a little too close to the road, with the effect that a neat rectangular shape is maintained by constant pruning from large vehicles. Staffordshire, England.

young trees trapped as shapeless bushes, if not killing them outright. Larger trees can have their lower growth pruned parallel to the ground by browsing animals such as rabbits, cattle and deer. Mechanical help is also to blame. Several roads near my home have beautiful lollipop-shaped hornbeams lining the verges, and most have a rectangular shape clipped out (Figure 7.22) showing where high-sided vehicles have repeatedly pruned back the young growth! Mechanical help can also kill. Witness 'Sheffield blight' – damage done to bark by lawnmovers, named in memory of Sheffield steel – which can do much to hasten the early demise of a tree.

## 🍂 *Further Reading*

Ball, P. (2009) *Branches*. Oxford University Press, Oxford.

Berninger, F., Mencuccini, M., Nikinmaa, E., Grace, J. & Hari, P. (1995) Evaporative demand determines branchiness of Scots pine. *Oecologia*, 102, 164–168.

Cline, M.G., Bhave, N. & Harrington, C.A. (2009) The possible roles of nutrient deprivation and auxin repression in apical control. *Trees*, 23, 489–500.

Dewit, L. & Reid, D.M. (1992) Branch abscission in balsam polar (*Populus balsamifera*): characterization of the phenomenon and the influence of wind. *International Journal of Plant Science*, 153, 556–564.

Fisher, J.B. & Stevenson, J.W. (1981) Occurrence of reaction wood in branches of dicotyledons and its role in tree architecture. *Botanical Gazette*, 142, 82–95.

Hallé, F., Oldeman, R.A.A. & Tomlinson, P.B. (1978) *Tropical Trees and Forests: an Architectural Analysis*. Springer, Berlin.

Hirons, A. (2012) Straightening out the Askenasy curve. *Arboriculture & Urban Forestry*, 38, 31–32.

Horn, H.S. (1971) *The Adaptive Geometry of Trees*. Princeton University Press, Princeton, New Jersey.

Jones, M. & Harper, J.L. (1987a) The influence of neighbours on the growth of trees: the demography of buds in *Betula pendula*. *Proceedings of the Royal Society of London*, B232, 1–18.

Jones, M. & Harper, J.L. (1987b) The influence of neighbours on the growth of trees: the fate of buds on long and short shoots in *Betula pendula*. *Proceedings of the Royal Society of London*, B232, 19–33.

Neel, P.L. & Harris, R.W. (1971) Motion-induced inhibition of elongation and induction of dormancy in *Liquidambar*. *Science*, 173, 58–59.

Porter, B.W., Zhu, Y.J., Webb, D.T. & Christopher, D.A. (2009) Novel thigmomorphogenetic responses in *Carica papaya*: touch decreases anthocyanin levels and stimulates petiole cork outgrowths. *Annals of Botany*, 103, 847–858.

Putz, F.E., Parker, G.G. & Archibold, R.M. (1984) Mechanical abrasion and intercrown spacing. *American Midland Naturalist*, 112, 24–28.

Suzuki, A.E. & Suzuki, M. (2009) Why do lower order branches show greater shoot growth than higher order branches? Considering space availability as a factor affecting shoot growth. *Trees*, 23, 69–77.

Tsiouvaras, C.N. (1988) Long-term effects of clipping on production and vigor of Kermes oak (*Quercus coccifera*). *Forest Ecology and Management*, 24, 159–166.

von Casteren, A., Sellers, W.I., Thorpe, S.K.S., *et al.* (2012) Why don't branches snap? The mechanics of bending failure in three temperate angiosperm trees. *Trees*, 26, 789–797.

Walker, D.J. & Kenkel, N.C. (2000) The adaptive geometry of boreal conifers. *Community Ecology*, 1, 13–23.

Watt, M.S., Downes, G.M., Jones, T., *et al.* (2009) Effect of stem guying on the incidence of resin pockets. *Forest Ecology and Management*, 258, 1913–1917.

Westing, A.H. (1965) Compression wood in the regulation of branch angle in gymnosperms. *Bulletin of the Torrey Botanical Club*, 92, 62–66.

# Chapter 8: The next generation: new trees from old

In Chapter 5 we followed the processes of reproduction through to the arrival of the seed on the ground. Here we will look at germination and early survival of the seedling, and ways of producing new trees without resorting to seed.

## The seed

Seeds remind me of spaceships: they contain everything they need to colonise new worlds given favourable conditions and water once they arrive. The outside is covered by the seed coat (the testa), designed to protect the contents. At the centre of the seed is the embryo, consisting of little more than a miniature root (the radicle) and shoot (the plumule). The rest of the seed is taken up with the food supply to keep the embryo alive before it germinates and to sustain early growth before photosynthesis takes over. This food is stored usually in the cotyledons (the seed leaves, part of the plumule), although some store it outside the cotyledons in the endosperm (which can be thought of as a short-lived half-brother of the embryo; all flowering plants have endosperm but in most it is used up quickly). In ash (*Fraxinus excelsior*), for example (Figure 8.1), the cotyledons are small, surrounded by endosperm, but in oak (*Quercus* spp.) the cotyledons are bloated and fill the seed with no remaining sign of the endosperm. The best example of endosperm is in the coconut (*Cocos nucifera*): part liquid (the milk) and part solid (the flesh). Wherever the food is stored, it is usually in the form of starch but oils are not uncommon especially in small wind-dispersed seeds. Oils contain more calories in a given mass and volume and so allow the seeds to travel light (having said that, large seeds can also contain oil: think of walnut oil for cooking and cocoa fat in chocolate). There is a price to pay: seeds with fats cost more to produce and are fairly short-lived (they go rancid).

## Seed dormancy

Most tropical species and a few temperate ones (notably elms, which produce their seeds in the spring; Chapter 5) have seeds which are ready to germinate as

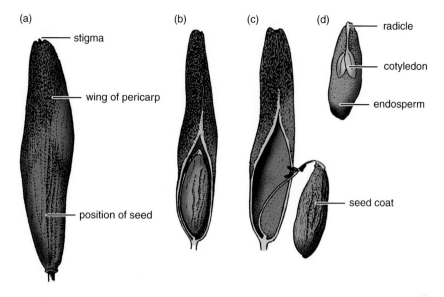

**Figure 8.1** Ash (*Fraxinus excelsior*). (a) a single fruit; (b) half the fruit removed to show the enclosed seed and (c) the seed displaced to show the attachment; (d) the seed split open to show the embryo with two cotyledons (seed leaves) embedded in the endosperm food supply. From: Priestley, J.H. & Scott, L.I. (1938) *An Introduction to Botany*. Longmans, Green & Co., London.

soon as they fall. But the seeds of most temperate trees are produced in the autumn and show some sort of dormancy to ensure they germinate in the spring rather than producing their delicate seedlings in the dying days of a warm autumn. In some, the seeds lie dormant for much longer and either germinate in a gentle dribble or await some signal to trigger mass germination.

The question then arises, what breaks the dormancy? Commonly, it is a period of winter chilling (but see fire and heat in Chapter 9). Bear in mind that not all parts of the seed need chilling. In oaks and viburnums, for example, the root starts growing as soon as the seed falls in the autumn (protected from the harshest conditions by the soil) but the shoot only grows after chilling. This gives the seedlings a head start in the spring. As with bud opening (Chapter 6), a need for chilling can be circumvented; after a mild winter, germination is eventually induced by spring warmth.

Dormancy can also be caused by the embryo being too immature to grow immediately after seed fall. The classic example is ash. The seeds fall in the autumn and after a winter chilling a few (less than 5%) may germinate but in the rest the embryo must mature over the following summer to germinate the next spring, around 18 months after the seed fell. Intuitively this would seem to put the seed at great risk of predation so there must be some overriding benefit that is not obvious.

**Figure 8.2** Seeds of gorse *(Ulex europaeus)* that have a hard shiny seed coat that prevents germination until it is cracked or removed.

A third cause of dormancy comes with the hard impermeable seed coat typical of the pea family and a number of others (Figure 8.2). The seed coat prevents the entry of water and germination cannot take place until the coat is ruptured in some way. This might be by fluctuating temperatures of hot summer days and cool nights, or the intense heat of a fire (e.g. gorse, *Ulex* spp., and acacias), rotting by fungi over time, or partial digestion when passing through an animal's gut. The seeds of yew *(Taxus baccata)* and juniper *(Juniperus* spp.), for example, germinate promptly once they have been through a bird's gut (which increases the permeability of the seed coat to water) and then chilled but if just dropped on the ground the thick wooden coat takes 1–2 years to rot enough for germination to be possible. Elephants, rhinos and bats[1] have been implicated in improving germination of tree seeds from typically 1–2% before eating to over 50% after, although in some cases this is due to the killing of insect predators that would otherwise eat the seed. But don't get the idea that hard seed coats are only found

---

[1] In the first edition I wrote that the extinction of the dodo on Mauritius in 1681 had left the tree *Calvaria major* (now called *Sideroxylon grandiflorum*) with no seedlings because the hard seeds needed to pass through the dodo to germinate. Temple (1977) fed the fruits to turkeys and supposedly produced the first seedlings for probably 300 years. Alas, this has proved to be false; the seeds do not need to pass through a bird to germinate, although the tree is endangered for a variety of other reasons. See Cheke & Hume (2008) for more details.

in exotic plants; even acorns will germinate quicker and more completely if the seed coat is removed first.

Dormancy can be complicated. For example, people are often disappointed with the germination of the European holly (*Ilex aquifolium*). In exasperation they cut open a few seeds and finding no embryo conclude that the seeds are sterile. What is really preventing germination is a two pronged dormancy. The hard coat prevents the entry of water (passage through a bird's gut helps). Once this is solved, the minute embryo develops and matures only after a prolonged warm period.

Gardeners have been imitating these natural processes for years. Seeds are given cold treatments ('stratified') by being mixed with damp soil and placed in a refrigerator. Others are scarified by abrasion, nicked with a knife or soaked in acid to break the hard coat to induce these seeds to germinate.

A few woody species will not germinate until given light. These are usually small seeds that need the physical cue that they are not deeply buried or heavily shaded. And it's not just any old light they need. Light is detected in the seeds by the pigment phytochrome (see Chapter 6). Red light (wavelength around 660 nm) tends to break dormancy whereas far-red light (c. 760 nm) induces it. Light passing through leaves has a larger proportion of far-red light, so seeds that need light may be inhibited from germinating beneath other trees. This makes ecological sense: it can be far better to sit and wait for a gap than face almost certain death as a small seedling in heavy shade. This switching effect of phytochrome may also explain why some trees have a light requirement for germination in the first place. Two ecologists, Cresswell and Grime (1981) working with herbaceous plants have shown that seeds surrounded by green fruits get a large dose of far-red light and a light requirement is induced; as the seed dries and becomes inactive this need is set. But where the fruit loses its green colour before the seed dries, seeds will have the phytochrome set in the active form and will not need light to germinate. Does this explain why most tree seeds have no light requirement because the thick fruit prevents far-red light reaching the seed?

In the same way that the need for cold can be circumvented, so also can the need for light. Birch will germinate in the dark if given heat.

## Soil seed bank

If seeds falling to the ground remain dormant and do not germinate in the first spring, they will become progressively buried by leaves and other debris and so be incorporated into the reservoir of seeds stored in the soil. How long they remain viable depends upon how long their food reserves last (all living things respire and burn up energy) and how long they survive the attention of predators and pathogens. Generally it is pioneer species that tend to be most prominent in the seed bank – those species that invade open conditions and are missing from the mature forest. These tend to have either small inconspicuous seeds (such as birches and tropical rainforest pioneers), easily overlooked by predators but which tend to be short-lived (2–5 years is probably average) or larger seeds with very good protection (such as cherries, gorse, brambles and raspberries) which may last 150–200 years. The oldest reliably dated tree seed to germinate is a date palm (*Phoenix dactylifera*) excavated from near the Dead Sea that was around 2000 years old. Reports of older seeds (up to 20 000+ years old) have been erroneous (see Zazula *et al.* 2009).

Densities can be high; in Europe, gorse (*Ulex europaeus*) and broom (*Cytisus scoparius*) – both with hard seeds as described under *Seed dormancy* above – have been found at densities of more than 30 000 and 50 000 seeds in a square metre, respectively (although herbaceous species can approach half a million seeds in a square metre). Many large-growing trees, however, do not form soil seed banks, probably because the seeds are so large that they are likely to be found and eaten. Others will store seed but only if they are deeply buried and so well-hidden. For example, seeds of pines that use birds to move the seeds (such as the whitebark pine, *Pinus albicaulis*, in the mountains of western N America) tend to stay viable in the soil for up to 3 years until suitable conditions for germination come along. They are fairly safe since the birds bury them well below the surface. This may be a mechanism evolved by the birds to iron out the variability between the infrequent seed years.

As described above, once the dormant seeds are buried, they need a trigger to tell them when to germinate. This may be fungal rotting of the thick seed coat that eventually lets in water, or it may be something much more specific. For example, seeds that germinate only after a fire will need a trigger to tell them when a fire has passed overhead. This may be the heat shock of a fire itself breaking open the seed coat through to chemicals in smoke that are washed down through the soil by rain (more details on fire-induced germination can be found in Thomas & McAlpine 2010).

## Other seeding strategies

As seen above, remarkably few of the dominant woodland trees in Europe (such as oaks, beech, ash, etc.) have seed in the soil seed bank. A side effect of this is that many of these trees do not fare well if we store the seeds for too long. Indeed, British acorns are readily killed by drying (termed 'recalcitrant' seeds) and therefore germinate quickly or not at all (as described above, the root grows out as soon as possible after seedfall and the shoot appears above ground the following spring). Acorns come off the tree at around 45%

**Figure 8.3** Seedling bank of balsam fir (*Abies balsamea*). These seedlings do not get enough light to grow quickly and survive as young seedlings until a tree in the canopy falls and allows light in. In this case part of the forest has been felled exposing the seedlings. The advantage to the balsam fir is that their seedlings are ready and waiting to grow into a gap while other species have to start from seed and are less likely to win the race to fill the gap. New Brunswick, Canada.

moisture and will cease to germinate if dried to 25% moisture (this can be reversed just a little by soaking in cold water for 48 hours).

Lack of soil storage is undoubtedly because they have alternative strategies for optimising the chances of producing new trees. They may sprout from the base (elms) or roots (cherries), or be masting species (predator satiation is inconsistent with seed longevity; see Chapter 5). Also, many shade-tolerant trees, such as beech and the N American balsam fir (*Abies balsamea*) and sugar maple (*Acer saccharum*) produce a mass of seedlings which can't grow to maturity in deep shade but can persist for years as stunted seedlings. This 'seedling bank' (Figure 8.3) gives a competitive edge over species starting from seed when a small gap opens up in the tree cover by a tree dying.

## Fire and storage on the tree: serotiny

Not all trees drop their seeds when ripe, rather they store the seed up in the canopy (officially called 'serotiny'). Getting on for a thousand or so species around the world do this including the eucalypts and she-oaks (*Casuarina* spp.)

of Australia, the giant sequoia (*Sequoia sempervirens*) of N America and many pines. Typically, the seeds are released by some sort of environmental trigger. This might be rainfall in desert shrubs and succulents but by far the most common trigger is fire. Just how and why serotiny works, and why it can be better than storing seeds in the soil, is discussed in detail in Chapter 9.

## Seed size

The smallest seeds in the world belong to orchids. The Scottish orchid, creeping lady's-tresses (*Goodyera repens*), for example, has individual seeds weighing just 0.000002 g, so small that there are 50 million to the kilogram. Tree seeds are altogether bigger and tend to be at the larger end of the seed size range. As shown in Box 8.1, one of the smallest seeds of a woody plant is the heather of the British uplands at 33 million to the kilogram. At the other end of the scale there are less than a hundred horse chestnuts in a kilogram, culminating in the huge Seychelles double coconuts that weigh 18–27 kg *each* in a fruit up to 45 cm long (Figure 8.4). (Incidentally, the reputation of this palm for sea dispersal derives from the discovery of dead seed washed up on the Maldive Islands 2000 km (1200 miles) away, but fresh seeds sink in sea water! The coconut we buy to eat (*Cocos nucifera*) *does* float, however; see Chapter 5.)

Figure 8.4 Seychelles double coconut (*Lodoicea maldivica*) growing in Singapore Botanic Gardens. (a) A semi-mature palm showing the large, fan-shaped leaves; (b) a young seedling growing from the double coconut (which is really one large seed).

| Box 8.1 Average number of cleaned seed (i.e. without any of the fruit) per kilogram | | | |
|---|---|---|---|
| Seychelles double coconut | *Lodoicea maldivica* | 18–27 kg each! | |
| Coconut | *Cocos nucifera* | 4 | |
| Walnut | *Juglans regia* | 85 | |
| Horse chestnut | *Aesculus hippocastanum* | 90 | |
| Sweet chestnut | *Castanea sativa* | 250 | |
| English oak | *Quercus robur* | 290 | |
| Sessile oak | *Quercus petraea* | 400 | |
| Hazel | *Corylus avellana* | 1200 | |
| Sycamore | *Acer pseudoplatanus* | 3300 | |
| Animal-dispersed pines | Swiss stone pine | *Pinus cembra* | 4400 |
| | Whitebark pine | *Pinus albicaulis* | 5700 |
| Beech | *Fagus sylvatica* | 4500 | |
| Rowan | *Sorbus aucuparia* | 5000 | |
| Ash | *Fraxinus excelsior* | 14 000 | |
| Hornbeam | *Carpinus betulina* | 24 000 | |
| Wind-dispersed pines | Ponderosa pine | *Pinus ponderosa* | 26 500 |
| | Weymouth pine/ Eastern white pine | *Pinus strobus* | 58 400 |
| | Scots pine | *Pinus sylvestris* | 200 000 |
| Small-leaved lime | *Tilia cordata* | 32 000 | |
| Ivy | *Hedera helix* | 49 000 | |
| Wych elm | *Ulmus glabra* | 73 000 | |
| Holly | *Ilex aquifolium* | 125 000 | |
| Gorse | *Ulex* spp. | 150 000 | |

| Box 8.1 (cont) | | |
|---|---|---|
| European larch | *Larix decidua* | 160 000 |
| Norway spruce | *Picea abies* | 195 000 |
| Giant sequoia | *Sequoiadendron giganteum* | 200 000 |
| Coastal redwood | *Sequoia sempervirens* | 260 000 |
| Alder | *Alnus glutinosa* | 770 000 |
| Silver birch | *Betula pendula* | 5 900 000 |
| Aspen | *Populus tremula* | 8 000 000 |
| Dwarf birch | *Betula glandulosa* | 8 450 000 |
| Rhododendron | *Rhododendron ponticum* | c 11 000 000 |
| Heather | *Calluna vulgaris* | 33 000 000 |

Based on data from: Salisbury, E.J. (1942) *The Reproductive Capacity of Plants*. George Bell, London; Grime, J.P., Hodgson, J.G. & Hunt, R. (2007) *Comparative Plant Ecology*, 2nd edn. Unwin Hyman, London; and Burns, R.M. & Honkala, B.H. (1990) *Silvics of North America. Vol I, Conifers; Vol 2, Hardwoods*. USDA, Forest Service, Agricultural Handbook 654, Washington, DC.

## Germination

Once any dormancy has been broken and germination conditions are met (warmth, water and sometimes light), germination follows. As water is absorbed, the swelling seed splits the seed coat. The young root (the radicle) is first to emerge in search of a stable water supply, followed by the young shoot. As the shoot emerges above ground two things can happen to the cotyledons (seed leaves). In small seeds, as in pines and beeches, the portion of shoot below the cotyledons usually expands rapidly, pulling the cotyledons, often at first still in their seed case, above ground where they will expand, turn green and start photosynthesising (Figure 8.5a, b). The young seedling then gets the best of both worlds; it uses the stored food and what the cotyledons can produce themselves. The cotyledons continue this dual function for normally 1–3 months until the first true leaves develop and the stored food is exhausted, whereupon the cotyledons wither and fall. This type of 'epigeal

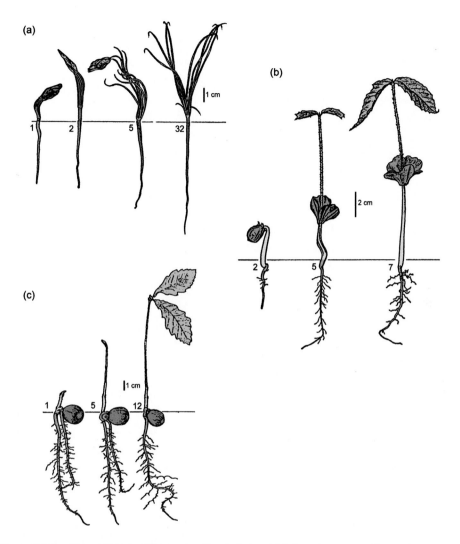

**Figure 8.5** Seedlings of (a) pine (*Pinus palustris*) at 1, 2, 5 and 32 days after germination; (b) beech (*Fagus grandifolia*) at 2, 5 and 7 days; and (c) oak (*Quercus macrocarpa*) at 1, 5 and 12 days. Germination in pine and beech involves bringing the cotyledons above the ground surface ('epigeal') whereas in oak the cotyledons stay below ground ('hypogeal'). Pines have up to a dozen or even more cotyledons while the dicot hardwood plants (see Chapter 1 for an explanation) have two, and the advanced monocots have just one. Mistakes are not uncommon though; sycamore will frequently have three or four cotyledons and in a handful of crab apples seeds I once germinated, several had an extra cotyledon. From: Anon. (1948) *Woody-Plant Seed Manual.* USDA, Forest Service, Miscellaneous Publication No. 654.

germination' ('epi' means above) is found in many trees including maples, beeches, ashes, most conifers and small-seeded tropical species.

The alternative way of doing things is 'hypogeal germination' ('hypo', below). Here the shoot *above* the cotyledons grows, drawing the baby leaves above ground, leaving the cotyledons underground (as in oaks, see Figure 8.5c).

The cotyledons cannot, of course, photosynthesise and the plant has to wait until its first true leaves are functional before it grows any of its own food. This would seem to put hypogeal seeds at a great disadvantage in the race to the sun. The clue to resolving this paradox lies in seed size. Hypogeal germination is found in larger seeds: oaks, walnuts, cherries, horse chestnut (*Aesculus hippocastanum*), hazel (*Corylus avellana*) and the rubber tree (*Hevea brasiliensis*). It would be risky to bring all that food above ground where it would be readily sought by herbivores, especially as it may last the whole growing season, and anyway such cotyledons are bulky objects to pull out of the ground (with possible damage to the delicate roots) and heavy to support in the air.

Despite their size, cotyledons carefully hidden below ground may not be as crucial as we might think to seedling survival. In Europe, acorns are primarily dispersed by jays (Chapter 5) which cache the acorns in the soil for winter fodder. Those that are not eaten will germinate and produce the next generation of oaks. But the jays can still find the seedlings in the spring. When they find one, they give it a yank upwards exposing the acorn containing the cotyledons, which are nipped off and eaten. Small seedlings may be uprooted but most survive without damage. Experiments have shown that seedling survival and growth is unaffected by such removal, even on nutrient-poor soil. It seems that once the first leaves are out, the seedling does not need the cotyledons. So why do oaks have such large acorns? The answer can be found below!

## The significance of seed size

Seeds are often remarkably constant in size. So much so that the seed of the Mediterranean carob tree (*Ceratonia siliqua*) was used as the original measure of gold, the carat (200 g). Having said this, in most carob trees there can be just as much variation in seed size as in most other trees. It seems that in the case of the carob, humans would have weeded out the smallest and biggest seeds, using just the middle-sized seeds. In carob samples it has been seen that seeds vary by 26% in weight but the seeds used to measure gold varied by just 5%. Similar records exist for a number of other trees. For example, the saga or red sandalwood, native to India and S China but widely planted in the tropics, has bright red seeds (Figure 8.6) that were used to weigh gold in India. Seed size does tend to be fairly constant in most trees. Growing conditions will alter the number of seeds produced much more than their size. This indicates that size must play a pretty crucial role in plant establishment. First of all though, seed size is not necessarily related to the eventual size of the tree: the world's biggest trees (coastal redwoods and giant sequoias; see Chapter 6) are well down the order of seed size (Box 8.1).

**Figure 8.6** Saga or red sandalwood seeds (*Adenanthera pavonina*) that have been used as weights for gold (4 seeds weigh about 1 g). Saga is derived from the Arabic word for goldsmith. Note, however, that the seeds are not completely uniform in size and when used as weights seeds of uniform size are selected. Singapore Botanic Gardens.

Generally, seed size is a delicate balance between on the one hand being small enough to maximise the number produced and (for wind-dispersed trees) maximise the distance they will spread, and on the other hand being large enough to give seedlings a good start. Not surprisingly, in pioneer species which invade open areas (like birch and aspen), the emphasis is on small mobile seeds to ensure the greatest spread of seed. Conversely, trees that establish under dense shade (like beeches and some oaks) do better with the bigger supplies of energy, nutrients and, sometimes, water found in large seeds. Larger seeds produce taller seedlings which can more readily get above the shade of the woodland litter and herb layer (oak seedlings can grow 5–10 cm before the first leaves appear). Also, once growing, larger food reserves enable seedlings to be generally more tolerant of low light, nutrients and water availability, and are physically more resistant to disturbance of the soil. All this can enable a seedling to persist longer while waiting in the seedling bank for a light-gap. Seeds may also be lucky enough to escape the worst of the competition and attention by predators by germinating higher above ground on, for example a rotting 'nurse log' (Figure 8.7). This doesn't just jack the seedling above ground but the rotting wood can provide a ready source of water and nutrients. It's not uncommon in rainforests to see lines of young trees marking their success growing along the top of a large fallen tree which is now long gone.

But coping with shade is not the whole story. Large seeds are also of value in dealing with the deep, loose layers of undecomposed leaves found in

**Figure 8.7** Rotting nurse logs are often the site of successful seed germination and establishment since they provide a good source of water and help lift the seedling above competing vegetation. In this case, a western hemlock (*Tsuga heterophylla*) has started life on a large rotting stump in the temperate rainforest of Quinault Valley, Washington State, USA. Note that the old stump has a springboard notch; a shallow notch used to hold a plank on which the tree feller would have stood.

woodland which are notorious for rapid drying, leaving a shallowly rooted seedling high and dry. A large seed can grow a root rapidly giving it a better chance of reaching the more constant water supply of the underlying soil before the litter dries to a crisp. As an example, the acorns of the cork oak (*Quercus suber*) can vary from 3 to 8 g, and the larger seeds are better able to produce seedlings under drought conditions. Small-seeded pioneers tend to invade open areas bare of litter. And it is noticeable that where the climate is more constantly wet there are more smaller-seeded woodland species. Fast root growth also goes some way to explaining why maritime trees (like coconuts) are often large seeded: they need to grow roots quickly down into water layers below that influenced by salt water. Indeed early growth from the huge Seychelles double coconut (*Lodoicea maldivica*) can put the young embryo a half metre or so underground and up to 3 m sideways from the coconut: no wonder they're difficult to cultivate in pots (but see Figure 8.4). The downside of large seeds is the expense of growing them. The Seychelles double coconut can bear only 4–11 fruits at a time and these take 10 years to develop.

This sinking of the seeds contents deeper into the ground has been termed 'crytogeal germination' and is found in some African trees as a way of getting the seed out of the way of frequent fires. It is also seen in the monkey puzzle (*Araucaria araucana*) and a relative from Brazil, the parana pine (*A. angustifolia*).

As it germinates, food is moved from the seed into the base of the young root (the hypocotyl) making it tuberous. This seems to be a way of getting valuable resources in the large seed sitting on the soil surface underground as quickly as possible and before they are needed for growth.

We haven't finished with seed size yet because there are still several other factors which influence size. Large seeds are attractive bundles of food to herbivores and are not easily hidden. So large tree seeds generally germinate more quickly than small seeds because they are more likely to be eaten once on the ground. This puts them at a disadvantage if their germination window happens to be unsuitable; they may all fail. In this case, smaller-seeded species would have the advantage in spreading their germination over a longer time. Other large-seeded species use costly defences to protect their investment in each seed. Eventually, however, there comes a point where it is 'cheaper' to be smaller. This is nicely illustrated by a group of woody legumes in Central America which are plagued by the larvae of bruchid beetles which live inside the seed. They divide neatly into those species with large seeds (an average of 3.0 g per seed) heavily protected by toxins, those with small seeds (average 0.26 g) which are readily attacked and rely on some being missed, and one species with very small seed (0.003 g) which is too small for a beetle larva to grow in and so goes unmolested. It is also true, however, that larger seeds can suffer less from seed-eating insects simply because they can afford to lose more flesh before they are killed. This was found to be the case with the Oriental white oak (*Quercus aliena*, a common tree in eastern Asia) infested by weevils; large seeds were more likely to survive weevil infestations to germinate and grow. Plus, there is evidence that partially eaten acorns of Pyrenean oak (*Quercus pyrenaica*) produce seedlings just as well as intact seeds.

Soil fertility also has potentially interesting evolutionary consequences for seed size. On poorer soils, where the vegetation cover is slow to develop, seeds of pioneer species are smaller because they need fewer reserves to compete with neighbours and smaller seeds are more likely to spread wider and be more successful. For those invading dense forest the opposite is true; heavier seeds, with a greater supply of nutrients, will tend to win the battle to dominate. This could explain the production of huge seeds typical on the impoverished soils of dense tropical forests, so much so that the heavy fruits have to be borne on the trunk rather than thin branches (see Chapter 5).

This is complicated by the fact that the very largest seeds are probably an adaptation to attracting animal dispersers rather than anything to do with establishment. Here, we return to the jay. In Europe, oak is almost exclusively dependent on jays for dispersal. Jays tend to go for large acorns but only so large (usually no more than 17–19 mm wide) because they swallow 1–5 per journey and they need to fit in the jay's throat. Having said this, jays will take the occasional larger acorn which they carry in the beak. Since jays mostly leave acorns that are too large or small, it's the midsize ones that are taken away from the shade of the mother-tree and are therefore most likely to produce the next generation of trees. So oak appear to have evolved big seed for dispersal rather than establishment (explaining why cotyledon removal has little effect on the seedling; see *Germination* above). Similarly, many pines around the world have evolved to be spread by other corvid birds in the same manner; they have large wingless seeds (4000–6000 per kg; see Box 8.1) compared to 200 000 per kg of the wind-dispersed Scots pine.

From all this it can be seen that there is a trade-off between seed size and seed number. Sometimes this is simple and readily understood (such as with the size of coconuts). But in most cases, the seed size that each species ends up with is a complex compromise between a number of competing factors. As importantly, seed size is by no means completely fixed between years. Every tree species will be constantly changing their position on this size-number scale, even if by small amounts, as their surroundings hone the seed size for maximal survival.

## The odds against success

An oak can produce more than five million acorns over its life, and some birches can easily reach hundreds of millions of seeds in their short life. For the most part, we are not over-run by these trees so the mortality rate must be high. As you would probably predict, the highest mortality is amongst the seeds and tender young seedlings, declining with age into adulthood. As a broad-brush figure the mortality of seed alone can often be around 95%. Of the 5% that germinate, another 95% may well die within the first year. These are very much general figures; the mortality of seeds and seedlings will obviously vary tremendously with species, year, site and chance events (like landslides and volcanic eruptions) but figures in the order of one in a thousand acorns becoming a seedling and one in a million acorns becoming a mature oak are probably worth considering when thinking of survival rates. Why such tremendous wastage, especially when we can plant a few seeds in a plant pot and most of them will grow? The answer lies with a fundamental limitation of plants: they can't move. Seeds have to be planted using animals or physical agents like wind as intermediaries, which is inevitably a hit and miss process. How many seeds would we need if we planted our gardens by throwing handfuls of seed out of a window? Add in the ravages of disease and herbivores, and the need for such large numbers of seeds to get just a few trees becomes a pragmatic reality.

Periodic establishment of new seedlings is by no means uncommon. This may be because seedlings can only establish when a tree dies creating an opening in dense woodland, or only after rain following a long drought. Or establishment may need some sort of environmental disturbance such as fire, avalanches, wind storms, earthquakes and volcanic eruptions to play a part. The worrying lack of some southern beeches (*Nothofagus* spp.) in Chile, which live for four or five centuries, was found to be because they establish after periodic devastating earthquakes. We must make sure that we look at seedling establishment on the time scale of the trees. For an oak living more than 500 years, a few decades of no success is a drop in the ocean. At the same time, we must not be complacent about lack of new seedlings if we humans have had a part to play. The lack of oaks in the UK in the latter part of the last century was claimed to be due to the persecution of jays (thankfully largely a thing of the past) and to the removal of large native herbivores which would have created good seedbeds. Certainly, the 'recent' lack of seedlings in oak and beech woods has been blamed by some on the cessation of woodland grazing by cattle and pannage of pigs.

In temperate forests dominated by just a few evenly spread species, the survival of seedlings is often fairly random (and thus independent of seedling density; see Shabel & Peart 1994). In others, and especially tropical forests, survival can be linked to how many other seedlings are nearby (density-dependent survival; see Boudreau & Lawes 2008). In some cases it is those packed together that survive best and in others it is those furthest apart. The former – survival in clumps – works on the same principle as survival of seeds in masting: predator satiation (Chapter 5). Large numbers of seedlings in a small area will overwhelm the ability of local herbivores to eat them all while they are young and vulnerable and some will therefore survive. Since most seeds do not travel very far, the seedlings end up clumped around the parent tree. The opposite – seedlings surviving when away from their fellows – is found where herbivores roam large areas searching for specific types of seeds or seedling and will feed voraciously when a clump is found. Thus it is the odd isolated seedlings at the outer edge of the dispersal area that are most likely to be missed and survive to grow. This 'escape hypothesis' may go some way to explain why individuals of most tropical trees are usually well spaced through a forest and not clumped as in many temperate forests. These ideas have been formalised into the Janzen–Connell hypothesis (named after the two promin- ent tropical forest ecologists who independently came up with the idea in the 1970s). They say that the number of seedlings will peak at a certain distance from the parent tree. Close to the tree, seed and seedling mortality is high because they are dense enough to be easily found by animals, further away from the tree, there are fewer seeds and so fewer seedlings, but in the middle the numbers peak. It certainly works in lowland tropical and temperate forests but for some reason not in the montane forests of Central and S America (see Swamy *et al.* 2011 for more detail).

# New trees without seed: vegetative propagation

There are several ways that new trees can be produced from pieces of an existing tree. Suckers (which are shoots from roots; Figure 8.8) are common on a number of temperate trees such as alders, elms, cherries and plums, false acacia/black locust (*Robinia pseudoacacia*), sweet gum (*Liquidambar styraciflua*), aspens and poplars. Suckering, especially in the last two, happens fairly readily and allows them to invade sites not suitable for establishment from seed. In this way great stands of aspen and poplar have developed on the floodplains of N America, covering many hectares each (Chapter 6) with up to half a million shoots, all clones (i.e. genetically identical) to the original tree(s). Some of the other trees listed are less prone to suckering unless damaged, and even trees that readily sucker do so more when damaged (for example, aspen by fire) or if the roots are exposed to light. When a cherry tree outside my garden was heavily pruned and at the same time some of the roots were disturbed by digging, the tree began throwing up suckers over a radius of 10 m from the trunk. These sucker shoots initially depend on the roots from which they are growing but quickly develop roots of their own. The whole lot usually stays connected to the original roots for some time but the connection may eventually be broken by disturbance, disease or death of the parent, and they then become truly independent. If you're interested in finding out which was the original trunk in a colony of suckers (and it is not always obvious from trunk size), you can try looking for a

Figure 8.8 A sucker on a monkey puzzle (*Araucaria araucana*). A sucker is a new shoot coming up from a root and so can be some way away from the main trunk (the tree on the right). The top of the root can be seen to the left of the sucker. Co. Wicklow, Ireland.

tap root. In poplars at least, the initial tree normally has a tap root while the suckers do not. Since the new shoots are backed up by a mature root system, they can grow very fast, up to a metre in height a year.

Branches touching the ground may root and produce new upright stems by 'layering'. This may be when the bottom branches of an upright tree bend under their own weight to touch the ground (or even the ground in mossy areas grows up to engulf the bottom branches). Like suckering, this is rare in the tropics and most common in high latitudes and alpine areas where damp peaty soil is common. Conifers seem to be best at it: spruce, juniper, fir, yew, hemlock, false cypresses in the north (*Chamaecyparis* spp.) and podocarps in the south (but only rarely in pines). Hardwood trees can also layer naturally; this is commonest in shrubs (e.g. *Forsythia* and the lowest shoots of coppiced hornbeam, *Carpinus betulus*) but can occur in large forest trees. It is also possible to induce it artificially since layering is a handy means of propagation in those trees hard to root from cuttings. The new shoots develop adventitious roots at the same time as they start growing and they soon become quasi-independent. As Figure 8.9a shows, even a fairly small branch can produce a new tree which can reach a large

(a)

Figure 8.9 (a) A branch (shown by the white arrow) of the large sweet chestnut (*Castanea sativa*) at the back has touched the ground, rooted and given rise to another trunk. The branch has carried on growing (black arrow) and is capable of doing the same again further away from the main trunk to produce a grove of trees. (b) The resulting new trees can reach a large size while still attached to the parent. Weston Park, England.

(b)

Figure 8.9 (cont)

size while still attached to the parent (Figure 8.9b) and in extreme cases (often it seems in beech *Fagus sylvatica* or hornbeam) a number of the outer branches sag to touch the ground around the tree, root and produce a ring of new 'trees' like a palace guard around the outside of the venerable parent.

This can be even more dramatic when a tree is partially uprooted and falls over with some of its roots still intact. Where the branches are pushed hard onto the ground (preventing movement by wind) they will readily root. These new shoots and branches on top of the trunk now have full sunlight and grow upwards into 'new' trees. They will eventually produce their own adventitious roots and as the old trunk dies and rots (often dying first back from the top to the last upright stem), the new trees are left standing in a line to confuse future generations who wonder why someone planted all those trees in a straight line in the middle of nowhere. This can happen in many different trees, and can even be seen in conifers such as European larch (*Larix decidua*) and Douglas fir (*Pseudotsuga menziesii*).

New trees can also be produced by twigs breaking from or being shed by the tree (see Chapter 7), sticking in the ground and growing as natural cuttings. This obviously works best in wet places and may explain why poplars and willows along a river tend to be either all male or female rather than the mixture you would expect from seed. Poplars, willows and the tree of heaven certainly drop branches that are in peak health, full of nutrients and with plenty of live buds: perfect propagation material. Normally when a branch or a leaf is shed, the nutrients are scavenged back by the tree before it is shed, but, not surprisingly, shoots designed to be used for reproduction are shed with all their nutrients intact so that they have every chance of growing.

As we have seen before, most hardwoods and a handful of conifers can regrow from the stump if the canopy is damaged, removing apical regulation. Old or newly formed buds (epicormic and adventitious buds, respectively; Chapter 3) grow into new stems using the stored food and roots of the old tree. However, since this just replaces the previous tree it is debatable whether you would call this propagation. Nevertheless it can be impressive: Willis Linn Jepson, a renowned Californian botanist, described a colony of 45 large coastal redwoods (*Sequoia sempervirens*) that formed a third generation 'fairy ring' 17 m by 15 m in diameter (Del Tredici 1998; Figure 8.10).

New trees can be produced artificially by micro-propagation (or tissue culture). Individual cells taken from a tree can be grown in the laboratory into

**Figure 8.10** A partial ring of coastal redwoods (*Sequoia sempervirens*) at John Muir Woods, California. There was originally a tree in the middle that was snapped off or killed above ground and then produced new stems around the outside of the stump. The old stump is now gone but has left the two trees joined at the base.

whole new seedlings, each genetically identical to the original tree (cloning). In this way, good trees can be reproduced without the gamble of genetic variation inherent in seed production. Such techniques have been used to propagate timber trees such as cherries and fruit trees such as dates, as well as being vital in the genetic modification of trees (see below).

## Producing new *types* of tree

Cuttings (especially from young growth) can be rooted to create new trees. Although this cannot produce new species, it can create new growth forms (or cultivated varieties, shortened to cultivars). This relies on the way a cutting 'remembers' its way of growing (called topophysis; see Chapter 7). Thus cuttings taken from the top of a tree produce fast, upright growth while side branches once rooted will often carry on growing sideways giving a prostrate shrub or at least a slow-growing shrubby specimen. By taking cuttings from the tops of mature hollies or false acacias where there are few spines (temperate herbivores can't reach that far and nature doesn't waste effort) we can create thornless varieties.

New varieties are also produced by watching for genetic mutations or 'sports'. Very occasionally a seed will produce an unusual specimen (a seedling sport) due to a mutation in its genetic code. Copper beech (*Fagus sylvatica* 'Purpurea') arises this way and a number have been found growing wild in continental Europe (though most seedling sports will only survive in the cosseted nursery bed and garden). If you want one of these lovely beeches, however, it is easier to take a cutting than having to plant thousands of seeds in the faint hope of striking lucky (even seed from a copper beech produces a mix of green or just faintly coloured seedlings).

More commonly, the odd unusual branch can be found on a tree (a bud or branch sport) that, if rooted as a cutting or grafted (see below), will produce a whole tree of the sport type. Pink grapefruits, seedless grapes, navel oranges and Delicious apples all come from single such branches. If you consider that a large tree may have between 10 000 and 100 000 growing points (buds), even a relatively low mutation rate of around 1 in 10 000 per year produces a large number of sports. Most mutations are detrimental and the branch quickly dies. But occasionally one will be benign enough to merely lead to an interesting variation which we find worth propagating. Similar mutations can happen over and over again, resulting in confusingly similar varieties hitting the market. Mutations can produce dwarfness, odd shapes, colour variation (think of the large number of variegated and golden conifers) and juvenile fixation where juvenile foliage (mostly on conifers; see Chapter 2) is kept right through to maturity (that's how we get, for example, the 'Squarrosa' and to a lesser extent the 'Plumosa' varieties of the sawara cypress, *Chamaecyparis pisifera*).

Note that trees propagated from bud mutations tend to be unstable and revert back to produce vigorous foliage of the original tree that should be cut out to prevent it taking over.

Occasionally a cutting from a 'witch's broom' in a hardwood will produce dwarf trees. The causes of witch's brooms (the ball-like dense tangle of small twigs looking almost like a round bird's nest) are largely unknown but see Chapter 9.

Hybridisation (reproduction between usually closely related species), is another way of getting new types of tree. Often this happens naturally but we are not beyond encouraging it. Perhaps one of the most famous deliberately produced hybrids is the hybrid larch (*Larix* x *eurolepis*[2]), a hybrid between the European and Japanese larches *(L. decidua* and *L. kaempferi)* that is beloved by foresters because of its hybrid vigour (called heterosis), growing more vigorously than either parent at least while it is young. But probably the most famous hybrid, if the large number cluttering up urban Britain is any indication, is the Leyland cypress (x *Cupressocyparis leylandii*), a cross between two species in different genera, the Nootka cypress (*Chamaecyparis nootkatensis*) and the Monterey cypress (*Cupressus macrocarpa*). The original Leyland cypress hybrid occurred at Leighton Park in Wales in 1888 where both parents happened to be growing near each other (in their natural range along the western seaboard of N America they don't come within 300 miles of each other). Six plants were raised from seed collected from a Nootka cypress by C.J. Leyland. In 1911 the reverse cross happened: seeds collected from a Monterey cypress produced two seedlings. Leyland cypress shows hybrid vigour and the tallest is over 35 m and growing strongly. At this rate it has been suggested that they have the potential to become the tallest trees in the world and we'll all be pale troglodytes! Unlike animals, hybrids in plants are often fully fertile. In tree families where hybridisation is common (the birch, pine, rose and willow families) this has led to backcrosses with the parents to produce a wide range of intermediates (a hybrid swarm) creating a nightmare for field botanists and students facing an identification exam!

Grafting can also be used to create new types of trees, or at least new sizes. In grafting, the top of one tree (the scion) is grafted onto the bottom of another (the stock). This allows a weak plant to benefit from the strong roots of another, or a vigorous tree (such as an apple) to be kept small by growing on 'dwarfing root stock'. A big problem with grafts is the need to cut back vigorous

---

[2] A hybrid between two species in the same genus is denoted by a cross before the new species name (e.g. *Larix* x *eurolepis*). A hybrid between two species in separate genera is shown by a cross in front of the new genus name (e.g. x *Cupressocyparis leylandii*). In the same way a + sign is conventionally used to indicate a plant that is a graft hybrid.

Figure 8.11 Problems with grafted ash trees (*Fraxinus* spp.). In the foreground the root stock is growing in diameter faster than the grafted-on scion, and in the background the reverse is happening. Royal Botanic Gardens, Kew, England.

shoots from the stock which might swamp a delicate scion. One also needs to ensure that the stock and scion will grow in diameter at the same rate otherwise strange looking trees result (Figure 8.11).

Whether you consider grafting to be a way of producing new types of tree is arguable but grafts can give rise to something far more interesting. The scion is joined to the stock only at the point of grafting. But occasionally a bud arising from the union point grows into a strange mixture of the two plants which literally has the core of one plant and a thin skin of the other: a graft-hybrid or graft chimaera (from the Greek mythological fire-breathing monster with the head of a lion, body of a goat and tail of a dragon). Probably the best know woody graft-hybrid is the laburnum-Spanish broom hybrid + *Laburnocytisus adamii*. The original specimen (from which all subsequent trees have come as cuttings) arose in France in 1825 in the garden of a nurseryman called Adam who grafted the Spanish broom (*Cytisus purpureus*) onto a laburnum (*Laburnum anagyroides*). The tree is a laburnum at its core, a fact which is reflected in its overall shape, but

**Figure 8.12** Leaves of the cut-leaf beech *Fagus sylvatica* 'Asplenifolia' at Keele University, England. This beech is a graft-hybrid or graft chimaera such that the cut-leaf beech is a thin layer over a core of ordinary beech. The leaves on the top and right are the cut-leaf but the two lower left-hand branches are ordinary beech that has broken through. The photograph was taken in late autumn so all the leaves are past their best!

the skin, just one cell thick in this case, is Spanish broom and the leaves and twigs are broom with a hint of laburnum in the shape. The flowers can be broom purple or laburnum yellow or a mixture, sometimes divided down the middle into half yellow-half purple! In general the flowers of graft-hybrids are either sterile or produce seeds of the parent at the core (e.g. laburnum). These trees are usually stable but some buds may develop into branches of the core species. Thus in the cut-leaf or fern-leaf beech (*Fagus sylvatica* 'Asplenifolia'), where a core of ordinary beech has a skin of the cut-leaf form, branches with ordinary beech or intermediate shapes are not uncommon (Figure 8.12).

To make life more interesting, there are other types of chimaeras. Variegated plants are chimaeras formed naturally within one plant by mutation in the growing point. In most plants the growing point or meristem is composed of discrete layers of cells each of which arises from one initial cell. A mutation in an initial cell will be spread through the resulting layer which in turn produces layers in all parts of the plant. Bud mutations which result in sandwiches of colourless and green layers produce the white-edged variegated leaves commonly seen in pelargoniums and a variety of trees from maples to hollies.

# Tree breeding and genetic engineering

A long established way of producing new trees has been by tree breeding. This has often been a matter of finding a 'good' tree (perhaps one that grows fast, has a particular shape such as short branches, or is resistant to disease), collecting seeds from it, growing the seeds and selecting the best seedlings. A step beyond this is taking pollen from a good tree and using it to fertilise the flowers of another good tree to attempt to get even better trees amongst the resulting seedlings. The trouble with these approaches is that it can take decades for each generation to be produced. This has led to a lot of work on whether you can predict how well your tree will eventually grow from some feature that appears early and is easy to measure such as leaf size or number. Much of this work has been done using poplar trees: they are fast growing, have lots of uses and are easy to hybridise. As an example of the results, in some poplar hybrids it has been found that the amount of wood grown each year is tightly linked to the area of the largest leaf, so looking at the leaf size of young saplings tells you which trees are worth planting out to grow as a crop (see Marron & Ceulemans 2006). Nevertheless, it is still a slow process and, perhaps more importantly, it can still be very hit and miss; if selecting for rapid growth, for example, it might turn out that the best trees for growth have poor disease resistance, so you are probably no better off than you were with the original trees.

A solution to these problems is to tinker directly with the genes inside the cells, and produce new trees by genetic manipulation. However, first of all we need to know how much of what a tree does is determined by the genes rather than the environment in which the tree is growing. On poorer soils or colder sites, trees will grow more slowly, determined by the environment. But if individuals of that same species from different areas are grown together on the same site the growth rate will almost certainly still vary due to genetic differences. (This is the difference between the genotype and the phenotype of an organism; see Box 8.2.) But what is the relative effect of genes and environment? Following a lot of work, it can be said that generally such things as phenology (the timing of yearly events; see Chapter 6) and wood density are highly controlled by genes whereas growth is not (Neale & Kremer 2001). So, giving a tree better growing conditions is likely to improve growth much more than it will lead to earlier flowering. To put this the other way round, by altering the genetic make-up of the tree it is easier to alter flowering time than to improve growth, much to the annoyance of foresters. This 'low heritability' of wood production is why the breeding of poplars for fast growth has been so difficult and laborious in the past (Marron *et al.* 2007). To add another twist in the story, it is also known that some things are controlled by single genes (for

| **Box 8.2** Words that go with genetics |
| --- |

### Genome

The complete set of DNA in a cell (all the genes along all the chromosomes). Every cell contains the same genome. Working this out involves cataloguing all the genes (gene discovery) and where they are along the chromosomes (gene mapping).

### Proteome

The complete set of proteins produced in the tree by the genome. Since different parts of a tree will produce different proteins, this is much more complex than the genome.

### Genotype

The genetic make-up of a particular organism; all the genes that make up my genome can be referred to as the genotype. My genotype will be different from yours.

### Phenotype

How a tree looks resulting from the interaction of the genotype with the environment. For an oak the genotype may code for a 40 m high tree but in unsuitably windy areas the phenotype will be a tree just a few metres high. My genotype gives me brown hair but after the environmental consequences of raising two boys my phenotype is premature grey hair.

### Genetic engineering

The general term for modifying the genetic content of a cell to make it capable of performing in a new way. When these cells are grown into whole genetically modified (GM) trees, the effect might, for example, be greater resistance to disease or to stop flowering (quite important in stopping the modified genes escaping into wild plants via pollen).

### Transgenic

A major form of genetic engineering that involves moving one or more genes from one species to another. Laigeng Li from the USA did this to produce the transgenic poplars with less lignin described in this chapter. The modified genes were incorporated into the DNA of bacteria which were allowed to infect leaf fragments of a poplar. Bacteria work by incorporating their own DNA into that of their hosts, so the new genes were inserted into the DNA of the poplar. The infected cells were then used to grow complete trees (this is possible since every cell has the whole DNA of the tree, and so a perfectly good tree can be grown from one cell by tissue culture).

example, resistance to white pine blister rust in a variety of pines) while others, including many aspects of growth, are controlled by a whole set of genes working together. This is like trying to bribe corrupt officials: if there is only one to deal with it may work, but if there is a whole committee of them it becomes much more delicate and complicated to get the desired outcome. So it is with genes since the more genes you tinker with, the more likely it is that there will be other consequences than the one you want.

Leaving these problems aside, to genetically modify genes there is a need to know where these genes are. Every cell of every tree has a complete copy of the tree's DNA which consists of strings of genes that control the development of each cell and thus the whole tree. The strings are the chromosomes inside the cell: poplar has 19 pairs of chromosomes in each cell, we have 23 pairs. This complete set of DNA in a cell (all the genes along all the chromosomes) is referred to as the genome, and since every cell has the same set of DNA if you can work out the genome of a cell, you have the genome of the tree. The first complete genome of a tree, the black cottonwood (*Populus trichocarpa*), was published in 2006 by Gerald Tuskan and his 108 colleagues from the USA, Canada and Europe. They used this tree because it is widely used commercially and it has a small genome for a tree. The genome of the cottonwood is a sixth the size of us humans; amongst trees that of eucalypts (*Eucalyptus grandis*) is twice as big, that of oak (*Quercus robur*) four times as big, and spruces and pines have a genome 80 times the size of the cottonwood. I should note here that size of the genome does not match how complex the organism is: the humble single-celled freshwater amoeba has a genome more than 1300 times larger than that of the black cottonwood and is a comparatively simple organism!

Once the genome is known, and the position of specific genes along the chromosomes is known, work can begin to see what happens if these genes are manipulated. Again, it is research on poplars that has led to much of our understanding of how this works. There are several ways of working out which gene does what. First, you can use a tree with naturally occurring mutations that result in such things as unusual flower colour and investigate how the genome varies from a normal individual (this is called 'forward genetics'). Or 'reverse genetics' can be used where the function of a gene is learnt by chemically suppressing or interfering with it and seeing what happens when a tree is grown from the modified cells. Heather Coleman and colleagues published just such a study in 2008. By interfering with a particular gene in a poplar hybrid they found that less lignin was produced in the wood and so showed that this gene must be involved in the process. Moreover, just as a blockage in a production line leads to a build-up of half finished products, so by looking at the compounds that built up in the poplar cells, they deduced which part in the production line, from raw materials to finished lignin, this particular gene controlled. As we work out which bits of the genome controls such things as wood formation, flowering, and response to cold and frost, we can progressively 'engineer' new trees.

There are two main objectives in tree breeding. The first is to improve trees as a crop. Thus trees that have been genetically modified include eucalypts in Brazil that grow faster (to produce biomass for power plants), aspen in Europe to grow faster in height and be drought tolerant, and rubber trees in India to have higher rubber yields. Poplars have been extensively modified to reduce the amount of lignin in the wood. As noted above, this can be done by suppressing one or more genes involved in lignin production. In this way it has proved possible to produce wood with less than 50% of the normal lignin and with 30% extra cellulose (Li *et al.* 2003). Less lignin means that it is easier and cheaper to make paper (lignin is a difficult molecule to remove from the useful cellulose) and greater yields of bioethanol can be produced from the wood (in some cases more than 80% increase compared to non-modified poplars). Trials are also underway to grow trees that fix more carbon, to help remove and lock-up extra carbon from the atmosphere.

The second objective of plant breeding is to help trees cope with a new or changing environment. For example, trees are being produced that can cope with drier, warmer or more toxic environments. This gives us trees that can be used to reforest parts of the world where past removal of trees has resulted in open, dry growing conditions. Or trees that can grow where the climate is changing for the worse, and so prevent wholesale loss of forest. Other trees can be used to help clear up hostile environments. Engineered poplars have been produced that can break down volatile hydrocarbons such as trichloroethylene, vinyl chloride, carbon tetrachloride, benzene and chloroform, which are quite nasty but common pollutants. This new poplar does it by being engineered to produce a large amount of a particular enzyme which helps break the pollutants down.

The next step in manipulating plants comes under the realm of proteomics. Genes work by producing proteins that control what goes on inside a cell. While every cell in a tree has the same genome (set of genes), different types of cell have different roles in a tree (think of cells in leaves which are photosynthesising and cells in the root controlling water uptake). These different cells have different genes switched on and off and so produce different proteins. Thus the 'proteome' (the sum of all proteins in the tree) is infinitely more complex than the genome and will vary hugely depending upon the environmental stresses felt by the tree. But to fully understand how the tree ticks, the role and control of all these proteins needs to be understood.

Whether you agree with the production of GM trees, which have been referred to as 'frankentrees', and their release into the environment, involves complex arguments of pros and cons that are unfortunately beyond the remit of this book.

## Further Reading

Abril, N., Gion, J.-M., Kerner, R., *et al.* (2011) Proteomics research on forest trees, the most recalcitrant and orphan plant species. *Phytochemistry,* 72, 1219–1242.

Bossema, I. (1979) Jays and oaks: an eco-ethological study of a symbiosis. *Behaviour,* 70, 1–117.

Boudreau, S. & Lawes, M.J. (2008) Density- and distance-dependent seedling survival in a ballistically dispersed subtropical tree species *Philenoptera sutherlandii. Journal of Tropical Ecology,* 24, 1–8.

Bradshaw, H.D., Jr, Ceulemans, R., Davis, J. & Stettler, R. (2000) Emerging model systems in plant biology: poplar (*Populus*) as a model forest tree. *Journal of Plant Growth and Regulation,* 19, 306–313.

Cheke, A. & Hume, J. (2008) *Lost Land of the Dodo.* Poyser, London, UK.

Coleman, H.D., Park, J.-Y., Nair, R., Chapple, C. & Mansfield, S.D. (2008) RNAi-mediated suppression of *p*-coumaroyl-CoA 3'-hydroxylase in hybrid poplar impacts lignin deposition and soluble secondary metabolism. *PNAS,* 105, 4501–4506.

Cresswell, E.G. & Grime, J.P. (1981) Induction of a light requirement during seed development and its ecological consequences. *Nature,* 291, 583–585.

Daws, M.I., Crabtree, L.M., Dalling, J.W., Mullins, C.E. & Burslem, D.F.R.P. (2008) Germination responses to water potential in neotropical pioneers suggest large-seeded species take more risks. *Annals of Botany,* 102, 945–951.

Del Tredici, P. (1998) Lignotubers in *Sequoia sempervirens*: development and ecological significance. *Madroño,* 45, 255–260.

Dewit, L. & Reid, D.M. (1992) Branch abscission in balsam poplar (*Populus balsamifera*): characterization of the phenomenon and the influence of wind. *International Journal of Plant Science,* 153, 556–564.

Dillenburg, L.R., Rosa, L.M.G. & Mósena, M. (2010) Hypocotyl of seedlings of the large-seeded species *Araucaria angustifolia*: an important underground sink of the seed reserves. *Trees,* 24, 705–711.

Doty, S.L., James, C.A., Moore, A.L., *et al.* (2007) Enhanced phytoremediation of volatile environmental pollutants with transgenic trees. *PNAS,* 104, 16816–16821.

Foster, S.A. (1986) On the adaptive value of large seeds for tropical moist forest trees: a review and synthesis. *Botanical Review,* 52, 260–299.

Galloway, G. & Worrall, J. (1979) Cladoptosis: a reproductive strategy in black cottonwood? *Canadian Journal of Forest Research,* 9, 122–125.

Gosling, P.G. (1989) The effect of drying *Quercus robur* acorns to different moisture contents, following by storage, either with or without imbibition. *Forestry*, 62, 41–50.

Harper, J.L., Lovell, P.H. & Moore, K.G. (1970) The shapes and sizes of seeds. *Annual Review of Ecology and Systematics*, 1, 327–356.

Jansson, C., Wullschleger, S.D., Kalluri, U.C. & Tuskan, G.A. (2010) Phytosequestration: carbon biosequestration by plants and the prospects of genetic engineering. *BioScience*, 60, 685–696.

Kelly, C.K. (1995) Seed size in tropical trees: a comparative study of factors affecting seed size in Peruvian angiosperms. *Oecologia*, 102, 377–388.

Kinloch, B.B., Jr (2003) White pine blister rust in North America: past and prognosis. *Phytopathology*, 93, 1044–1047.

Koop, H. (1987) Vegetative reproduction of trees in some European natural forests. *Vegetatio*, 72, 103–110.

Krasny, M.E., Vogt, K.A. & Zasada, J.C. (1988) Establishment of four Salicaceae species on river bars in interior Alaska. *Holarctic Ecology*, 11, 210–219.

Li, L., Zhou, Y., Cheng, X., *et al.* (2003) Combinatorial modification of multiple lignin traits in trees through multigene cotransformation. *PNAS*, 100, 4939–4944.

Marron, N., Dillen, S.Y. & Ceulemans, R. (2007) Evaluation of leaf traits for indirect selection of high yielding poplar hybrids. *Environmental and Experimental Botany*, 61, 103–116.

Marron, N. & Ceulemans, R. (2006) Genetic variation of leaf traits related to productivity in a *Populus deltoides* × *Populus nigra* family. *Canadian Journal of Forest Research*, 36, 390–400.

Neale, D.B. & Kremer, A. (2001) Forest tree genomics: growing resources and applications. *Nature Reviews Genetics*, 12, 111–122.

Perea, R., San Miguel, A. & Gil, L. (2011) Leftovers in seed dispersal: ecological implications of partial seed consumption for oak regeneration. *Journal of Ecology*, 99, 194–201.

Pons, J. & Pausas, J.G. (2007) Not only size matters: acorn selection by the European jay (*Garrulus glandarius*). *Acta Oecologica*, 31, 353–360.

Rae, A.M., Tricker, P.J., Bunn, S.M. & Taylor, G. (2007) Adaptation of tree growth to elevated $CO_2$: quantitative trait loci for biomass in *Populus*. *New Phytologist*, 175, 59–69.

Ramírez-Valiente, J.A., Valladares, F., Gil, L. & Aranda, I. (2009) Population differences in juvenile survival under increasing drought are mediated by seed size in cork oak (*Quercus suber* L.). *Forest Ecology and Management*, 257, 1676–1683.

Shabel, A.B. & Peart, D.R. (1994) Effects of competition, herbivory and substrate disturbance on growth and size structure in pin cherry (*Prunus pensylvanica* L.) seedlings. *Oecologia*, 98, 150–158.

Sonesson, L.K. (1994) Growth and survival after cotyledon removal in *Quercus robur* seedlings, grown in different natural soil types. *Oikos*, 69, 65–70.

Swamy, V., Terborgh, J., Dexter, K.G., *et al.* (2011) Are all seeds equal? Spatially explicit comparisons of seed fall and sapling recruitment in a tropical forest. *Ecology Letters*, 14, 195–201.

Temple, S.A. (1977) Plant-animal mutualism: co-evolution with dodo leads to near extinction of plant. *Science*, 197, 885–886.

Thomas, P.A. & McAlpine, R.S. (2010) *Fire in the Forest.* Cambridge University Press, Cambridge.

Thomas, P.A. & Wein, R.W. (1985) Delayed emergence of four conifer species on postfire seedbeds. *Canadian Journal of Forest Research*, 15, 727–729.

Tilney-Bassett, R.A.E. (1986) *Plant Chimeras.* Edward Arnold, London.

Tomback, D.F., Anderies, A.J., Carsey, K.S., Powell, M.L. & Mellman-Brown, S. (2001) Delayed seed germination in whitebark pine and regeneration patterns following the Yellowstone fires. *Ecology*, 82, 2587–2600.

Turnbull, L.A., Santamaria, L., Martorell, T., Rallo, J. & Hector, A. (2006) Seed size variability: from carob to carats. *Biology Letters*, 2, 397–400.

Tuskan, G.A., DiFazio, S., Jansson, S. *et al.* (2006) The genome of black cottonwood, *Populus trichocarpa* (Torr. & Gray). *Science*, 313 (5793), 1596–1604.

Vázquez-Yanes, C. & Orozco-Segovia, A. (1992) Effects of litter from a tropical rainforest on tree seed germination and establishment under controlled conditions. *Tree Physiology*, 11, 391–400.

Whitham, T.G., Bailey, J.K., Schweitzer, J.A., *et al.* (2006) A framework for community and ecosystem genetics: from genes to ecosystems. *Nature Reviews Genetics*, 7, 510–523.

Yi, X.F. & Yang, Y.Q. (2010) Large acorns benefit seedling recruitment by satiating weevil larvae in *Quercus aliena*. *Plant Ecology*, 209, 291–300.

Zazula, G.D., Harrington, C.R., Telka, A.M. & Brock, F. (2009) Radiocarbon dates reveal that *Lupinus arcticus* plants were grown from modern not Pleistocene seeds. *New Phytologist*, 182, 788–792.

# Chapter 9: Age, health, damage and death: living in a hostile world

It's a tough world. Trees face a constant battle in competing for light, water and minerals with surrounding plants. As if that were not enough, they also have to fend off the attention of living things that view trees as good to eat and places to live. Insects chew away on all parts of a tree and are quite capable of completely defoliating it. Larger leaf-eating animals (which are usually on the ground since a belly full of compost heap is a heavy thing to carry around; leaf-eating monkeys are an exception) chew away at the lower parts of the tree, although giraffes can reach up around 5.5 m. Whole armies of animals that can climb and fly will feed on the more nutritious flowers, fruits and the sugar-filled inner bark (see Chapter 3). The grey squirrel, introduced to Britain from N America in the 1880s, is a prime example. This rodent does extensive damage to hardwoods by stripping bark in spring to get at the sweet sap. It seems that dense stands of self-sown hardwoods have little sap and are largely immune (which may be why it does not cause problems in its native home) but well-tended planted trees have thin bark and high sap content and are mercilessly attacked. So big is the problem that ash, lime and wild cherry may become more common in Britain because of their relatively low palatability to squirrels at the expense of palatable beech and sycamore.

Other animals are not averse to making their homes from or in trees. Many birds, including various eagles, can tear off sizeable branches for nest making. Others live completely off the tree. Gall wasp larvae stimulate their host plant to grow galls on whichever part each species specialises in (leaves, flowers, fruits and stems). The nutritious tissues swell up around the grub and give it a home and a food supply. Witches' brooms, the mass of densely branched small twigs that resemble a besom lodged in the canopy, are grown in the same way, induced by a range of different organisms: sometimes by fungi (e.g. *Taphrina betulina* in birches), mites, or even in N America by dwarf mistletoes (such as *Arceuthobium* spp.). To this list of parasites you can add a number that normally live on the tree's roots and are only seen when they flower, such as the broom-rapes (*Orabanche* spp.) and toothwort (*Lathraea squamaria*) of Europe. European mistletoe (*Viscum album*), in contrast, is only partly parasitic; it just takes water from the tree's plumbing and grows its own sugars using its green leaves.

(a) (b)

Figure 9.1 (a) A line of roadside oaks (*Quercus robur*) in winter showing the dense growth of evergreen ivy through the canopy. (b) A section of trunk cleared of ivy stems to kill the ivy above; is this necessary? Market Drayton, England.

Epiphytic plants such as lichens and ivy merely use the tree as a place to grow up near the sun. In theory they take nothing from the tree. However, there is evidence that tropical epiphytes, including a range of bromeliads and orchids, intercept nutrients washed from the tree's leaves and branches by rain which would otherwise eventually reach the tree's roots. This 'nutritional piracy' may be significant to tree health and explain canopy roots, particularly in the tropics (see Chapter 4). A word about ivy (*Hedera helix*), a common woodland plant in Europe (Figure 9.1). It has been debated long and hard whether ivy kills trees. I stand in the camp that says ivy poses no problem for a healthy tree. The problems come with old weak trees where the sail area created by the evergreen ivy can make the tree vulnerable to windthrow in the winter when winds tend to be strongest. Likewise, it is rare for the ivy to swamp the leaves of a healthy tree except on very old slow-growing trees with a sparse canopy. The light intensity inside the canopy – ash (*Fraxinus excelsior*) is an exception – is normally low enough to prevent the flowering ivy shoots from growing (see Chapter 7); and it is these flowering shoots growing upwards like bushes that are most likely to compete with new growth.

As if all this was not enough, diseases also take their toll. Fungal diseases, especially of the roots and stems, are particularly important in tree health (more on this later). Bacteria and possibly viruses also play a role. For example, 'wetwood' is a bacterial rot creating pockets of moist rot with plenty of methane.

## The age of trees

Some trees can be very short-lived. For example, papaya trees (*Carica papaya*) in forest openings in Mexico can live for just 8 years, and the huge palm *Tahina spectabilis* in Madagascar which has fan-shaped leaves 5 m wide, flowers just once before dying at less than 50 years old. Nevertheless, most trees are long-lived by the standards of other plants and animals (hence the Bible passage in Isaiah 65:22 'As the days of a tree are the days of my people'), and, of course, the oldest known living things in the world are trees! The life span of most trees is measured in centuries, most being comfortably seen out in less than 500 years (see Box 9.1). Tropical trees are probably no exception (but given the normal lack of annual rings it can be hard to tell; Martínez-Ramos & Alvarez-Buylla (1998) give a good review of the methods for ageing tropical trees). Work on the revegetation of Krakatoa after the devastating eruption of 1883 suggests the average life span of pioneer tropical trees there to be 80–120 years, about the same as temperate birches. Richards, in his acclaimed book *The Tropical Rainforest*, says that the average maximum lifespan of only two species is known: *Shorea leprosula* at 250 years and *Parashorea malaanonan* at 200 years with a maximum age of perhaps 300–350 years. It certainly seems true that the majority of tropical trees live for less than 400 years although there is evidence that some tropical trees in the Amazon Basin can live for 1200–1400 years but this is probably exceptional.

Around the world there are a number of trees that live significantly longer (Box 9.1). Many of the redwoods are long-lived with the coastal redwood (*Sequoia sempervirens*) and giant sequoia (*Sequoiadendron giganteum*) living over 2000 and 3000 years, respectively. The oldest living trunks belong to the bristlecone pines (*Pinus longaeva*)[1] growing at more than 3000 m (10 000 ft) in California and Nevada (Figure 9.2). The oldest known living specimen was a short, gnarled tree on Wheeler Peak, Nevada, called Prometheus, just 2 m (6 ft 3 in) in diameter. This was cut down on 7 August 1964 for a geologist, Donald Currey, who was studying glaciology of the area and using the trees to date the moraines. Up to this point he had been taking thin straw-like cores from trees to count the rings but his corer broke in this tree and a Forest Service crew

---

[1] There is some difference of opinion over the correct scientific name. Americans have liked to call these old bristlecone pines *Pinus longaeva* while others have used the older name, *Pinus aristata*. The *World Checklist and Bibliography of Conifers* (Farjon 2001) now uses the former for these old trees.

**Box 9.1** The expected life span of a range of trees in years

| | | |
|---|---|---|
| Birches | *Betula* spp. | 80–200 |
| Red maple | *Acer rubrum* | 110 |
| Ash | *Fraxinus excelsior* | 200–300 |
| Beech | *Fagus sylvatica* | 200–400 |
| Scots pine | *Pinus sylvestris* | 500 |
| Oak | *Quercus* spp. | 700–1000 |
| Ponderosa pine | *Pinus ponderosa* | 1000+ |
| Douglas fir | *Pseudotsuga menziesii* | 1200 |
| Limber pine | *Pinus flexilis* | 2000 |
| Coastal redwood | *Sequoia sempervirens* | 2200 |
| Huon pine | *Lagarostrobos franklinii* | 2500 |
| Giant sequoia | *Sequoiadendron giganteum* | 3200 |
| Alerce/Patagonian cypress | *Fitzroya cupressoides* | 3600 |
| Bristlecone pine | *Pinus longaeva* | 4900+ |
| Yew | *Taxus baccata* | possibly 5000+ |

felled the tree for him to allow him to count the rings. Currey counted 4844 rings, which was later recounted at 4862 rings, giving a birth date of 2898 BC, the oldest known bristlecone pine. This tree would be 4911 as I write in 2013 and so a life span of over 5000 years seems possible. When the Egyptians were building the pyramids these trees were already well established.

Yew trees in Britain can also live for thousands of years. Certainly the largest are impressive (Figure 9.3a). The crumbling shell of the Fortingall yew on Tayside, Scotland has a staggering diameter of 5.4 m and is suggested to be up to 5000 years old (Figure 9.3b). A stumbling block in deciding precisely *how* old yews are has been the mechanics of ageing them. The oldest specimens are hollow, and even if they were not, it would be difficult to get access to the rings even with a corer due to their size. Felling a venerable tree would be a little drastic just to see how old it was. Mitchell's rule (see Chapter 6) of 1/2 to 1 inch (1.2 to 2.5 cm) growth in girth (circumference) per year can be a useful way of estimating age in a non-destructive way. This can be made more accurate using

Figure 9.2 Bristlecone pine (*Pinus longaeva*) growing in the White Mountains, California, USA.

girth data from trees where the rings *have* been counted or the planting date is known to construct a graph of girth against age (Thomas & Polwart 2003 give an example for yew). Yews, however, do not seem to necessarily put on annual rings (see Chapter 3) and trees are known which show no apparent growth in girth for over 300 years (though loss of part of the trunk or different accuracies/ heights of measuring need to be considered). Repeated measurements have shown that the average growth in girth of yews may be just 5 mm per year, reducing to 0.3 mm per year in old trees (corresponding to ring-widths of 0.8 mm and 0.05 mm, respectively). This is much slower than Mitchell's 1/2

(a)

**Figure 9.3** (a) An old yew (*Taxus baccata*) in Church Preen, Shropshire, England thought to be planted in 457 AD. (b) The Fortingall yew on the north side of Loch Tay in Perthshire, Scotland, possibly 5000 years old. Although just two pieces of the trunk are now alive, the wooden posts outline the position of the original trunk. Old drawings from the 18th and 19th century depict this as a single tree.

to 1 inch, so the Fortingall yew *could* be 5000 years old but it will never be known for sure. The long life span of yews is perhaps the answer to the age-old question of why yews are traditionally planted in British churchyards: the churches were built around the yews revered by pagans.

Others types of tree *may* be even older. The dragon trees (*Dracaena draco*) on Sumatra are thought by some to be up to 10 000 years old, and some cycads are speculated to be 14 000 years or older. Since neither of these produce annual rings or have a solid centre that can be carbon dated, I suspect that these estimates are tales that have grown in fond telling.

(b)

Figure 9.3 (cont)

Stories occasionally emerge of incredibly old trees being discovered. For example, a creosote bush (*Larrea tridentata*) in California has been dated as 11 700 years old, a Huon pine (*Lagarostrobos franklinii*) in Tasmania is possibly 10 000 years old, and a shrub called King's holly (*Lomatia tasmania*) growing in southwest Tasmania may be up to 40 000 years old. It has been suggested (but with no direct evidence) that an aspen clone in Utah called Pando (meaning 'I spread': it covers 43 ha with around 47 000 stems; see Chapter 6) could possibly have lived for an impressive million years (Barnes 1966). Are these for real? Well, yes and no. In these cases it is the plant clump that is so ancient, not the individual stems. The Huon pine and King's holly are clones which have probably been resident on the site for thousands of years but the current stems are all fairly young. In the case of the Huon pine there is no trunk older than around 2000 years. Similarly with the Pando aspen clump the average age of the stems is 130 years. Leif Kullman carbon-dated wood found in the soil beneath a Norway spruce (*Picea abies*) in the mountains of northern Sweden in 2004 and found it to be 9550 years old. This suggests that the tree (called Old Tjikko) has been on the site for all this time (see http://www.kullmantreeline. dinstudio.se). The oldest trunks are no more than 600 years old and the tree has probably survived for millennia by branches touching the ground and rooting (layering), providing new trunks when the old ones die. All these examples are rather like having an old broom that has done 50 years' service with only five new heads and two new handles. Since these trees survive by cloning themselves (producing genetically identical trunks with new roots; see Chapter 6), perhaps the best way of looking at this is to call these trees the oldest clonal trees and the bristlecone pine the oldest non-clonal tree: accurate if somewhat cumbersome! Ultimately, all these trees are impressive but, personally, to touch a bristlecone pine trunk that has been growing for over 4000 years takes some beating.

## Why do trees live for so long?

The most obvious answer is to say that it is an evolved survival strategy. Since trees are usually such large things, the woody framework represents a huge investment which is cost effective only if it is kept for a long time. A less obvious answer, perhaps, is that the longevity allows a tree to endure periods of poor growing conditions while taking advantage of good periods. This is especially true if the conditions are poor for seed production since a few decades of few or no seeds is probably of little consequence compared to centuries or millennia of good seed production. This is something that would trouble short-lived plants much more. In the same way, in environments where successful production of seedlings is rare, such as in very dry or cold habitats, long-lived species are more likely to still be alive and producing seeds when good conditions come along. Petit & Hampe (2006) explore these arguments in more detail.

Another way of looking at this question is to say that trees live for so long because they can. This is discussed below under *Starvation and old 'age'*.

## Defences

Being firmly rooted in the ground, a tree cannot move to escape harsh conditions or the unwanted attentions of animals and diseases (seeds are the only parts with an option for movement). And there are no two ways about it, because trees are so long-lived they face a tremendous number of problems over their lives. They have therefore developed an impressive array of defences to protect themselves where they stand. The living skin of a tree is normally no older than two or three decades, and the leaves, flowers and fruits are usually even shorter lived. Consequently, the defences of these parts are similar to those in non-woody plants. The real specialisation of defence comes in maintaining the woody skeleton which may persist for centuries or millennia. We'll consider these defences in turn.

## First-line defences: stopping damage

### PHYSICAL DEFENCES: SPINES, THORNS AND PRICKLES

Large herbivores are capable of eating copious quantities of leaves along with the odd twig. Indeed the low nutritional value of leaves *requires* large volumes to be eaten. The chief initial defences against these large herbivores are spines, thorns and prickles. Spines and thorns (which botanically are the same thing) are usually modifications of either a leaf, part of a leaf or a whole stem. In *Berberis* species (as in cacti) it is the leaf which is turned into a spine (Figure 9.4). The branch is then left with no green leaves so new ones are produced from the bud in the leaf axil growing out this year rather than next as is normal. Holly (*Ilex aquifolium*) could be regarded as a halfway house with just the margin of the leaf growing spines. In false acacia/black locust (*Robinia pseudoacacia*) it is just the stipules (outgrowths of the leaf base) that are modified into persistent thorns (see Figure 9.4c). Hawthorns (*Crataegus* spp.) and firethorns (*Pyracantha* spp.) have thorns *above* the leaf showing that the thorn is a modified branch, grown out from the bud. Growth next year can still take place because these trees cunningly grow extra buds beside the thorns. Again there are halfway houses; blackthorn (*Prunus spinosa*), some brooms (*Cytisus* spp.), crab apples (*Malus* spp.) and other *Prunus* species produce thorns at the ends of normal leaf-bearing branches. Thorns are also possible on the trunks of trees (Figure 9.5).

The prickles of roses (*Rosa* spp.) and blackberries (*Rubus* spp.) are quite different from thorns (Figure 9.4), being merely outgrowths of the bark or skin (and so with no vascular tissues) that can be easily broken off. However, while of use for defence, their main use is undoubtedly to help these scrambling plants hook onto whatever they're climbing over.

Spines and thorns are expensive things to produce and will only be grown where they are needed. Since most leaf-eaters stand on the ground to eat, simply because they are too heavy to climb or fly, it is perhaps not surprising that trees like holly produce thorns mostly on the lower 2 m of the canopy and the leaves above that have few or no thorns. But why do thorns appear in some trees but not others? A number of years ago Peter Grubb of Cambridge University proposed a solution to what appears to be a haphazard distribution. First of all, physical defences are found in habitats where nutritious growth is scarce. This explains why thorny plants are found in deserts and heathlands (e.g. gorse, *Ulex* spp.). But it also explains why thorns are found on plants that invade gaps in forests,(e.g. hawthorns, apples, honey-locust *Gleditsia triacanthos*, and false acacia/black locust *Robinia pseudoacacia*): food is scarce because other vegetation is out of the reach of large herbivores. The same principle applies to European holly which, being an evergreen in a

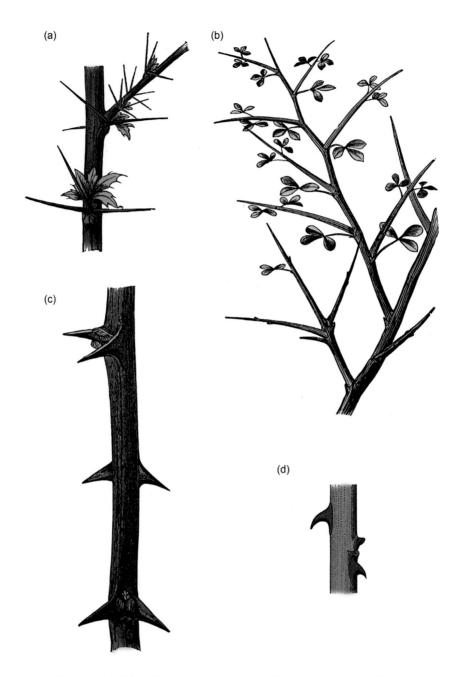

**Figure 9.4** Thorns and prickles. (a) Barberry (*Berberis* spp.); thorns made from modified leaves. (b) A spiny broom (*Cytisus spinosus*); thorns made from modified branches. (c) False acacia/black locust (*Robinia pseudoacacia*); thorns made from the two stipules at the leaf base, which remain on the twig when the leaf falls. (d) Prickles as found on the stems of a rose (*Rosa* spp.). From: (a) – (c) Oliver, F.W. (1902) *The Natural History of Plants.* Vol. 1: *Biology and Configuration of Plants.* Gresham, London. (d) Troll, W. (1959) *Allgemeine Botanik.* Ferdinand Enke, Stuttgart.

Figure 9.5 Thorns on the trunks of trees. Perhaps the best known is the floss-silk tree (*Ceiba speciosa*, used to be called *Chorisia speciosa*) native to Brazil and Argentina and a relative of the kapok tree (see Figures 4.16d and 5.19). (a) The branches and trunk of young trees are densely thorny, giving protection to the easily damaged thin bark. (b) As the tree ages it usually loses most or all of the thorns. Photographs from (a) Brisbane Botanic Gardens, Australia, (b) Singapore Botanic Gardens.

background of deciduous trees, provides scarce fodder in the winter. (Indeed around the southern Pennines in England, shepherds would cut the upper branches of holly (no prickles!) as fodder, particularly in spring when other foods were not available.) Finally the spines common around the growing points of palms and tree ferns are well worth the investment because they have only one growing point which, if damaged, means death to the plant (see Chapter 3).

## OTHER PHYSICAL DEFENCES

Trees have not been slow in evolving other physical defences. Hairs help deter insects attracted to young vulnerable growth, especially if they are dense and branched. In some cases this is a physical barrier (like us having to chew through a pillow to get food); in others the hairs contain chemical deterrents (as in the nettle-like hairs of the Australian rainforest stinging trees, *Dendrocnide* spp.).

Other physical defences can be quite subtle. For example, because of the arrangement of their mouths, caterpillars have to eat a leaf from the edge, so holly has evolved a thickened margin to prevent the caterpillars getting a start.

So why haven't more trees evolved a similar mechanism? Probably because the cost is only worthwhile in long-lived evergreen leaves and these are usually protected by chemicals instead. Physical defence can be yet more subtle. One example of many is the subterfuge indulged in by passion flower vines (*Passiflora* spp.) which are eaten by caterpillars of heliconid butterflies. The butterflies rarely lay their yellow eggs where there are already plenty (this would be pointless competition for food amongst the caterpillars) so the vine grows imitation yellow eggs on its leaves!

## ANTS, OTHER BEASTIES AND FUNGI

Around the world, there are many examples of trees that use ants as their main defence, particularly amongst tropical trees but also some temperate trees such as the black cherry (*Prunus serotina*) of N America. The ants run around the tree preventing birds and animals from nesting, dissuading herbivores, cutting away epiphytes and lianas, and in some cases killing anything growing within a 10 m radius of their tree. These ants usually have vicious stings as is rapidly found out if you lean against their tree! In return the tree provides food and often a home as well. For example, the 'ant acacias' of the New and Old World have swollen thorns in which ants hollow out nests (Figure 9.6). Nectar is delivered from leaf stalks ('extrafloral nectaries', i.e. nectaries not in flowers) and protein, fats and carbohydrates from small knobs (called Beltian bodies) produced at the tips of the leaflets. Ant leaves are less bitter and tough than in other trees, containing fewer internal deterrent chemicals: presumably the cost of looking after ants is cheaper than producing chemical defences. But if the ants are experimentally removed then a tree will usually produce more chemical defences to protect itself.

A similar sort of arrangement is seen in temperate areas where trees such as the N American aspen (*Populus tremuloides*) and the European wild cherry (*Prunus avium*) have extrafloral nectaries visited by parasitic wasps and predatory flies, which presumably do a similar job for the trees.

In some cases the tree also gains nutrition from the ants.[2] This was reported by Daniel Janzen in 1974 who found four such shrubs in Sarawak, Malaysia. These grow as epiphytes on other vegetation which in turn are growing on very poor soils. Ants are housed on the shrubs in a swollen stem base, swollen roots or rolled-up leaves and instead of the ants jettisoning their waste (dead ants and bits of their prey) they store it inside their homes where it decomposes, feeding the host plant, especially with nitrogen. In the tropical tree

---

[2] Trees that have a relationship with ants are referred to as myrmecophiles (Greek *myrmex* meaning ant and *philos*, loving). Those that gain nutrition from the ants are also myrmecotrophic, *trophia* meaning to nourish.

Figure 9.6 Thorns on the whistling thorn (*Acacia drepanolobium*) in E Africa. In amongst ordinary thorns are those with bulbous swellings 2–3 cm across and into which ants burrow and make their nests. Photograph by Nevit Dilmen, reproduced under Creative Commons Attribution-Share Alike 3.0 Generic, courtesy of Wikimedia Commons.

*Cecropia peltata* of Central and S America, which can reach impressive size, up to 90% of its nitrogen can come from ant debris.

Employing guards can work on a much smaller scale. It is common to find little tufts of hair or pockets of tissue at the junctions where veins join together on the underside of leaves. A study on the European evergreen shrub *Viburnum tinus* found 10 species of mite (mostly predators and microbivores, i.e. eating microbes) living in the crevices (or 'domatia') between these hairs. In other words, the plant is providing homes for mites that help keep other harmful organisms at bay.

Tree defences are also helped by the presence of soil bacteria and particularly mycorrhizal fungi (Chapter 4). This help was long thought to be via these organisms helping the tree to get more nutrients from the soil, making the plant healthier and so better able to defend itself. But the interaction is more subtle than that. These organisms appear to increase the sensitivity of the tree's cells to hormones, and thus be more sensitive to herbivore attack in producing plant defences. And it is not just the soil since fungi inside tree leaves have been shown to be very beneficial in reducing damage. A study of the cocoa tree (*Theobroma cacao*) found that fungi inside leaves reduced infections by the pathogen *Phytophthora palmivora*, probably by the fungi being directly antagonistic to the pathogen.

Some plants use primarily physical defences, while others use the chemical defences described below, but most trees, of course, use a mixture of both.

## CHEMICAL DEFENCES

Woody plants produce a great variety of chemical compounds to provide protection against other plants, diseases and herbivores big and small. The main groups of chemicals are alkaloids (such as caffeine, strychnine, quinine, morphine and nicotine; think what they do to *our* physiology!), terpenes (the active ingredient of latex and resins) and phenolics (such as tannins). There are also a number of others including steroids, cyanide producers (cyanogenic glycocides) and lectins (including ricin, one of the most poisonous chemicals known, that comes from the castor oil plant *Ricinus communis*, admittedly not a tree!). Chemical defences also include enzymes that digest chitin (the main ingredient of fungal cell walls and insect skeletons). Collectively, these chemicals can be found in just about any part of the plant (for example, the waxy surface of apple leaves contains toxins to repel certain aphids). They can also be emitted into the air in considerable quantities which explains why pine woods smell so nice and the eucalypt-dominated bushlands of Australia have a blue haze (too much of this can be considered as pollution: see Chapter 10). These chemicals seep out through the bark, the holes in the leaf (the stomata) or ooze out of special glands, often on teeth around the leaf edge (as found in pines, alders, willows, hornbeam, maples, ashes, elms and viburnums) usually when the leaf is young but sometimes (as in crack willow, *Salix fragilis*) even late in life. Are these emissions of expensive chemicals just inadvertent leaks or do they have a purpose? Certainly in oaks and many other trees it allows neighbours to communicate with each other: just how is revealed below.

Chemical defences work in several ways. Some are highly toxic and will kill attackers in small doses (referred to as 'qualitative' defences since they work by being what they are – their quality – rather than how much there is). These include the alkaloids which often give a bitter taste, such as those in rhododendron (honey from its flowers is poisonous to humans but is so bitter it's uneatable), and nicotine in tobacco plants. Cyanide (hydrocyanic acid) comes into the same category produced, for example, by crushing leaves of cherry laurel (*Prunus laurocerasus*; Figure 9.7). A few crushed leaves put into a bottle makes a very effective insect killing-bottle. Other chemicals work by

Figure 9.7 Cherry laurel (*Prunus laurocerasus*). Crushed leaves produce cyanide (hydrocyanic acid). The building chemicals (cyanogenic glycosides) and the necessary enzymes to produce cyanide are kept separate in the plant cells until they are damaged by browsing or frost and the two mix together. Keele University, England.

building up in the herbivore (called 'quantitative' defences since they become more effective the more there is). Classic examples are resins in creosote bushes (*Larrea tridentata*) and tannins found in a number of plants. Tannins are a 'protein precipitant' that bind onto proteins making food less digestible so herbivores end up starving and stunted, and they may also work by damaging the gut of the eater. This binding ability explains why adding milk to tea makes it less bitter; the milk proteins bind to the tannins of the tea and we no longer taste them. These quantitative defences often have the effect of causing the herbivore to move elsewhere. Incidentally, it is possible that the endangered British red squirrel is declining because of acorns and the introduced N American grey squirrel. Acorns contain digestion inhibitors that greys can disarm but reds can't, so red squirrels do less well in oak woods when in competition with greys.

Plants are not immune to the effects of chemical defences. Spanish moss (*Tillandsia usneoides*) that festoons trees in southern USA and S America (and is not a moss but a sophisticated relative of the pineapple) never grows on pines presumably because of chemicals in the resins. Perhaps the best known example of anti-plant chemical defences was first reported by Pliny the Elder in the first century AD who wrote that 'The shadow of walnut trees is poison to all plants within its compass'. Walnuts (and all trees in the genus *Juglans*) contain the chemical juglone which, seeping from the roots, reduces the germination of competitors, stunts their growth and even kills nearby plants (a process referred to as allelopathy), resulting in open-canopied walnuts having very little growing under them. Tomatoes, apples, rhododendrons and roses are very susceptible but many grasses, vegetables and Virginia creeper (*Parthenocissus* spp.) will happily live under walnuts. These allelopathic chemicals are also produced by the tree of heaven (*Ailanthus altissima*), some eucalypts and acacias.

Which defensive chemicals you find in a tree partly depends upon what sort of area the tree normally grows in. What this really comes down to is the balance between the availability of sugars and of nutrients such as nitrogen (referred to as the carbon/nutrient balance). On poor soils, trees will be short of nutrients but will still be able to produce sugars by photosynthesis and so defences tend to be carbohydrate based (terpenes, phenolics). Under deep shade, photosynthesis is limited and so sugars are in short supply, so providing nutrients are readily available, trees tend to have nitrogen-based defence compounds such as alkaloids and cyanide.

Defensive chemicals are costly to produce and store (many are toxic to the plant producing them). Oaks may put up to 15% of their energy production into chemical defences, which explains why stressed trees are most prone to attack: they have less energy available to produce defences. Growth and seed production can also be reduced in plants producing large quantities of chemical defences. Thus, although there may be large pools of defences swilling around (preformed 'permanent' defence) where attacks are common and ferocious, it makes evolutionary sense in lesser situations to produce the chemicals in earnest only when they are needed (induced defence). Oaks and a number of other trees including conifers will tick along with low concentrations of tannins in their leaves but once part of the canopy is attacked, the whole tree will produce tannins in large quantity. More than that, the chemicals released into the air (particularly jasmonates including methyl jasmonate involved in the fragrance of jasmine) are detected by surrounding oaks and they in turn will produce more tannins in preparation for the onslaught. There is also evidence that these chemical signals can attract the enemies of the attackers. Communication appears to be normally through the air, but can also happen underground as seems to be the case in conifers (see Dicke *et al.* 2003). These induced defences may have an effect for some time; birches can remain unpalatable for up to 3 years after being browsed.

Expensive defences are saved not just *when* they are not needed but also *where*. The N American aspen (*Populus tremuloides*) has been shown to produce fewer tannins in trees growing on fertile soils: it is cheaper to replace lost material than defend it. This may explain why white oak (*Quercus alba*), used by native people of Massachusetts for food, has less tannins than red oak (*Q. rubra*). Tannins were leached from white oak acorns by putting baskets of acorns in streams for several days. Acorns of emory oak (*Q. emoryi*) in southwest USA have so few tannins that they can be eaten straight from the tree.

In the same way, more is invested in chemical defences where herbivores and diseases are more common. This is certainly true of tropical forests where 7–8% of annual growth is consumed by herbivores compared to 3.5–4% in temperate forests. This explains why so many interesting chemicals come from tropical trees, from which we are not immune! Some can be particularly unpleasant, doing to us what they are designed to do to the animals that venture to eat bits of them. Strychnine, from the bark of various *Strychnos* lianas (primarily *Strychnos nux-vomica*, native to Burma and India) causes twitching and muscle spasms that can cause the head to be forced back to the buttocks, breaking the spine. Only death from exhaustion releases the victim from indescribable agony. Curare from the bark of a number of lianas (including *Strychnos toxifera* and *Chondodendrum tomentosum*) paralyses muscles (a healthy herbivore defence!) which can cause death by asphyxiation and heart failure but which fortunately is also a boon in helping surgeons to operate on relaxed patients. Even something as familiar as nutmeg (*Myristica fragrans*) packs a punch. I'm told that 1 teaspoon (5 ml) of ground nutmeg causes hallucinations and 5 teaspoons worth is fatal. But the vicious arsenal is not restricted to the tropics. Seeds of various members of the rose family (apples, plums, peach, apricot and almond) contain doses of cyanide as hydrocyanic acid. Leaves of the California bay (*Umbellularia californica*; Figure 9.8) contain cyanic acid which when handled give off an unpleasant smell, causing headaches and even unconsciousness!

Other chemicals are less toxic to us humans and we even find them quite pleasurable. Where would we be without the defensive chemical caffeine found in coffee (*Coffea* spp.) and tea plants (*Camellia sinensis*), or the chemicals in the bark of the cinnamon tree (*Cinnamomum verum*)? Indeed some have saved lives. In 1535 Jacques Cartier was stranded in the ice-bound Canadian St Lawrence River near modern-day Montreal. Twenty-five men had already been lost to scurvy when passing Iroquois Indians taught them to boil the bark and foliage of what we now call white cedar (*Thuja occidentalis*) to make a tea. The high vitamin C content of the tea apparently worked miracles. Cartier named the tree 'arbor vitae', tree of life. There are many similar stories of using other woody plants as antiabscobants, from Captain Cook using spruce beer, to Captain Vancouver using Winter's bark (*Drimys winteri*) from S America. (Incidentally, unpasteurised cider has a high vitamin C content, a good excuse for imbibing a little of the pale fluid if considering a sea voyage!) Vitamin

**Figure 9.8** Leaves and fruits of the California bay (*Umbellularia californica*) at Point Reyes, California. The smell from the leaves can cause headaches.

C seems to do the same job for plants as for animals; it is an antioxidant keeping plants safe from their own aerobic metabolism and a range of pollutants. I can't pass without mentioning that the ancient Greeks used finely ground pine cones mixed with herbs as a cure for haemorrhoids.

## The evolutionary battle

No defence is impenetrable. Over evolutionary time herbivores find ways around a defence and the plant counter-attacks with new defences. And so the battle goes on. This has been called the Red Queen theory, from *Through the Looking-Glass* where Alice is told by the Queen that it takes all the running you can do to keep in the same place. In the arms race of evolutionary battles, both sides have to evolve as fast as possible in order just to survive. Yew is poisonous to cattle and horses and yet deer will happily eat it. Red and fallow deer eat holly despite the prickles. Deer and rabbits don't eat sycamore bark but squirrels love the inner bark. Caterpillars can eat though a whole tree of young oak leaves despite the tannins (the caterpillars have an alkaline gut which reduces the tannins' effectiveness). The tree responds by shedding the damaged leaves to prevent disease entry and the waste of further nutrients on substandard leaves. Willow leaf beetles not only survive the salicin (raw aspirin) from their hosts, they use it themselves to produce a defensive secretion. The mountain pine beetle (*Dendroctonus ponderosae*), currently devastating western N American pines, goes a step further. Normally, beetles are trapped and killed by the pine's resin. However, the mountain pine beetle carries a fungus

*Grosmannia clavigera* from tree to tree which is capable of breaking down the toxins in the resin, protecting the beetle as it burrows through the tree.

It is puzzling how a tree can survive in the battle of attacks and counter-attacks. A tree living for centuries with a genetically fixed array of defences will be faced with insects that produce thousands of generations during this time with thousands of opportunities for producing new ways of getting around the tree's defences. Even for a tree such as the European sycamore (*Acer pseudoplatanus*), which is fairly quick off the mark and starts producing flowers at 5–20 years old, it will be attacked by around 400 generations of aphids before it can produce any seeds. And so trees seem to be at a distinct disadvantage.

Part of the solution is that tree populations accumulate a large number of genes that give resistance to pathogens such as viruses, bacteria and fungi. For example, the black cottonwood (*Populus trichocarpa*; whose genome was published in 2006, see Chapter 8) has around 400 of these genes which are responsible for making about 1% of all proteins in the plant, twice what have been found in many herbaceous plants (see Veluthakkal & Dasgupta 2010). So when a population of trees is attacked by newly improved insects or fungi, there is hopefully an individual somewhere in the population that already has suitable defences.

Another part of the solution may well lie in a similar accumulation of genetic diversity within an individual tree, possibly due to the sheer size of the tree. A large oak or pine carries something between 10 000 and 100 000 buds (meristems). A naturally occurring genetic mutation, or other genetic change, in one of these growing points will result in a whole branch which is genetically distinct from the rest of the tree. Even though rates of mutation are low in trees and most mutations are likely to be deleterious (and will be shed as dead twigs and branches), genetically distinct branches should accumulate over the years turning the tree into a giant genetic mosaic. This may explain why some branches on a tree tend to lose their leaves earlier in autumn. It is likely that some of these distinct branches will be better resistant to the evolving insect attack than others, and so prosper: natural selection operating within a tree. As these resistant branches grow most vigorously they will make up a larger proportion of the canopy. They will also be the branches to produce most seed which will be genetically best-adapted to cope with the current set of pests. In this way, long-lived trees can go some way towards decoupling their rate of evolution from their generation time and give them a chance against rapidly multiplying pests. The main problem with this part of the answer is that trees have fairly slow rates of mutations, so this is likely to be a slow process. But given the longevity of trees they still have time to accumulate a large number of genetically distinct branches.

The final part of the solution lies in getting help from other living things. As we've noted above, trees often maintain colonies of the enemies of things that might attack them, such as ants kept by acacias and fungal endophytes described above. These presumably undergo as rapid rates of evolution as the pests and, as noted by the Red Queen, both sides are then running equally fast.

## Defending the woody skeleton

At first sight it may seem that wood does not need much defending: it's pretty inedible stuff. The large quantities of cellulose (40–55%), hemicellulose (25–40%) and lignin (18–35%) are all tough carbohydrates that are quite hard to digest or decompose. Moreover, wood is incredibly poor in protein and hence nitrogen (typically 0.03–0.1% nitrogen by weight compared to the 1–5% found in green foliage). Just how poor wood is as a food is illustrated by the goat moth whose caterpillars burrow through willow wood and take up to four years to grow to maturity, and yet they can mature in just a few months if fed on a good rich diet. Despite the starvation diet that wood offers, there are many insects, fungi and bacteria that are capable of living off wood.

### Keeping things out: resins, gums and latex

Bark itself is very resistant to attack by fungi and many insects. Suberin in the bark is very rot resistant and is often backed up by polyphenols, terpenes and other antimicrobial chemicals. The bark is often the last part of the tree to disappear after death; this can be seen in rotting logs where a tube of fairly sound bark holds together a mass of powdery formless wood. Nevertheless, for any number of reasons, the bark can be broken, potentially allowing all-comers entrance to the wood, so trees have a first line of defence using resins, gums and latex. These fluids have a primary role in rapidly sealing over wounds and in deterring animals from forcing their way in. Any animal rash enough to burrow is physically swamped and trapped, and may be overcome by chemical toxicity (though bark beetles are seen swimming through resin apparently unharmed, if a little hindered). The insects found trapped in amber (fossilised resin) show that this battle has been going on for at least 100 million years. A secondary role of these fluids is to defend the wood against fungi. Cedar resin is so good at this that Egyptians used the resin in the embalming of mummies and used cedar wood to build sarcophagi.

**Resins** If you have leant against an old pine or handled a cone you will be well aware of the ability of conifers to produce copious quantities of resin. The typical 'pine' smell of conifer foliage comes from the resin. Yews (*Taxus* spp.) do not contain resin and consequently do not smell in the same way.

In most conifers, including pines, Douglas fir (*Pseudotsuga menziesii*), larches (*Larix* spp.) and spruces (Picea spp.), the resin is contained in ducts which run through the bark and wood, tapering off into the roots and needles. Others such as the hemlocks (*Tsuga* spp.), true cedars (*Cedrus* spp.) and true firs (*Abies* spp.) have resin restricted more or less to the bark, although, like other conifers, they are capable of producing 'traumatic resin canals' in the wood after injury or infection. In the true firs the resin is contained in raised blisters on the bark. Cells along the ducts or blisters secrete resin, creating a slight positive pressure, so if the tree is damaged the resin oozes out (Figure 9.9). Once in the air, the lighter oils evaporate leaving a solidified scab of resin over the wound.

All resins are based on terpenes, but different species of tree produce resins of different composition. At the beginning of the American Civil War, Union forces in the Californian foothills were cut off from their normal supply of turpentine so they distilled it from the resin of the ponderosa pine (*Pinus ponderosa*). But at slightly higher altitudes there grows the very similar Jeffrey pine (*Pinus jeffreyi*) whose resin contains high levels of heptane, a highly inflammable liquid found in raw petroleum. Firing up a primitive turpentine still full of Jeffrey pine wood was like building a fire under a petrol tank!

Figure 9.9 The scar on a Japanese red pine (*Pinus densiflora*) left by tapping the tree for resin. The resin oozes out under a slight positive pressure, and successively lower strips of bark have been removed to keep the wound open and the resin flowing. This is a 200-year-old tree at Samcheok, South Korea.

Trees other than conifers produce resins. Members of the Burseraceae family contain resins particularly in the bark. This includes frankincense (*Boswellia carteri*), used in incense and chewed by Arabs as a breath freshener (and now becoming an endangered species), and myrrh (*Commiphora* spp.) used in incense and perfumes.

**Gums** Fulfilling a similar function, a wide range of woody plants produce gums. The family of Anacardiaceae is notable for gum-producing trees including the varnish tree (*Rhus verniciflua*), a native of China whose gum is used as the basis of lacquer. Gums are also found oozing from wounds in a variety of temperate trees such as those in the genus *Prunus* (the cherries, plums, etc.; Figure 9.10) and others such as the kauri (*Agathis australis*) in New Zealand. These gums are carried in ducts, which, as in the conifers, are in the bark and often the wood where they follow the rays and grain. Traumatic canals can be formed in the wood of some hardwoods, for example, sweet gum (*Liquidambar styraciflua*) and cherries.

Figure 9.10 Gum from a wound in a cherry tree that has solidified into a bright orange plug. Devon, England.

**Latex** Latex (a milky mixture of such things as resins, oils, gums and proteins) is found in different plants from some fungi to dandelions to trees. Many types of trees and shrubs have had their latex collected for making rubber (including the 'rubber tree', *Ficus elastica*, now grown commonly as a house plant). Around one third of rainforest trees have latex. The best commercial supply comes from *Hevea brasiliensis* in the spurge family (Euphorbiaceae) native to the Amazon and Orinoco river valleys of S America. The rich latex of this tree is about 33–75% water and 20–60% rubber. Latex is found in special ducts (lactifers) running through the bark in concentric circles. As with resin in conifers, the latex is under slight pressure, ensuring that any wound is sealed by coagulating latex (including those made deliberately to collect the latex, and which are treated with an anticoagulant to ensure a good collection).

## Callus growth

The production of resins, gums and latex is often insufficient to seal over large wounds such as the breakage of large branches and the removal of areas of bark by, for example, squirrels. If the wound is kept artificially moist (by covering with plastic, lanolin or other non-toxic substance) so that the living cells of the rays do not dry out, a new bark will regenerate in the same season in many species (Figure 9.11a). (Note that in pruning off branches it is only the younger parenchyma cells of the sapwood around the edge of a large wound that are capable of doing this, leaving a hole in the centre.) But if the newly exposed parenchyma cells are killed by toxic materials in paint or allowed to desiccate, then the wound can only be covered by the slow growth of callus tissue (sometimes called wound wood) from the living cambium and rays around the edge. As the callus grows, it starts off uniformly around the wound but the sides tend to grow more rapidly (Figure 9.11b) resulting in a circular wound becoming a spindle-shaped scar (just why this is so is considered below under *Wind*). Once the sides of the callus meet, the cambium joins so new complete cylinders of wood are again laid down underneath the scarred bark.

## Healing wounds

While callus tissue does not stop rot getting into a wound before it is covered over, it has been noted that wood decay stops spreading, or at least slows down, once callus tissue covers the wound and cuts off the main supply of oxygen. However, it is important to note that, unlike animals, trees cannot heal wounds; once wood starts to rot it cannot be repaired. Trees can only ever cover wounds up. New wood can be grown over the top and look healthy but the rotting wood underneath is still there. Hence the wise saying quoted by

(a)

(b)

Figure 9.11 (a) New bark being grown from the ends of rays in Oriental beech (*Fagus orientalis*). (b) New callus from the sides growing over a wound in an oak (*Quercus robur*) is concentrated at the sides of the wound, producing an oval shape over the years, because this is where most stress is placed on the tree as it flexes in the wind. Photographs from (a) Westonbirt Arborteum and (b) Keele University, England.

William Pontey in *The Forest Pruner* (1810), 'An old oak is like a merchant, you never know his real worth till he be dead'.

## Internal defences

Although resins, etc. and new wood help to seal off wounds there will always be some fungal spores and other damaging agents that slip by these first defences. So the tree must have other internal defences to deal with destructive agents once they get in. These defences cannot always stop the spread of destructive agents but they will slow them down. Before looking at those defences, it is worth looking at the biggest cause of the problem.

### Fungal rot

If we are thinking of large structural decay, we are dealing primarily with fungal rots that are capable of decomposing the cell walls of wood for food.

At an early stage of decay a number of fungi and bacteria will live on the nutritious cell contents but do little to structurally weaken wood. These are a nuisance to foresters since they cause staining of the wood, decreasing its value for timber and paper production. Structural decay comes from two main groups of fungi. Brown rots attack the hemicellulose and cellulose of the wood leaving the lignin untouched as a brown mass usually cracked into cubes. White rots attack all components of the wood, reducing it to a light-coloured spongy mass. Speed of rot varies tremendously with fungal species and conditions but as a rule of thumb, even virulent white rots such as *Kretzschmaria deusta* (formerly called *Ustulina deusta*) and giant polypore *Meripilus giganteus* take at least 20–30 years to weaken a tree enough for it to fall.

Decay progresses most rapidly when the wood is moist, that is, above 20% moisture (furniture and timbers in a house are generally drier than this and hence don't rot; dry rot fungus gets around this by making its own water from sugars and by moving water considerable distances). Wood that is *too* wet is also safe from rot. Logs have traditionally been stored for long periods in ponds. And after the 1987 hurricane in southern England (when 15 million trees were brought down) some timber was stored in huge piles constantly wetted by sprinklers to be sold when prices had picked up. Wood has been stored in this way for up to 4 years with no appreciable deterioration in timber quality. And 'bog wood', oak, pine and many other trees preserved in acidic, wet peat, can be as sound as freshly cut wood even after 4000–7000 years of preservation in peat bogs (Figure 9.12).

You may have noticed black lines running through a piece of well-rotted wood, especially in hardwoods, referred to by wood-turners as 'spalted wood'. These lines are really double lines formed of masses of gnarled dark-coloured fungal strands (hyphae) and are usually laid down by two different fungi species (or different strains of the same species) when they meet as a sort of double garden fence between them (officially called a 'zone of antagonism' with resulting 'demarcation lines'). Some fungi will, however, produce a black line around themselves even when no other fungus is present, perhaps marking where the fungus intermittently stops growing. Much more detail on the interaction of wood and fungi can be found in the excellent book by Schwarze *et al.* (2000).

## *Defence of wood: sapwood and heartwood*

As shown in Chapter 3, the wood inside a tree can be divided into a central core of heartwood surrounded by sapwood. Sapwood is responsible for conducting water and storing food in the living tissue of the rays. Heartwood is completely dead and filled with chemicals designed to repel all boarders. Over time the innermost sapwood is gradually converted into heartwood. The

(a)

(b)

**Figure 9.12** (a) Bog oak, preserved in a peat bog for around the last 6500 years, still structurally sound apart from the outer edge which would have started rotting before the trunk was covered by peat. The only substantial change is to change the colour of the wood to black. The base is 17 cm long. Using the dendrochronology techniques shown in Figure 6.7, the ring at the centre of this piece of wood has been dated as growing in 4596 BC. (b) Whole oak trees are still being uncovered in the UK. Wood and trunk are from Fenns & Whixall Mosses on the Wales–England border.

(a)

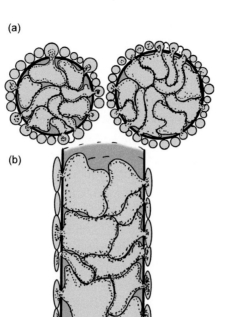

(b)

**Figure 9.13** The balloon-like outgrowths (tyloses – tylosis singular) found in some hardwoods that burst into the water-conducting tubes of the wood from surrounding living cells to form a plug: (a) looking from above and (b) looking from the side. From: Shigo, A.L. (1991) *Modern Arboriculture*. Shigo & Trees, Durham, NH.

creation of heartwood is a deliberate process and is not, as is sometimes said, just old living tissue gradually fading away. It has also been argued that heartwood is just a very useful dumping ground for waste products that would otherwise be hard to get rid of. This may be true but many of the compounds incorporated into heartwood are specifically and expensively produced. The energy cost per unit mass of these compounds is twice that of wood.

One of the first processes towards the formation of heartwood is to block the water-carrying tubes, although this is primarily in response to air in the system and may happen sometime before heartwood is formed. In conifers the pits irreversibly close (see Chapter 3) when air enters the tubes, and resins may add further blockage. In hardwoods, where living cells in sapwood are more numerous, vessels can be plugged by the cells either exuding gums or producing balloon-like outgrowths (called a tylosis or plural, tyloses) into the vessel (Figure 9.13): the original air-bags! Trees with small vessels (less than 80 µm wide) tend to have gum plugs while those with larger vessels have tyloses (although there are exceptions: magnolias with narrow vessels have tyloses and the tree of heaven (*Ailanthus altissima*) with wide tubes uses gums; and some trees have both). Tyloses stick to each other and the vessel wall, and are often impregnated with phenols, making them a physical and chemical barrier.

Blockage of the tubes by closing of pits and adding gums and tyloses is normally complete by the time the heartwood is formed, but not all trees do block up the tubes: red oak (*Quercus rubra*) has very few tyloses and it is possible to blow through a piece of heartwood (which also explains why it is useless for making whisky barrels!). These blockages will inevitably slow the movement of rot through the wood since it must digest its way through rather than growing unhindered through the tubes.

As sapwood is turned into heartwood, the living cells on the edge of the heartwood die and any food they are storing is either used in making the heartwood or is moved elsewhere. Cell death is accompanied by the formation of a variety of compounds (typically lignins, polyphenols, gums and resins containing terpenes) known collectively as 'extractables'. These include some compounds that we find pleasurable but which are toxic to wood-deteriorating organisms. Think of camphor tree (*Cinnamomum camphora*), used to make chests which repel insects, sandalwood (*Santalum album*) used for centuries as incense in the Orient, and the distinctive oils in cedar wood (*Cedrus* spp.), which has been used as a mosquito repellent. Others are less inviting. Volatile oils give fresh elm (*Ulmus* spp.) a distinctive and unpleasant aroma; my colleagues once complained that something had died in the service ducts when I was storing a fresh elm slice in my office!

The extractives are added to both the walls and the bore of the former water-conducting tubes in small amounts, typically making up less than 1% of the mass of dry wood. Although so little, they account for the greater density and durability of heartwood. Woods with few extractives (e.g. alder, ash, beech, lime and willow) may rot in less than 5 years when in contact with the ground, whereas extractive-rich woods (e.g. oak, sweet chestnut, yew and western red cedar, *Thuja plicata*) are usually durable for more than 25 years. Having said this, elm is usually considered non-durable and indeed makes short-lived fence posts, but, as with many woods, keep it saturated with water and it lasts for hundreds of years. The Romans knew this and used hollow elm logs as water pipes (apparently it was a long 'boring' job drilling these out, hence our use of this term). Elm in particular was probably used since it's very resistant to splitting and will cope with being bored out. Elm water pipes unearthed in 1930 in London were still sound after more than 300 years underground.

## Compartmentalisation of decay in trees (CODIT)

Decay-resistant heartwood goes a long way to keeping a tree full of wood but wood is not just a lump of hard-to-decay material; protection is much more

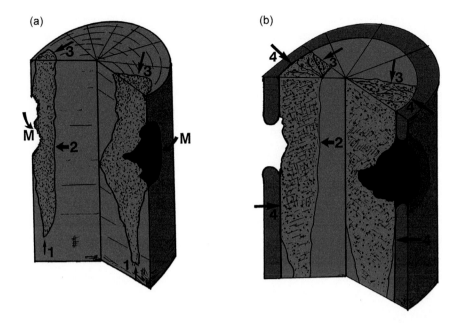

Figure 9.14 Compartmentalisation of decays in wood. (a) When a tree is wounded (marked M) three walls (1 to 3) prevent the spread of rot vertically, towards the centre and tangentially. These correspond to the tube cross-walls, the growth rings and rays, respectively. (b) New growth after the wounding is isolated from the damaged centre by the formation of a strong fourth wall – the 'barrier zone' – (marked 4). As the rot spreads past walls 1 to 3 to fill the centre of the tree, it does not readily penetrate through to the new wood. Thus a tree with rot in the centre is not necessarily immediately doomed. From: Shigo, A.L. (1991) *Modern Arboriculture*. Shigo & Trees, Durham, NH.

subtle. Alex Shigo, an American forester who spent decades looking at the insides of trees, has championed the idea that the wood is divided into compartments. He saw that rot does not always spread through the whole tree even when it has had sufficient time. There appear to be three 'walls' preventing the spread of disease (Figure 9.14a). Wall 1 resists vertical decay and not surprisingly given the structure of wood, is the weakest, allowing fairly rapid spread of rot up and down the tree; tyloses and gums presumably play a part in this 'wall' since wood without them tends to be least durable. Wall 2 corresponds to the ring boundaries and resists inward radial decay; cells here have thick walls that are rich in lignin. In some trees dense parenchyma cells at the end of the growth ring can produce the same corky suberin found in bark (here called 'interxylary cork'). Wall 3, represented by the rays, resists tangential spread around the tree. The rays don't form a continuous barrier and much of the strength of this 'wall' comes from the lack of cells running round the tree; fungi spreading sideways is literally going across the grain, having to go through or around millions of cells.

These three walls are often referred to together as the 'reaction' zone. Thus, in the heartwood which has no living tissue, these dead defensive walls forming the reaction zone are laid down as the wood is formed and left behind to divide the wood into compartments. These chemical walls are similar to a regular grid of anti-tank road blocks that might be laid down in a military retreat. When the enemy arrives, there is no living soul around but their tanks meet a resistant road block at every street corner.

The same walls are found in the sapwood, but because there is living tissue here, and the water-conducting tubes are mostly open and full of water, the reaction zone is not entirely passive. When damage occurs, changes in pressure and drying of tissues causes the normal reactions already discussed such as pit closure and the release of gums and tyloses, forming Wall 1. The living cells around the damage also react by producing antifungal compounds, insoluble, often gummy polyphenols, and, in some cases, corky suberin deposits. These dark-coloured reaction zones between healthy and damaged wood are commonest supplementing Wall 1, almost like thin layers of heartwood, actively produced in response to damage. Since this response in sapwood starts when the wood is first damaged, and is really a response to air getting in and so happens even before it starts to decay, some have argued that CODIT should stand for compartmentalisation of *damage* in trees.

After a tree is wounded, by for example being hit by a lawn-mower, these three walls, the *reaction zone*, will act to slow down the spread of rot that enters. But that is not all. In most trees, in the next growing season after the damage, the cambium lays down a fourth wall, rich in parenchyma cells (up to 30 cells thick) and chemically reinforced by corky cells (especially in oaks, less so in willows and conifers), resins and other defensive compounds between the wood present at the time of wounding and the wood grown afterwards (Figure 9.14b). This *barrier zone* may stretch right around the tree or be restricted to a small arc around the circumference, particularly at the wound site, like a sticky plaster on a child's knee. This reaction after the damage has occurred provides the strongest wall yet that isolates any rot in the centre from the new wood grown after the damage. Strong though it is, the wall is not infallible and rot can eventually break through. If this happens, the tree responds by laying down another barrier zone under the bark. The cambium may not be touched by the break-away rot but it receives the message (presumably via the living tissue in the rays). While this is a superb way of defending the wood, there is a chink in its armour. Many of the major wood-decaying fungi invade from the roots up through the heartwood. Since this is literally dead wood, the tree has no way of detecting the rot and much of the tree may rot away before the sapwood is touched and the cambium is notified to produce a barrier zone. This is undoubtedly what led to the evolution of walls 1 to 3 (the reaction zone); a set of traps to delay the spread of undetected fungi in dead wood. Nevertheless, the value of the barrier zone is that a tree can become hollow without being compromised and imminently doomed; the tree is compartmentalised to protect the outer living skin of the tree.

## What kills a tree?

Throughout their lives woody plants are subjected to damage from adverse environmental conditions and from other living organisms which can affect the growth and which will sometimes lead to death if the tree is stressed too far. A tree may readily succumb to some deficiency (such as drought) or excess (e.g. frost, road salt or pollution) in its environment; see Box 9.2. But this can take time with trees: damage to roots by digging may not show until a dry year several years later. In subalpine forests of the American Rocky Mountains, mortality after a drought can still be showing up 11 years later in subalpine fir (*Abies lasiocarpa*). Moreover, the ability to survive one sort of stress may not help with others; for example, in temperate areas shade-tolerant specialists are usually not very tolerant of drought or waterlogging.

## Damage from the environment: harsh conditions and pollution

The environment can throw up any number of problems – extreme temperatures, drought, flooding, lightning (see Chapter 3), soil impoverishment, wind, fire – and the list goes on. Trees can be mechanically destroyed by devastating events: wind, earthquakes and volcanic activity. Yet these events are often less injurious to trees than they seem and death is by no means always certain. For example, a ginkgo (*Ginkgo biloba*) almost at the epicentre of the 1945 Hiroshima nuclear explosion regrew from the base after its trunk was completely destroyed.

We are also merrily adding to these problems by simple things such as digging around trees (which damages roots), extensive use of herbicides and large-scale pollution: this list could also go on.

Pollution is a perennial concern. Fortunately, a number of trees are relatively tolerant (Box 9.2). London plane (*Platanus* x *hispanica*) is particularly tolerant of grimy city atmospheres with a noted ability to root in compacted and covered soil and an ability to shed bark clogged with soot. A number of trees are relatively resistant to sulphur dioxide including Lawson cypress (*Chamaecyparis lawsoniana*), junipers (*Juniperus* spp.), Corsican pine (*Pinus nigra*), western red cedar (*Thuja plicata*), ginkgo, beeches, oaks, hornbeam

**Box 9.2** Susceptibility of trees to different problems

### Frost

| Susceptible | | Tolerant | |
|---|---|---|---|
| Walnuts | *Juglans* spp. | Birches | *Betula* spp. |
| Ash | *Fraxinus excelsior* | Hazel | *Corylus avellana* |
| Sweet chestnut | *Castanea sativa* | Hornbeam | *Carpinus betulus* |
| Oaks | *Quercus* spp. | Limes | *Tilia* spp. |
| Beech | *Fagus sylvatica* | Elms | *Ulmus* spp. |
| Grand fir | *Abies grandis* | Many poplars | *Populus* spp. |
| Sitka spruce | *Picea sitchensis* | Scots pine | *Pinus sylvestris* |
| Norway spruce | *Picea abies* | Monterey cypress | *Cupressus macrocarpa* |
| Larches | *Larix* spp. | | |
| Western hemlock | *Tsuga heterophylla* | | |
| Western red cedar | *Thuja plicata* | | |

### Sulphur dioxide

| Susceptible | | Tolerant | |
|---|---|---|---|
| Walnuts | *Juglans* spp. | Field maple | *Acer campestre* |
| Birches | *Betula* spp. | Hornbeam | *Carpinus betulus* |
| Apples | *Malus* spp. | Beech | *Fagus sylvatica* |
| Italian poplar | *Populus nigra* 'Italica' | Planes | *Platanus* spp. |
| Willows | *Salix* spp. | Eastern cottonwood | *Populus deltoides* |
| Larches | *Larix* spp. | Oaks | *Quercus* spp. |
| | | Lawson's cypress | *Chamaecyparis lawsoniana* |
| | | Junipers | *Juniperus* spp. |

| **Box 9.2** (cont) | | | |
| --- | --- | --- | --- |
| Sulphur dioxide | | | |
| Susceptible | | Tolerant | |
| | | Corsican pine | *Pinus nigra* var. *maritima* |
| | | Western red cedar | *Thuja plicata* |
| | | Ginkgo | *Ginkgo biloba* |
| Salt | | | |
| Susceptible | | Tolerant | |
| Field maple | *Acer campestre* | Elms | *Ulmus* spp. |
| Norway maple | *Acer platanoides* | Holm oak | *Quercus ilex* |
| Sycamore | *Acer pseudoplatanus* | American plane | *Platanus occidentalis* |
| Horse chestnut | *Aesculus hippocastanum* | Alder | *Alnus glutinosa* |
| Grey alder | *Alnus incana* | False acacia/ black locust | *Robinia pseudoacacia* |
| Green alder | *Alnus viridis* | Golden willow | *Salix alba* var. *vitellina* |
| Beech | *Fagus sylvatica* | Tamarisk | *Tamarix* spp. |
| Limes | *Tilia* spp. | Stone pine | *Pinus pinea* |
| Hawthorns | *Crataegus* spp. | Maritime pine | *Pinus pinaster* |
| Norway spruce | *Picea abies* | Sitka spruce | *Picea sitchensis* |
| | | Ginkgo | *Ginkgo biloba* |
| Honey fungus | | | |
| Susceptible | | Tolerant | |
| Lilac | *Syringia vulgaris* | Oaks | *Quercus* spp. |
| Apples | *Malus* spp. | Box | *Buxus sempervirens* |
| Privet | *Ligustrum* spp. | Hawthorns | *Crataegus* spp. |

| Box 9.2 (cont) | | | |
|---|---|---|---|
| Honey fungus | | | |
| Susceptible | | Tolerant | |
| Willows | *Salix* spp. | Ivy | *Hedera helix* |
| Walnuts | *Juglans* spp. | Holly | *Ilex aquifolium* |
| Cedars | *Cedrus* spp. | False acacia/ black locust | *Robinia pseudoacacia* |
| Cypresses | *Cupressus* spp. | Oregon grapes | *Mahonia* spp. |
| Western red cedar | *Thuja plicata* | Tree of heaven | *Ailanthus altissima* |
| Monkey puzzle | *Araucaria araucana* | Cherry laurel | *Prunus laurocerasus* |
| Giant sequoia | *Sequoiadendron giganteum* | Blackthorn | *Prunus spinosa* |
| | | Tamarisk | *Tamarix* spp. |
| | | Yew | *Taxus baccata* |

These are not complete lists. For further information, refer to Philips, D.H. & Burdekin, D.A. (1982) *Diseases of Forest and Ornamental Trees.* Macmillan, London; Strouts, R.G. & Winter, T.G. (1994) *Diagnosis of Ill-health in Trees.* Forestry Commission, HMSO, London; or a good gardening manual.

(*Carpinus betulus*) and planes (*Platanus* spp.). Nevertheless, there is plenty of evidence that pollutants do affect trees:

- Acid rain kills roots and can adversely affect mycorrhizas through soil acidification, and it can strip large amounts of nutrients from the leaves as it falls through the canopy.

- Excess nitrogen from burning fossil fuels and too many cattle (which release huge quantities of ammonia) has led to soil acidification and potential forest decline around the world. Some extra nitrogen can be beneficial; in Finland, rowan (*Sorbus aucuparia*) has become more common in urban forests, most likely due to extra nitrogen. The problem comes with too much.

- Too much ozone disrupts photosynthesis and, with other pollutants, can sterilise pollen and so reduce seed production. Too little overall in the upper atmosphere will expose trees to unhealthy quantities of ultraviolet light.

- Heavy metal contamination can be directly lethal, and indirectly through changing soil acidity.

- Particles of soot on leaves can also be harmful by shading. A 2500-year-old fig tree in north India under which Buddha attained enlightenment is under threat from smoke from pilgrims' candles: soot is blackening the leaves and preventing photosynthesis.

In urban areas you can add a whole gamut of extra stresses:

- Deicing salt on roads.

- Tarmac over roots.

- Weed and grass competition.

- Untold quantities of dog urine (which provides excess potassium) and faeces.

- Soil compaction (Chapter 4).

- Vandalism.

Threats can come from quite unexpected sources. For example, there is some evidence that WiFi transmitters (using radio waves to move data over a computer network) may reduce the health of trees. A study at Wageningen University in 2010 looked at ash trees grown within 50 cm of transmitters producing radiation at around 2.4 GHz at 100 mW EIRP (Effective Isotropic Radiated Power), which is a common frequency used and at the maximum power allowed in most European countries. After a few months the ash leaves developed a metallic sheen due to the disappearance of the outer cell layer. This was followed by desiccation and death of parts of the leaves. Trees would not normally grow so close to transmitters and whether trees at more normal distances would suffer any effects is open to speculation. Nevertheless, a survey of trees in urban areas of the Netherlands showed that 70% of deciduous trees had similar symptoms of decline, compared to only 10% 5 years ago, and in rural wooded areas trees were unaffected.[3] It is possible that such transmitters are providing another weakening stress that contributes to a tree's demise.

Although we know at least partially the effects of individual pollutants and other stresses on the physiology and performance of trees, it is often difficult to pinpoint the precise cause behind a sick tree or forest. This is sometimes because problems share the same symptoms. For example, natural gas leaking from underground pipelines is not directly poisonous but displaces oxygen and produces similar symptoms to waterlogging or soil compaction. Moreover, oxidation of natural gas (methane) by bacteria produces water so gas-leak sites

---

[3] see http://www.physorg.com/news/2010–11-dutch-wi-fi-possibly-trees.html

are often wetter! To make life more complicated the lethal fungal-like disease phytophthora (a cause of sudden oak death) is closely associated with wet soil conditions. So what *appears* to be killing the tree may be masking a much more deadly cause. Trees weakened by one thing, such as disease or pollution, are more susceptible to another, such as drought, windthrow or another disease. The obvious cause of death then gets the blame letting the initial cause of decline off the hook. In a similar way, small stresses can add up to one serious problem. Forest decline which became 'an ecological crisis throughout the developed world' at the end of the last century (Klein & Perkins 1988) does not have a single cause but is due to an amalgam of a number of stresses. In the same way, the disease called 'sudden oak death' has been blamed primarily on a set of phytophthora fungi (particularly *Phytophthora ramorum*, *P. kernoviae* and *P. cinnamoni*) but general oak decline in Europe is undoubtedly contributed to by drought, air pollution, soil compaction, waterlogging and insect damage.

Is it any wonder that urban trees have a reputation for being short lived? Isolated trees, such as those in hedges, are more prone to damage because they filter out more pollution than woodland trees which protect each other. In New York, the average life span of trees is 7–40 years.

## Climate change

In terms of pollution, carbon dioxide is an interesting case. Although it is controversial, an increasing amount of carbon dioxide in the atmosphere does seem to be affecting the climate. As carbon dioxide has increased from 280 ppm[4] before the industrial revolution to the current 394 ppm (as of July 2012), global air temperature has risen by 0.74°C, and is predicted to increase by another 1.1–6.4°C by 2100 depending upon how we regulate carbon emissions in the future (atmosphere content could reach 650 ppm). Although these seem very small temperature changes, these are annual averages and they hide a lot of unpleasant changes.

Climate has never been static and there have been huge changes over the history of the earth (see Chapter 1). Some past changes are easy to see: the few trees of Saharan cypress (*Cupressus dupreziana*) remaining in dried up wadis of the southern Sahara are a remnant of a wetter climate 4000 years ago and only survive because the big trees are able to reach the water table. There are no seedlings so this is a population waiting to die. But plants are by no means unchanging and can react to a changing climate. For example, as described in Chapter 2, the number of stomata on leaves has varied as the carbon dioxide

---

[4] ppm is parts per million. 1 ppm is the same as 0.0001%, so 394 ppm is 0.0394%.

level in the atmosphere has changed in the past. And this is still happening. So what's the problem?

Climate change presents several conflicting issues for trees. The increase in carbon dioxide itself can be a benefit since it is used in photosynthesis and more of it will surely lead to more sugar being produced. In many trees this is true but not in all. In some, other things may limit sugar production such as light or water supply, and the extra carbon dioxide is of no value. Moreover, in those that do show increased growth with extra carbon dioxide, in many it looks like the increase is temporary and after a few years sugar production returns back to its previous level. On a more positive note, increased carbon dioxide can lead to a mature tree using up to 10% less water. The extra gas means that stomata can be partially closed and still let in enough carbon dioxide, with the consequence that less water escapes by evaporation.

Increased temperature can also be beneficial to plants growing in cold areas that are limited by a short growing season. In some trees, seed production is increased in warmer conditions; more seeds are produced and, unlike in many non-woody plants, the seed quality (in terms of size and viability) is not reduced. But for many trees, higher temperatures mean more water loss to keep cool and possible heat stress. Moreover, increased temperatures have a large number of other ramifications since they often seem to lead to detrimental changes in weather patterns; long-term drought, flooding, hail damage and an increase in severe storms and hurricanes.

Climate change thus has a number of potential short-term problems for trees. But the biggest problems come from trees being so long-lived. Over a life span of hundreds or thousands of years, the climate may now be changing so rapidly that they will not be able to cope. Already, forests around the world are dying, attributed to climate change (see Allen *et al.* 2010 for a global review). Trees can move, of course, by seed spreading so maybe if enough seeds move polewards, the species will survive. Loarie *et al.* (2009) calculated that bands of temperature are moving north (and south in the southern hemisphere) at an average rate of 0.42 km per year (0.26 miles per year). While this doesn't look much, even over the life of a 1000-year-old oak, this is 420 km (260 miles). To put this into context, it is calculated that trees may be able to move northwards by themselves at an average rate more like 100 m per year (0.1 km, 0.06 miles per year, or 320 ft) meaning that left to themselves they will be left far behind to face an uncertain future. A solution that is in our hands is to start planting trees from sources further south. If we're looking even just 100 years ahead, in southern England we should be planting acorns collected in northern France. This is easily done but is not without its own problems. Insects, other animals and plants in an oak woodland have their lives finely tuned to the oaks around them; they might fare less well if French oaks were planted on a UK site since subtle differences in such things as leaf chemistry, time of bud burst and acorn size might have important knock-on effects for the woodland inhabitants. Given the complexity of predicting what will happen and the even greater complexity in trying to solve the many problems, it would seem that by far the best solution is to control climate change. You may not be convinced that climate change is happening, but can we take the risk of widespread tree death?

## Cold

We saw in Chapter 3 that cold weather can cause frost cracks in trunks (Figure 3.13). Cold can also cause the frost heaving of young trees and death of small twigs. The latter is especially a problem with late spring frosts which catch the buds as they lose their cold hardiness. Some hardwood trees are more susceptible than others (see Box 9.2) and evergreens are not immune. The foliage of the 'Elegans' variety of Japanese red cedar (*Cryptomeria japonica* 'Elegans') and young western red cedars (*Thuja plicata*), amongst others, turns a most delicious bronze over winter, a sign of stress. More severe is the 'red belt' damage that occurs in conifers when the air is warm while the ground is still frozen: the leaves lose too much water and die, seen as a red belt across the landscape.

Despite such damage, many trees can survive low temperatures given the right 'hardening off'. A variety of mechanisms give different degrees of protection. Plants in fairly mild climates can survive purely by the insulation given by bark (like lagging on pipes). If living cells approach freezing, the first step is to prevent ice forming in cells because growing ice crystals will puncture the cells and kill them. The first line of defence is depression of the freezing point by accumulating sugars, organic acids or amino acids inside the cell, usually in response to short day length (e.g. white spruce, *Picea glauca*) or lowered temperatures (e.g. Nootka cypress, *Chamaecyparis nootkatensis* and western red cedar, *Thuja plicata*). This works in the same way that adding salt to roads prevents them freezing. Using this mechanism plants tolerate mild frosts down to −1 or −2°C. Indeed, some alpine conifers can carry on photosynthesising down to −35°C (by way of comparison, wood and root growth in temperate and northern trees is rare below 4–5°C). The second line of defence common in woody plants is provided by the cell contents supercooling without freezing. This gives protection down to −80°C but needs to be preceded by several days below 5°C (which explains why a cold spell in early winter can kill when the same temperature later does not, and why rapid fluctuations between freezing and thawing, as in alpine areas, can damage even the most hardy of trees). The third line of defence, needed in only the most severe climates by trees such as some birches, poplars and willows, is to take all the freezable water from the cells and allow it to freeze *between* the cells while tolerating the extreme desiccation caused inside the cells. This appears to be conditional on slow, continual cooling but if these conditions are met it allows dormant twigs of many northern woody plants, including conifers, to survive down to at least the temperature of liquid nitrogen (–196°C) without harm!

## Heat and fire

In temperate regions, temperatures much above 17–20°C slow down photo-synthesis and once they reach 50°C most living tissue is well on the way to dying. Under normal circumstances, a plant has a number of ways of staying cool with evaporation of water being foremost. Plants can thus cope with most environments, even deserts where water is in short supply. But how can plants cope with temperatures in a fire that are in the order of 800–1200°C whether we are talking about a candle flame or a raging forest fire?

The first point to appreciate is that fire is a naturally occurring event in many types of vegetation, including forests, around the world. A patch of prairie grassland may burn every 1–10 years while dry deciduous and conifer forests around the globe would naturally be burnt every 50–150 years on average. Moist broad-leaved forests tend to be less flammable but even seasonal tropical rainforest (where dry periods occur regularly) can burn every few centuries. Because fires have been burning around the world for over 400 million years (caused naturally mainly by lightning), we should expect that plants will not only be able to cope with high temperatures but even be able to take advantage of the situation. And that's exactly what you find.

Many trees, like aspen, have thin bark and are easily killed by even gentle fires. But the soil is a good insulator with lethal temperatures rarely found below 2–5 cm. The roots of aspen and many other woodland shrubs and herbs thus survive to resprout (Figure 9.15). However, this is a problem for grafted trees since sprouts will be from the root stock, and for conifers since they rarely sprout from the stump or roots (Chapter 3 lists the exceptions). Other trees have different strategies. Douglas fir (*Pseudotsuga menziesii*) and the giant sequoia (*Sequoiadendron giganteum*; Figure 6.2) of western N America, amongst other trees, have developed thick non-flammable bark to insulate the living tissue from the heat of the flames. In the Sierra Nevada Mountains of California, fires would naturally burn through the giant sequoia forest something in the order of every 25 years. Since the giant sequoias live for around 3000 years they will meet maybe 120 fires in their life. Because the fires are so frequent, comparatively little burnable material (dead needles, twigs, etc.) builds up, so

**Figure 9.15** New sprouts of mesquite shrubs (*Prosopis glandulosa*) from below ground after the stems above ground were killed two months ago by fire in Texas. From: Thomas, P.A. & McAlpine, R.S. (2010) *Fire in the Forest*. Cambridge University Press, Cambridge.

the fires are comparatively gentle. Add to this a bark which is thick (often 30 cm thick and up to 120 cm), full of insulating air, and virtually non-flammable, and you can see why the trees can survive so many fires. By contrast its thin-skinned competitors, such as white fir (*Abies concolor*), are more readily killed. Fire thus allows the giant sequoias to rule the forest. Heat-proof bark is also seen to a lesser extent in temperate deciduous forests where thin-barked American beech (*Fagus grandifolia*) and American plane (*Platanus occidentalis*) are more readily killed than the thicker-barked oaks and American chestnut (*Castanea dentata*).

The giant sequoia has another trick up its metaphorical sleeve. Along with hundreds of other species of tree around the world (see Chapter 8), the giant sequoia keeps its new cones tightly shut (called serotiny). In this way thousands of seeds are stored in the canopy for decades (pines can hold cones with viable seed for 20–50 years or more, *Banksia* species for 10–20 years, *Protea* species for just a few years). The seeds are most commonly released by fire. In serotinous conifers the cone scales are sealed shut with resin and few if any seeds are released until the cones are heated to 45–60°C (Figure 9.16a). As nature would have it, this sort of temperature in the canopy is likely to be reached only in those fires hot enough to kill the parents. Within hours or a

(b)

(a)

Figure 9.16 (a) The jack pine (*Pinus banksiana*) cone on the right was heated in an oven a few hours earlier to 60°C for 2 minutes and has subsequently fully opened while the unheated cone on the left stays resolutely closed. (b) A burnt Banksia cone; the two-lipped fruits buried in a mass of woody bracts are now opening, letting out the seed. Photographs (a) Canadian Rocky Mountains (From: Thomas, P.A. & McAlpine, R.S. (2010) *Fire in the Forest*. Cambridge University Press, Cambridge), (b) Perth, Australia.

few days of the fire, the scales open and a flurry of seeds drop to the ground, which by then is perfectly cool. In the southern hemisphere the seeds are equally well protected by woody fruits: the banksias have woody fruits (follicles) moulded together with woody bracts into the characteristic 'bottle' (Figure 9.16b), and the she-oak has a woody cone-like fruit. Characteristically in banksias the old flowers do not fall off and the heat produced by these catching fire (100–300°C for at least 2 minutes) is enough to induce opening. None of these serotinous fruits are remarkably heat resistant but they are capable of protecting the seeds from the brief high temperatures encountered. This is helped by dry, dormant seeds being very heat resistant. Seeds of jack pine (*Pinus banksiana*) have been exposed to 370°C for 5–20 seconds and even 430°C for 5–10 seconds without deleterious effects.

It just so happens that the seeds of serotinous species do best on a burnt soil (lack of competition, plenty of light, warmth and abundant nutrients in the ash). Thus over the next few years the forest becomes dominated by the next generation of these serotinous species. Like the phoenix, they emerge from the ashes of their own species. So good is the mechanism that many serotinous species have evolved to be highly flammable with peeling bark (as in eucalypts) or dead branches with flammable lichens (as in pines) to carry the fire into the canopy, aiding complete immolation of self and neighbours. The other side of the coin can be currently seen in Canada's forests where fire-prone areas are naturally dominated by fire-adapted jack pine (*Pinus banksiana*) and lodgepole pine (*P. contorta*). Where fires have been suppressed for many decades, the fire-created stands are old and frail and a new vigorous layer of fire-intolerant firs and hemlocks grow up underneath to take the pines' place.

A number of trees with small or hard seeds regenerate after fire from seeds stored in the soil (Chapter 8). The trick is to detect when the fire has burnt past. In some cases the fire itself is the trigger. Gorse seeds (*Ulex* spp.) in European heathlands have a hard impervious seed coat that is split by the high fire temperatures so allowing in water and thus germination. Others (e.g. heathers) detect not the fire but the fluctuating temperatures that result from the removal of the vegetation. A number of South African and Australian species are stimulated to germinate by chemicals in the smoke, and shrubs in the Californian chaparral are stimulated by the *removal* of inhibiting chemicals (put out by nasty neighbours) by the fire.

The clear advantage of seed storage in the canopy is the mass release of seed after a fire, which takes advantage of a good seedbed cleared of debris and competition, and satiates predators in the same way as masting (Chapter 5), producing a dense growth of the next generation. But why store seed in the canopy rather than the soil? Serotiny appears to put all seeds into one post-fire basket, and if the growing season following a fire is disastrous then everything is lost. The first part of the answer is that the seed supply is not as one-off as it sounds. Most serotinous species leak out a few seeds over the years in between fires, and once seeds are released by fire a small proportion stay in the cones to be released later and, of those that fall, some usually do not germinate till the following year or so, giving a small insurance policy against disastrous years. The second part of the answer is that fires can readily burn-off the top few centimetres of litter and organic debris which contain the bulk of the soil seed bank. Plus, as discussed in Chapter 8, large seeds stored in the soil are much more likely to be eaten than if stored in woody fruits in the canopy. So it is safer to store seeds in the canopy, providing a fire is likely to come along before the tree dies and the stored seed is lost.

Nevertheless, not all species in fire-prone forests are serotinous, and their abundance has a lot to do with how often a fire is expected in relation to the reproductive life span of a tree. If fires tend to be very frequent (every decade or less) or unpredictable, then trees that are tolerant of burning (by having thick bark) or are good at resprouting from the burnt stump are likely to dominate. When the normal interval between fires exceeds the life span of a tree, serotiny (canopy storage of seeds) will be of little use since once the tree dies the stored seed will not survive for long. In this case the next generation must come from seed stored in the soil or seed coming in from outside the area. In between these extremes lies the realm of serotinous species.

There are many other adaptations to fire which room does not permit to mention; see Thomas & McAlpine (2010) for a much fuller account. But I cannot pass without mentioning the longleaf pine (*Pinus palustris*) of southeast USA. It has a problem in that the young seedlings are readily killed by the frequent fires that sweep through its open grassy habitat. If they can survive to reach 2 m height, the bark is thick enough to resist fire and they are comparatively safe. To solve the problem, the newly germinated seedling goes through a 'grass-stage' where the shoot grows to several centimetres in a few weeks and then stops growing above ground for 3–7 years. During this time the bud is protected from fire by a dense tuft of needles, and all energy goes into the roots. When sufficient energy has been stored the shoot makes a mad dash upwards, growing rapidly to reduce the time spent as a vulnerable juvenile.

A final note on firewood. Ash (*Fraxinus* spp.) and holly (*Ilex aquifolium*) are renowned for making good firewood which will burn even when green. Is this related to their fire ecology? On the whole, no: these species rarely meet fire naturally. Rather it is the nature of their storing food as oils (Chapter 2). In the case of ash it is the large amounts of oleic acid that makes the wood so flammable (this is a fatty acid constituent of olive oil: ash is in the olive family, Oleaceae).

## Waterlogging and drought

The effect of waterlogging on roots is covered in Chapter 4, and drought effects on leaves in Chapter 2. Drought tends to increase the amount of roots in proportion to shoots by causing extra root growth. Trees at their maximum height are most prone to die during drought since they have narrow hydraulic safety margins (see Chapter 3). High temperatures and herbivory or disease can add to the effects of drought but generally, since the root system is so extensive, extreme drought years tend to have no more than short-term, reversible effects on trees. Much more damaging are long droughts such as seen in western Canada and large parts of Australia and the Amazonian rainforest.

## Wind

The storms of October 1987 in southern England (with 185 km per hour/ 115 mph gusts) broke or uprooted 19 million trees and killed 18 people.

Although unusual this storm was by no means unique and many other examples around the world could equally well have been quoted here. Whole woodlands of trees broken off or uprooted give a lasting impression of wind as a devastating agent that the tree can do little to defend against. Despite appearances, trees do have a variety of mechanisms for coping with wind.

Leaves can be torn apart and pulled off by high winds, especially when young and tender. But given their size and shape it is perhaps surprising that they remain undamaged at all. Big flat leaves cope mainly by being flexible: at lower speeds they flutter and as wind gets faster they tend to curl into cones which get tighter the stronger the wind. Vogel (2009) suggests that is best seen in leaves with a stalk (petiole) over 2 cm long and with lobes at the base that can curl over the leaf turning it into a streamlined cone (and basal lobes are often biggest in leaves in high light, which tends to be the windiest part of the tree). With smaller leaves, such as in birch and poplar, the whole branch tends to curl into a long cone, while pinnate leaves curl into cylinders.

Wind blowing against a whole tree acts like a hand pushing on the end of a lever acting to bend the tree (in the same way that a heavy branch or leaning tree bends under the pull of gravity, as discussed in Chapter 7). The taller the tree, the stronger the lever pushing against the tree (in the same way that a longer spanner more easily shifts a tight bolt). In the long term, as we have already seen, strong winds lead to trees staying shorter to reduce this lever. In the short term, such as in a particularly strong wind storm, trees can reduce the length of the lever by being flexible. When the wind blows, the branches bend and, in effect, reduce the height of the tree by 'putting its ears back'. As wind speed increases, smaller branches are snapped as a sacrifice to save the trunk by further reducing the sail area and height of the tree. The shorter the apparent height, the less force exerted by the wind at the base.

Leonardo da Vinci observed that where a tree splits into branches the branches above have the same cross-sectional area as the trunk below. While this may be true around one junction, it doesn't apply to the whole trunk or we would have parallel-sided trees, and most tree trunks are fatter at the bottom than the top. In reality, the shape of the tree is carefully adjusted to reduce weak areas where the tree might snap. As a tree is bent by the wind it will snap if one area has more stress on it than others. But the cambium can detect stressed areas and lays down more wood to strengthen them. Conversely, areas experiencing less stress will have less wood expended on them. This is Claus Mattheck's axiom of uniform stress: the safest, optimal structure has a uniform stress over the whole of its surface as it bends.

Branch junctions (as shown in Chapter 3) are also cleverly designed. Duncan Slater of Myerscough College has shown that there are often greater stresses in the junction between a branch and its stem than elsewhere on the tree. The bulge of the branch collar helps to spread this extra stress and the wood around the branch is also very dense, holding it firm. Even more well designed, the wood at the centre of the branch and at its very base is fairly low density which allows it to deform and flex in the wind, like a shock absorber. But this low density area is also a place of weakness so if the wind gets too strong the branch will break, sacrificing the branch but protecting the more valuable trunk from damage.

As the tree flexes in the wind, most stress is felt in the outer skin of the tree. In the same way that a rock in a river forces the water aside and increases the water speed at its sides, so a defect in a tree's surface will divert the stress lines around itself such that the stress is greatest at the sides of the defect. Any cracking of the trunk will start at the sides of a hole, and this is where 'repair' of damage is concentrated. The tree will detect the stress and concentrate new callus growth at the sides of the hole (see Figure 9.11b). In the same way, when a heavy object is leant against a tree, or a tree getting fatter meets an immovable object like a railing, the tree detects the stress and adds more wood at this point to spread the load over as large an area as possible to reduce the stress on any one point (Figure 9.17). This is analogous to leaning against the corner of a table and putting a cushion behind you to spread the painful load more evenly.

## Trees still fail in windy conditions: why?

Wood is expensive stuff to produce. A tree that puts wood where it is not needed will be wasting resources and will likely suffer in the competitive struggle for life. So the design of trees, like most things, is a compromise between risk and safety, as described in Box 6.1. Trees are not so overdesigned to protect against any eventuality. A natural consequence of this is that trees will sometimes break when under extreme stress such as hurricanes. So, if a tree falls over in high winds and flattens your car you can't necessarily blame or sue someone, you have to accept that this is how trees are designed! And is this so unacceptable? High winds blow down chimneys and cause roofs to shed tiles, yet we don't banish people's roofs and chimneys in the same way we do trees, or assume someone is at fault if this does happen.

## Hollow trees and snapping

Whatever the strategy of safety used by a tree, it is intuitively right that rot eating away the centre of a tree will make it more prone to being snapped in the wind. Or is it? Claus Mattheck looked at 1200 trees of a variety of species, some that had snapped, others that hadn't, and measured the amount of sound wood left around the edge in proportion to the width of the tree. He

(a)

Figure 9.17 When a tree grows fatter and meets an immovable object like (a) a railing or (b) a concrete wall, the tree will detect the stress at that point and grow extra wood to spread the load over as large an area as possible to reduce the risk of snapping at that point. Both in Blanes, Spain.

found that regardless of species or size, trees rarely snapped until more than 70% of the diameter was rotted away (i.e. the solid wood remaining was less than 0.3 of the radius). Thus a tree 50 cm in diameter (i.e. 25 cm radius), should have solid wood under the bark at least 7.5 cm thick (0.3 x 25 cm) to ensure safety, and a tree 100 cm in diameter should have solid wood 15 cm thick. That's not to say that trees with thinner walls than this can't be found standing; these generally have a small canopy and so are less affected by wind or are in very sheltered areas. This may seem like a remarkably small amount of sound wood to hold up a large tree, but if you think about a tree bending in the wind the greatest compression and stretching forces will be felt nearest the bark while the centre of the tree is comparatively stress-free. So it is this outer layer of sound wood that is most important for holding up the tree.

(b)

Figure 9.17 (cont)

These results clearly show that much of the wood in a tree trunk is redundant when it comes to holding up the tree. Indeed, hollownesss can be an advantage. After severe gales (such as those in southern England in 1987) many hollow trees are left standing while their solid neighbours fall. This may be partially explained by old hollow trees having a smaller canopy – a smaller sail area – but it seems that hollow trees are more flexible and better able to withstand the buffeting and swaying.

Hollowness is also beneficial in another way. It is well known amongst tropical trees but also temperate trees such as yews (*Taxus* spp.) and elms (*Ulmus* spp.) that roots can be produced from the sound wood (adventitious roots) into the rotting mass in the centre. This allows the tree to suck up the nutrients being released by the rotting and, in effect, recycle itself. What's more, any droppings or dead remains left by animals living in the hollow add to the soup, all of which gives the tree extra supplies in its battle for dominance. An elm I saw felled had several dozen pencil-thickness roots growing down the centre with fine feeding roots firmly embedded in the brown rot

**Figure 9.18** A felled Huntingdon elm (*Ulmus* x *hollandica* 'Vegeta') showing the pencil-like roots at the back growing down the hollow centre. The fine feeding roots would have been absorbing nutrients from the decomposing wood. An example of self-recycling!

around the inside of the hollow (Figure 9.18). The mystery in this is why more species don't have internal roots.

So when you see a hollow tree don't feel sorry; hollowness appears to be a deliberate part of its survival strategy, making it more flexible and better able to resist high winds, and allowing it to recycle itself.

## Windthrow: uprooting trees

As explained in Chapter 3, in most trees it is the big framework roots (which spread roughly the width of the canopy) that hold the tree up. These roots create a 'root plate' of soil welded together by roots. Sometimes the sinkers that grow down from the plate produce side roots that weld the soil into a deeper root plate, often called a root cage. Still other trees (e.g. birch, larch, lime and Norway maple, *Acer platanoides*) have a deep tap root or descending lateral branches to produce a 'heart-root' system akin to a root ball. See Figure 4.3 as a reminder of the differences.

How does a tree fall over? This depends upon whether the tree has a more rounded root ball (heart-root system) or a flat root plate (or root cage). In a tree with heart roots with or without a long tap root, as the tree sways in the wind the soil begins to crack and horizontal roots are pulled out or snapped on the windward side (Figure 9.19a & b). As these 'guy roots' are removed, the root ball begins to rotate in its 'socket' of soil and the tree starts to lean tearing out more roots as it goes (Figure 9.19c). Normally just 2.5–4.0 degrees of lean on the stem is enough to cause the failure of most of the roots. The tree forms less of a target for the wind but its weight, now being way off-centre, helps to bring the tree crashing down (Figures 9.19d and 4.3b). It is the friction between the root ball and the surrounding soil that really holds the tree up. Thus wet soils (where the water reduces friction) are likely to result in more windthrow, and the tree uprooting can be considered a failure of the soil rather than the tree.

A young tree with just a long tap root and few lateral roots, tends to stay in the ground in the same way as a fence post. As the wind blows and the whole tree bends, it is the stiffness of the root and the resistance of the soil to sideways movement that holds it upright.

Trees with very shallow roots, such as spruce and beech, fall in a slightly different way. The initial stages are the same as above with the soil cracking and windward roots being pulled out or broken (Figure 9.20a). But then the whole root plate hinges along the leeward side and falls, just like a hat stand or a wine glass toppling over (Figures 9.20b and 4.3a). The large lateral roots are strengthened by being figure-of-eight shaped in cross-section (resembling a steel I-beam) since it is the top and bottom of the root that are most compressed and pulled when the tree rocks. Nevertheless, the tree normally falls when there is a hinge failure; the lateral roots can't stand the bending force and break. This is obviously more likely in roots weakened by fungal rot and is like trying to keep a wine glass upright when most of the base has been broken away. What holds these trees up is firstly the weight of the tree and the soil carried on the root plate which must be lifted around the edge of the hinge, secondly the sinker roots acting like tent pegs into the soil, and thirdly the length of the roots (longer roots make the lever from the hinge to the centre of gravity longer and so harder work). Friction has little to do with the process. Thus the shallow roots that are inevitable on wet soils are the best type for holding the trees up. Nevertheless, the environmental dice are loaded against the tree; shallow-rooted trees are still less stable than deeper-rooted neighbours. Windthrow is common amongst conifer plantations in upland Britain

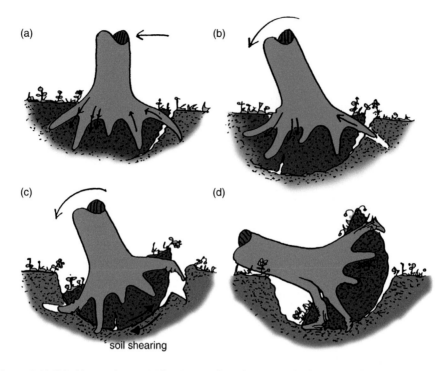

Figure 9.19 Windthrow of a tree I. The pictures show four stages in the process for a tree with comparatively deep heart roots (Figure 4.3) where the tree falls by the root ball rotating in an earthen socket. After Coutts from Mattheck, C. & Breloer, H. (1994) *The Body Language of Trees: a Handbook for Failure Analysis. Research for Amenity Trees, No 4.* HMSO, London. Reproduced under the Open Government Licence.

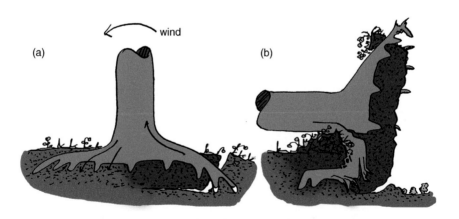

Figure 9.20 Windthrow of a tree II. A tree with shallow roots falls in the same way that a coat stand or wine glass topples over. The weight of the tree and root plate must be lifted as it pivots over the edge of the root plate. From: Mattheck, C. & Breloer, H. (1994) *The Body Language of Trees: a Handbook for Failure Analysis. Research for Amenity Trees, No 4.* HMSO, London. Reproduced under the Open Government Licence.

for this reason. A consequence of using a root plate is that as a tree gets bigger and heavier (pushing more into the soil) the root plate is proportionately smaller: small trees may have a root plate more than 15 times the stem radius while in larger trees it needs to be no more than three times the stem radius.

Trees are not passive in the face of wind. They usually grow more roots on the side towards prevailing winds and to a lesser extent on the lee side, with fewest at right angles to the wind. This is why strong wind from an unusual direction can more easily uproot trees. Winds are usually highest in winter when the trees are mostly dormant but there is evidence from Scots pine (*Pinus sylvestris*) that trees can retain information about mechanical forces acting on their stems during the winter and respond the following growing season. As discussed in the sections above, trees do not squander limited resources; they do not grow roots that are not needed. Thus, trees in the middle of groups often have less well developed roots compared to the individuals around the edges. This represents good economy until part of the stand is felled or blown over and the remainder is exposed to high winds before extra roots can be grown.

## Does wind kill trees?

It is recognised that a large proportion of the 19 million trees which snapped and blew over in the 1987 storms in southeast England were not as dead as they first looked. A tree left leaning after high winds is by no means finished. The roots regrow and the tree straightens itself by growing 'reaction wood' (Chapter 3). Leaning trees can also be winched back to the vertical and if held there for a few years will regrow roots to hold themselves upright. Too many leaning trees are given up as lost and cleared away by tidy-minded people. Even a completely prostrate tree with some roots intact (more likely in a shallow-rooted tree; Figure 9.20) can produce suckers from the surviving roots or create new trunks out of branches now pointing skywards (and so creating strange lines of trees to fox the unwary naturalist once the original stem is buried in leaf litter!). Similarly, the stump of a snapped tree can usually grow new shoots from stored (epicormic) buds. The exception, where winds *can* kill trees, is found in old weak trees, as described at the end of this chapter.

## Damage from living things: virulent diseases

Sudden death can come with diseases which are able to rapidly and catastrophically overcome the defences of the infected tree. Honey fungus is probably the most notorious (see Box 9.2). A number of other diseases are specific to certain types of tree and have caused consternation around the world. Chestnut blight in N America (caused by a fungus *Cryphonectria parasitica* and which killed over three *billion* American chestnuts, *Castanea dentata*) and Dutch elm disease around the world come readily to mind although a number of others could be listed. A number are waiting in the wings including the many phytophthora species (causing sudden oak death and which look set to devastate a number of other species) and the more recent ash dieback in Europe caused by the fungus *Chalara fraxinea* (also called *Hymenoscyphus pseudoalbidus*) which has been marching across Europe for more than a decade and has just reached the UK and looks set to eradicate large numbers of ash (*Fraxinus excelsior*). However, there is emerging evidence that some ash are genetically resistant to the disease and should survive, and can be used in a breeding programme to produce very resistant ash.

## Dutch elm disease

Dutch elm disease is worth looking at as an exemplary virulent disease. The fungus behind the disease, first investigated by the Dutch, hence the name, has caused extensive losses of elms in Europe, N America and western Asia. It was first reported in England in 1927, probably brought into the country on diseased logs from N America. The first epidemic in Europe in the 1920s to 1940s was fairly mild but sometime after this the original strain of fungus, *Ophiostoma ulmi* (previously called *Ceratocystis ulmi*) changed into a much more aggressive strain (*O. novo-ulmi*) which swept through Britain in the late 1960s killing 25 million of the UK's estimated 30 million elms. Many trees, particularly the wych elm (*Ulmus glabra*), survived to produce suckers from the roots but these have been hit by repeated cycles of the disease as they become large enough to be of interest to the bark beetles responsible for spreading the fungus.

The fungus is moved from tree to tree primarily by the large elm bark beetle (*Scolytus scolytus*). Young beetles leaving an infected tree (from May onwards) carry with them sticky spores produced by the fungus inside the beetles' tunnels. Once out they fly up to 5 km searching the wind for the specific chemicals released by elm trees, particularly weakened or diseased trees. Once a female beetle finds a suitable site she releases an odour (called an 'aggregation pheromone') to attract other searching females. As these beetles mate, the smell they release changes and puts off other beetles which fly on past, preventing too many competing for space on one tree. The happy females burrow into the bark (preferring the crotch of largish branches usually around 4 years old). As they eat the sugary inner bark the sapwood is scored and the spores carried by the beetle are introduced.

The fungus is classified as a 'vascular wilt fungus', meaning that it blocks the water-carrying tubes (the vascular tissue) causing wilting of the foliage above. As the fungal strands digest their way into the wood, air is sucked in (especially if the water is under great tension) stopping the tubes from working (see Chapter 3). Any left working will rapidly be blocked by fungal strands and spores being swept along and piling up against the perforation plates just like a drain in the road being bunged by debris caught in the grill. Lethal toxins are also produced by the fungus. The tree responds to these intrusions by blocking off the tubes above and below the disease with tyloses (Figure 9.13). Because elm has ring-porous wood (Chapter 3 may need to be consulted again) it moves most of its water only in the youngest ring just under the bark which makes the tree particularly vulnerable to Dutch elm disease. Trees show typical symptoms of yellowing, wilting leaves by June and can be dead within the year, maybe 2–3 years for older trees.

Whether a tree becomes infected once visited by beetles depends upon a numbers game and competition. Around 1000 fungal spores are needed for a successful invasion (just as a big army has more success in storming a fort than a lone soldier). Beetles lose spores as they travel and it is suggested that as many as 10 000 spores may be needed on a beetle leaving the bark to secure a new infection. Studies have shown that 60–90% of beetles leave the bark contaminated but only 10–50% arrive at a new tree still with enough spores, and only 3–5% of all feeding tunnels lead to infection. From this it should come as no surprise that in the UK the large elm bark beetle is better at transferring the disease than its smaller cousin, the small elm bark beetle (*Scolytus multistriatus*).

The fungus has its own natural virus-like diseases (called 'd-factors') which act to increase the number of spores needed to successfully invade a tree from around 1000 to around 50 000. This may be beyond the spore load deliverable by normal beetle densities and may explain the sudden and unexpected decline of the 1930s epidemic. Wych elm is less susceptible than English elm (*Ulmus procera*) to the disease, possibly because, although it is more susceptible to the fungus, it is much less favoured by the elm bark beetle for feeding, and a fungus *Phomopsis oblonga* (a rapid and common invader of the bark of newly dying wych elm, especially in the north and west of Britain) successfully competes with the beetle.

Various methods of controlling the disease have been tried:

- 'sanitation felling' and destruction of the wood and bark from infected trees;
- trenching to break root grafts between healthy and infected trees;
- pheromone traps and insecticides (mimicking the attractive pheromones produced by females to lure male beetles to sticky traps);

Figure 9.21 A scribbly gum (*Eucalyptus haemastoma*) named for the 'scribbles' on its bark. The scribbles are tunnels made by the caterpillar of the scribbly gum moth (*Ogmograptis scribula*). Its eggs are laid between layers of bark and the caterpillar burrows through the new bark and, as the old bark falls off, the tracks are exposed. There are 20 species of eucalypt that are prone to the attentions of some 6 species of moth, so scribbly gums are common in Australia. The result is primarily aesthetic (positively so) rather than damaging to the tree. Wallumatta Nature Reserve, Sydney, Australia.

- fungicides injected into trees as a curative or preventative treatment; and
- not pruning trees from early spring to late autumn since the beetles are attracted to the odour of fresh cuts.

The cost of these can be enormous with no guarantee of success. Fortunately, a large number of hybrid elms are now on the market using Asiatic elms as one of the parents due to their more consistent resistance (including the Siberian elm *Ulmus pumila*, Himalayan elm *U. wallichiana* and Japanese elm *U. japonica*). The new hybrids include *Ulmus* 'Nanguen' and *Ulmus* 'New Horizon' from Hillier Nurseries in the UK. But these are resistant not immune and may still present problems.

## Insects

Trees are host to many insects. Some are of aesthetic interest (Figure 9.21) and in most cases they do no serious harm. Even complete defoliation can be borne. In the UK, oaks regularly lose their entire set of leaves in May/June to the caterpillars of the oak leaf roller moth (*Tortrix viridana*) and the winter moth (*Operophtera brumata*) and, as discussed in Chapter 2, the oak will grow a new set of lammas leaves. Although the trees survive, such attacks do have an effect. So, for example, in the eucalypt Blakely's red gum (*Eucalyptus blakelyi*) found in southeast Australia, normal defoliation by insects over a season leads to the speed of water movement through the trunk being halved (4.4 as opposed to 9 cm per hour). Although this is likely to be no problem in itself, the cumulative effect of many such disturbances can weaken the tree and make it more prone to other problems.

There are many examples around the world where new introductions of insects can tip the balance from nuisance to death. In the UK, two moths from Europe, the gypsy moth (*Lymantria dispar*), which arrived in 1995, and the oak processionary moth (*Thaumetopoea processionea*), here in 2006, are adding extra pressure on oak trees since their caterpillars hatch a month or so later than the native moths described above and can catch the newly growing set of lammas leaves. As such they are major contributors to the recent oak decline.

Even native insects can be problematic if conditions change. Probably the worst example of this is the mountain pine beetle, *Dendroctonus ponderosae*, of western N America. This has always played a vital role in pine forests in helping to kill weak trees and so encouraging new trees to grow. However, a series of hot, dry summers and mild winters has led to population explosions of the beetle and wholesale death of pine forests through western Canada and down into Wyoming and Colorado. The prognosis is not good and 90% of western N American pine forests may succumb. Similar to Dutch elm disease, the beetle burrows into the bark of pines carrying with it a blue stain fungus *Grosmannia clavigera*. As noted above, this stains the wood without rotting it, but it does act to block up the water-conducting tubes which can obviously contribute to tree death.

## Starvation and old 'age'

Unlike animals, age as such is not a real problem for plants. Being modular, plants can grow new limbs when old ones die off: we would be more like plants if we could grow a new leg when one gets too old and arthritic to be useful! Moreover, in a tree which grows a new living skin every year, the oldest living bits are always young, rarely more than three decades old with maybe 50 years as the upper limit, and may form just 10% of the mass of the whole tree. The living parts of trees are eternally youthful.

What is more crucial to survival is the *size* of a tree. As a tree grows it gets to a point where the canopy reaches a maximum size. The tree cannot get taller, as described in Chapter 6, and the side branches cannot grow longer because they become too expensive to physically support. So the number of leaves a tree can hold becomes more or less fixed, which means that the tree's food production also becomes fixed. But each year the tree needs to add a new layer of wood under the bark, and as the tree gets bigger the amount of wood needed to coat the whole tree goes up each year, just like putting together a set of Russian dolls where each new doll on the outside has to be bigger. Moreover, as the tree gets bigger the amount of food needed for running the tree (respiration) increases, rising to two thirds of the sugary income in a mature tree. The tree then becomes like a bank balance where the income (food) is fixed but the outgoings (respiration and new wood) keep rising. The tree compensates for a time by producing narrower and narrower rings but there comes a point where they cannot get any narrower. At this point the tree is becoming 'senescent', defined as the irreversible accumulation of damage with age that leads to decline and eventually death. The small rings lead to water stress in the canopy, leading in turn to slower growth and less hormone production, fewer fruit, and poorer quality pollen and seeds.

Something has to give, which usually means the loss of the topmost branches which are under most water stress. The result is a stag-headed tree, so named for the antler-like dead branches sticking out of the top of the canopy (Figure 9.22). This is the start of the end because losing branches means fewer leaves and so less new wood, and the beginning of a downward spiral. But many trees can slow the process. A tree with epicormic buds can grow new branches from the trunk and the big branches that can hold enough leaves to go a long way towards making up for those lost higher up. And since these are borne on thin branches which do not need so much wood, the tree's outgoings are reduced. In effect they have kept the leaf area while cutting out the expensive-to-maintain upper trunk and its big branches. The resulting tree can look characteristically old with a less wide canopy and a general stubbiness to the trunk (look back at Figure 6.5).

These new branches (epicormic shoots) tend to be fairly short-lived (100 years in oaks, 60 years in hornbeam and beech and still less in birch and willow). Nevertheless, trees that have a plentiful supply of epicormic buds such as oaks and sweet chestnut (especially those with big burs) can keep producing new shoots and stave off death for centuries. As the old saying goes: oak takes 300 years to grow, 300 years it stays, 300 years it takes to decline. Perhaps we should think of a stag-headed oak as merely entering middle age and, like many humans, is just going a little bald on top! Others, such as ashes and beech, are not so good at this 'retrenchment' and decline rapidly to die relatively young. Although not many conifers can resprout from the base (see Chapter 3), a good many can produce epicormic shoots in the canopy, including various larches, giant sequoia, coastal redwood, Sitka spruce (*Picea*

Figure 9.22 An old Turkey oak (*Quercus cerris*) becoming stag-headed. Killerton Park, England.

*sitchensis*) and western hemlock (*Tsuga heterophylla*). It's not uncommon in old conifers to see the bottom half of the canopy made up of big dead branches with bright green epicormic shoots hugging the trunk. To put some numbers on this, Hiroaki Ishii and colleagues found that in 450-year-old Douglas firs (*Pseudotsuga menziesii*) in Washington State up to 83% of their branches are dead, and up to half of the live branches are epicormic, typically 3–15 cm in diameter and 2–7 m long.

If size is more important than age, then the life of a tree can be extended by keeping it small; in this way a tree can avoid entering senescence. Certainly, Lanner & Connor (2001) found none of the symptoms of senility in small bristlecone pines (*Pinus longaeva*) that were up to 4713 years old. Trees can be kept small by growing slowly: Schulman (1954) put it that 'adversity begets longevity'. Indeed, there is evidence that in trees there is only so much living to do, so you either do it quickly and not for long, or slower and for longer. The bristlecone pines grow on poor soil in a dry, cold environment (less than 30 cm of annual precipitation, most of which falls as snow) with a short growing season measured in weeks. One bristlecone at 3400 m (11 300 ft) in the American Southwest, just 1 m tall, 7 cm diameter, was recorded as 700 years old.

The other way of keeping the tree small and alive longer is, paradoxically, to keep cutting it down. This reduces the amount of wood needed each year and, in effect, rejuvenates the tree (but of course this will only work in trees capable of regrowing when cut). Thus in Britain the ash (*Fraxinus excelsior*) normally lives for 250 years and yet in Bradfield Woods, Suffolk there is a coppiced ash with a stump 5.6 m in diameter and at least 1000 years old.

The state of the bank balance is also influenced by savings in the form of food reserves. In Chapter 3 we saw that food is stored in the living cells of the sapwood. As a tree gets bigger and food production goes into the red, it has less spare food to store. At the same time, since less new wood is grown, the larder for storing food gets smaller. Moreover, as rot and infections accumulate in the structure, more food-storage capacity is lost behind the 'barrier zone' laid down by the cambium to seal off infected wood (see *Compartmentalisation of decay* above): in effect the living part of the tree is walled into a thinner and thinner layer under the bark. Respiration needs are somewhat reduced but it is not enough. As spare energy production and storage wanes, the tree becomes weaker. It is less able to grow barriers between central rot and the new wood and this, combined with the narrowness of the new wood, enables fungal rot to easily reach the bark. That part of the tree then dies. New epicormic branches can still save its life but large old trees are less good at producing new shoots perhaps because they are running out of stored epicormic buds or the buds are trapped below thick bark. Moreover, it is not uncommon for new branches on weak trees to die sometime later, usually just when people think the tree is going to live. This may be because the barrier zone is missing or very weak or because there are too few reserves left to grow a new strip of wood from the new branch down to the roots. Either way, the disease can easily take over the tree between the new branch and the roots, and the branch consequently withers away. Or the trunk becomes so weak that it literally collapses where it stands without resprouting. At this point the tired old tree bows out gracefully.

A consequence of the above is that a tree has a more variable life span than most animals. That is why if you look at survivorship for most animals, the death rate is higher in older age groups than in the young; as we get older, and nearer and beyond our three score years and ten, it is more likely we will die. For trees, however, it's a different shaped curve: there is a very high death rate when very young (amongst the vulnerable young seedlings and saplings) but once established, they have a good chance of living for a very long time, especially if they are kept small.

## 🍃 *Further Reading*

Allen, C.D., Macalady, A.K., Chenchouni, H., *et al.* (2010) A global overview of drought and heat-induced tree mortality reveals emerging climate change risks for forests. *Forest Ecology and Management*, 259, 660–684.

Arnold, A.E., Mejía, L.C., Kyllo, D., *et al.* (2003) Fungal endophytes limit pathogen damage in a tropical tree. *Proceedings of the National Academy of Sciences*, 100, 15649–15654.

Barnes, B.V. (1966) The clonal growth habit of American aspens. *Ecology*, 47, 439–447.

Bergstrom, C.T. & Lachmann, M. (2003) The Red King effect: when the slowest runner wins the coevolutionary race. *Proceedings of the National Academy of Sciences*, 100, 593–598.

Bigler, C., Gavin, D.G., Gunning, C. & Veblen, T.T. (2007) Drought induces lagged tree mortality in a subalpine forest in the Rocky Mountains. *Oikos*, 116, 1983–1994.

Chambers, Q.J., Higuchi, N. & Schimel, J.P. (1998) Ancient trees in Amazonia, *Nature*, 39, 135–136.

Coutts, M.P. & Grace, J. (1995) *Wind and Trees*. Cambridge University Press, Cambridge.

Cunningham, S.A., Pullen, K.R. & Colloff, M.J. (2009) Whole-tree sap flow is substantially diminished by leaf herbivory. *Oecologia*, 158, 633–640.

Damore, J.A. & Gore, J. (2011) A slowly evolving host moves first in symbiotic interactions. *Evolution*, 65, 2391–2398.

Dicke, M., Agrawal, A.A. & Bruin, J. (2003) Plants talk, but are they deaf? *Trends in Plant Science*, 8, 403–405.

Dujesiefken, D. & Liese, W. (2011) The CODIT principle: new results about wound reactions of trees. *Arborist News*, April 2011, 28–30.

Eloy, C. (2011) Leonardo's rule, self-similarity and wind-induced stresses in trees. *Physical Review Letters*, 107, article 258101.

Eyles, A., Bonello, P., Ganley, R. & Mohammed, C. (2010) Induced resistance to pests and pathogens in trees. *New Phytologist*, 185, 893–908.

Farjon, A. (2001) *World Checklist and Bibliography of Conifers* (2nd edn). Kew Publishing, Richmond.

Gasson, P.E. & Cutler, D.F. (1990) Tree root plate morphology. *Arboricultural Journal*, 14, 193–264.

George, M.F., Hong, S.G. & Burke, M.J. (1977) Cold hardiness and deep supercooling of hardwoods: its occurrence in provenance collections of red oak, yellow birch, black walnut and black cherry. *Ecology*, 58, 674–680.

Grace, S.L. & Platt, W.J. (1995). Effects of adult tree density and fire on the demography of pregrass stage juvenile longleaf pine (*Pinus palustris* Mill.). *Journal of Ecology*, 83, 75–86.

Groenendijk, P., Eshete, A., Sterck, F.J., Zuidema, P.A. & Bongers, F. (2012) Limitations to sustainable frankincense production: blocked regeneration, high adult mortality and declining populations. *Journal of Applied Ecology*, 49, 164–173.

Grostal, P. & O'Dowd, D.J. (1994) Plants, mites and mutualism: leaf domatia and the abundance and reproduction of mites on *Viburnum tinus* (Caprifoliaceae). *Oecologia*, 97, 308–315.

Grubb, P.J. (1992) Presidential address: a positive distrust in simplicity – lessons from plant defences and from competition among plants and among animals. *Journal of Ecology*, 80, 585–610.

Hamberg, L., Malmivaara-Lämsä, M., Lehvävirta, S. & Kotze, D.J. (2009) The effects of soil fertility on the abundance of rowan (*Sorbus aucuparia* L.) in urban forests. *Plant Ecology*, 204, 21–32.

Herre, E.A., Mejía, L.C., Kyllo, D.A., *et al.* (2007) Ecological implications of anti-pathogen effects of tropical fungal endophytes and mycorrhizae. *Ecology*, 88, 550–558.

Ishii, H. & Kadotani, T. (2006) Biomass and dynamics of attached dead branches in the canopy of 450-year-old Douglas-fir trees. *Canadian Journal of Forest Research*, 36, 378–389.

Issartel, J. & Coiffard, C. (2011) Extreme longevity in trees: live slow, die old? *Oecologia*, 165, 1–5.

Janzen, D.H. (1974) Epiphytic myrmecophytes in Sarawak: mutalism through the feeding of plants by ants. *Biotropica*, 6, 237–252.

Jungnikl, K., Goebbels, J., Burgert, I. & Fratzl, P. (2009) The role of material properties for the mechanical adaptation at branch junctions. *Trees*, 23, 605–610.

Klein, R.M. & Perkins, T.D. (1988) Primary and secondary causes and consequences of contemporary forest decline. *Botanical Review*, 54, 1–43.

Lanner, R.M. (2002) Why do trees live so long? *Ageing Research News*, 1, 653–671.

Lanner, R.M. (2007) *The Bristlecone Book*. Mountain Press, Missoula, Montana.

Lanner, R.M. & Connor, K.F. (2001) Does bristlecone pine senesce? *Experimental Gerontology*, 36, 675–685.

Loarie, S.R., Duffy, P.B., Hamilton, H., *et al.* (2009) The velocity of climate change. *Nature*, 462, 1052–1055.

Martínez-Ramos, M. & Alvarez-Buylla, E.R. (1998) How old are tropical rain forest trees? *Trends in Plant Science*, 3, 400–405.

Meier, E.S., Lischke, H., Schmatz, D.R. & Zimmermann, N.E. (2011) Climate, competition and connectivity affect future migration and ranges of European trees. *Global Ecology and Biogeography*, 21, 164–178.

Mencucci, M., Martínez-Vilalta, J., Vanderklein, D., *et al.* (2005) Size-mediated ageing reduces vigour in trees. *Ecology Letters*, 8, 1183–1190.

Ninemets, Ü. & Valladares, F. (2006) Tolerance to shade, drought, and waterlogging of temperate Northern Hemisphere trees and shrubs. *Ecological Monographs*, 76, 521–547.

Petit, R.J. & Hampe, A. (2006) Some evolutionary consequences of being a tree. *Annual Review of Ecology, Evolution and Systematics*, 37, 187–214.

Pineda, A., Zheng, S.-J., van Loon, J.J.A., Pieterse, C.M.J. & Dicke, M. (2010) Helping plants to deal with insects: the role of beneficial soil-borne microbes. *Trends in Plant Science*, 15, 507–514.

Radley, J. (1961) Holly as winter feed. *The Agricultural History Review*, 9, 89–92.

Rank, N.E. (1994) Host-plant effects on larval survival of a salicin-using leaf beetle *Chrysomela aeneicollis* Schaeffer (Coleoptera: Chrysomelidae). *Oecologia*, 97, 342–353.

Richards, P.W. (1996) *The Tropical Rain Forest*. Cambridge University Press, Cambridge.

Rico-Gray, V. & Oliveira, P.S. (2007) *The Ecology and Evolution of Ant–Plant Interactions*. University of Chicago Press, Chicago.

Schulman, E. (1954) Longevity under adversity in conifers. *Science*, 119, 396–399.

Schwarze, F.W.M.R., Engels, J. & Mattheck, C. (2000) *Fungal Strategies of Wood Decay in Trees*. Springer, Berlin.

Shigo, A.L. (1984) Compartmentalization: a conceptual framework for understanding how trees grow and defend themselves. *Annual Review of Phytopathology*, 22, 189–214.

Smirnoff, N. (1996) The function and metabolism of abscorbic acid in plants. *Annals of Botany*, 78, 661–669.

Stener, L.-G. (2012) Clonal differences in susceptibility to the dieback of *Fraxinus excelsior* in southern Sweden. *Scandinavian Journal of Forest Research*, 1–12, iFirst article.

Stephenson, N.L. & Demetry, A. (1995) Estimating ages of giant sequoias. *Canadian Journal of Forest Research*, 25, 223–233.

Strimbeck, G.R., Kjellsen, T.D., Schaberg, P.G. & Murakami, P.F. (2007) Cold in the common garden: comparative low-temperature tolerance of boreal and temperate conifer foliage. *Trees*, 21, 557–567.

Thomas, P.A. & McAlpine, R.S. (2010) *Fire in the Forest*. Cambridge University Press, Cambridge.

Thomas, P.A. & Polwart, A. (2003) Biological Flora of the British Isles: *Taxus baccata* L. *Journal of Ecology*, 91, 489–524.

Tilman, D. (1978) Cherries, ants and tent caterpillars: timing of nectar production in relation to susceptibility of caterpillars to ant predation. *Ecology*, 59, 686–692.

Veluthakkal, R. & Dasgupta, M.G. (2010) Pathogenesis-related genes and proteins in forest tree species. *Trees*, 24, 993–1006.

Vogel, S. (2009) Leaves in the lowest and highest winds: temperature, force and shape. *New Phytologist*, 183, 13–26.

Way, D.A., Ladeau, S.L., McCarthy, H.R., *et al.* (2010) Greater seed production in elevated $CO_2$ is not accompanied by reduced seed quality in *Pinus taeda* L. *Global Change Biology*, 16, 1046–1056.

Weber, K. & Mattheck, C. (2006) The effects of excessive drilling on decay propagation in trees. *Trees*, 20, 224–228.

Weisberg, P.J. & Ko, D.W. (2012) Old tree morphology in singleleaf pinyon pine (*Pinus monophylla*). *Forest Ecology and Management*, 263, 67–73.

White, J.E.J. & Patch, D. (1989) *Ivy – Boon or Bane? Arboriculture Research Note 81/90*. Forestry Commission Arboricultural Advisory and Information Service, Farnham.

Wooley, S.C., Donaldson, J.R., Gusse, A.C., Lindroth, R.L. & Stevens, M.T. (2007) Extrafloral nectaries in aspen (*Populus tremuloides*): heritable genetic variation and herbivore-induced expression. *Annals of Botany*, 100, 1337–1346.

# Chapter 10: Trees and us

Trees form the visual backdrop in most people's lives whether in an urban or rural setting. In England alone we have more than 90 million trees outside of woodlands in either small groups, hedgerows or as individual trees. But of what value are they to us apart from enhancing our surroundings?

## The value of trees

Over their long history, trees have played an important part in our lives that goes beyond just the supply of wood. Trees have been (and still are) sacred to many peoples; oaks were sacred to the European Druids, baobabs (*Adansonia digitata*) to African tribes, the ginkgo (*Ginkgo biloba*) to the Chinese and Japanese, sequoias (*Sequoiadendron giganteum*) to N American first people, and monkey puzzles (*Araucaria araucana*) to the Pehuenche people of Chile. Indeed, many of our words and expressions are derived from a close association with trees. Writing tablets were once made from slivers of beech wood (*Fagus sylvatica*), and 'beech' is the Anglo-Saxon word for book. Beech is still called 'bok' in Swedish and 'beuk' in Danish. Romans crowned athletes with wreaths of the bay laurel (*Laurus nobilis*); this was extended to poets and scholars in Middle Ages, hence Poet Laureate. Similarly, Roman students were called bachelors from the laurel berry (baccalaureus) leaving us with bachelor degrees (baccalaureate) and, since Roman students were forbidden to marry, unmarried bachelor males.

Despite modern technology, we are still very reliant on wood as a raw material. According to the Food and Agriculture Organisation (FAO) the world's annual production of timber in 2008 was more than 3800 million cubic metres (a well-grown conifer in a European plantation contains around 1–2 m$^3$ of wood at maturity). This is used for anything from building to paper making to the creation of chemicals including synthetic rubber. We get a million matchsticks from an average Canadian aspen. Cloth has historically been made from tree bark by the Polynesians and Africans and we now make rayon from wood. (Incidentally, wooden chopping boards show mild antibacterial properties and so are better than the seemingly more hygienic plastic ones!)

Numerous things we eat and drink come from trees. From the Old World comes citrus fruit, cinnamon (*Cinnamomum verum*), cloves (*Eugenia caryophyllus*), nutmeg (*Myristica fragrans*), coffee (*Coffea arabica* and *robusta*), tea (*Camellia sinensis*) and carob (*Ceratonia siliqua*). From the New World we get, amongst others, papaya (*Carica papaya*), avocado (*Persea americana*), cocoa (*Theobroma cacao*) and the brazil nut (*Bertholletia excelsa*). We also eat pine nuts collected from about 20 different species that produce seeds big enough to be worth collecting. In Europe these are from the stone pine, *Pinus pinea* (see Figure 7.4), and in N America from the pinyon pine, *P. monophylla* (named by Spanish explorers after the *P. pinea* seeds they ate at home). The main commercial source is now the Korean pine *P. koraiensis* grown in China; these are characteristically triangular in cross-section compared to the more rounded stone pine. Some people eat the trees themselves. North American first people have long eaten the inner bark (phloem) and cambium of pines and the western hemlock (*Tsuga heterophylla*). The Adirondack Aboriginals (whose name is used for the Mountains in New York State) are named after the Mohawk words for 'tree eaters'.

There are also oils such as olive oil (*Olea europaea*) and palm oil (*Elaeis guineensis*), the latter from tropical W Africa used for margarine, candles and soap. The artificial vanilla flavour of cheap ice-cream is a chemical derivative of wood. And cellulose extracted from wood is a common ingredient in instant mashed potato (and disposable nappies!). Medicinal compounds from trees are legionary including quinine (*Cinchona* spp., 'Jesuit's bark' from Peru and Ecuador) used to fight malaria (hence the colonial passion for gin and tonic, a palatable way of taking your bitter quinine medicine). Others are still being discovered: extracts from the ginkgo are advocated for improving blood flow to the brain to improve memory particularly for those suffering from Alzheimer's disease. Paclitaxel (which used to be called taxol), originally from the bark of the Pacific yew (*Taxus brevifolia*), is now used very successfully in chemotherapy of various types of cancer.

It would be remiss not to mention wine. Who can forget the unique taste of Greek retsina, white or rosé wine flavoured with pine resin? Neolithic pottery 5000 years old has been found that contained resinated wine, the resin undoubtedly being added to stop bacteria turning the wine into vinegar. Wine barrels made from oak are renowned for the flavour and colour they impart to wine. Three oaks species are traditionally used, sessile oak (*Quercus petraea*) and English oak (*Q. robur*) from France, and white oak (*Q. alba*) from N America. The interaction between wine and oak is complex, reliant on oxygen and chemicals from the wood to give the 'oaked' flavour to wine. The chemicals are primarily from the heartwood 'extractables' (Chapter 9), made up of lactones (a coconut flavour) and phenols (vanilla).

As is often the case, we find that the chemicals used to defend the plant, in this case the wood, in small doses are very palatable to humans.

Trees fulfil many other uses. Where would we be without rubber, most of which comes from one species, *Heavea brasiliensis*? Or brake linings in our cars (which can be made from lignin extracted from wood)? We can even run our vehicles on wood. One tonne of wood can produce 250 litres of petrol (27 imperial gallons). As well as trees replacing carbon usage, they also store a lot of carbon that would otherwise be in the atmosphere. In 2002 it was estimated that the total carbon storage in just urban trees in the USA was over 70 million tonnes, to which was being added an extra 22.8 million tonnes per year. A single mature beech tree can produce enough oxygen for 10 people every year and fix 2 kg of carbon dioxide per hour. This mounts up; it is calculated that urban trees in the USA produce 61 million tonnes of oxygen per year, enough for two thirds of the American population.

New uses are still being found for trees. Genes for a sweet-tasting but low-calorie compound called monellin have been taken from a tropical shrub and put into tomatoes and lettuces to make them sweeter.

In the natural world, trees and shrubs often act as nurse plants, aiding the establishment of other species. An example is seen in the Sonoran desert where most of the big saguaro cacti (*Carnegiea gigantea*) start life in the shade of a tree. Animals can also benefit; cows in the Midlands of England (and undoubtedly elsewhere) give more milk when they can shelter behind shrubby hedges. In ecological parlance, woody plants are often 'keystone' or 'foundation' species, those on which many other plant and animals depend.

Two areas are worth exploring further: the effect of trees on pollution and on our health.

## Pollution control

Trees, being such big things (Figure 10.1), are very good at absorbing gaseous pollutants and trapping particles of pollution. Up to 70% of gaseous pollutants such as ozone, sulphur dioxide and nitrogen dioxide can be filtered out by street trees. Trees also intercept small particles (such as from smoke and diesel engines), pollen and dust, especially those that have sticky hairs. One hectare of mixed woodland can remove 15 tonnes of particulate matter from the air each year while a pure spruce forest, with its greater surface area of needles, can remove 2–3 times as much.

Most trees and other plants produce volatile organic compounds (VOCs) made up primarily of monoterpenes with some isoprenes. Conifers in

Figure 10.1 Trees around buildings, being tall and large, have a large surface area to catch pollutants, both gases and particulate material. In this case the trees are in a leafy part of the large city of Sheffield, England.

particular produce large quantities (but so do many hardwoods) and it seems that the VOCs play a role in protecting leaves from overheating and damage from ozone. They also play an important positive role in our health (see below) but in large quantities are serious pollutants, contributing to ozone production and acting as a greenhouse gas. VOCs are also produced by motor vehicles, coal burning and other human activities and there has been much debate as to how much VOCs are produced by nature and how much by humans, and this led to an infamous quote by Ronald Reagan that 'approximately 80% of our air pollution stems from hydrocarbons released by vegetation' which led in turn to him saying 'so let's not go overboard in setting and enforcing tough emission standards from man-made sources' (Pope 1980). More recently, however, it has been found that trees such as poplars also reabsorb large amounts of VOCs and so their net production is remarkably low.

The net effect is that trees are efficient removers of pollutants. In the USA alone it has been estimated that urban trees remove more than 700 000 tonnes of pollution each year, a service valued at $3.8 billion.

# Health benefits of trees

Most people would, I think, intuitively agree that trees are good for us. And they are! On the physical side, the particulate matter they trap reduces respiratory problems, and the cooling they provide reduces problems during heat waves.

Trees are also very important to our psychological wellbeing (Figure 10.2). A concrete piece of evidence for this comes from a now famous study by Roger Ulrich at the University of Delaware, published in 1984. He found that being able to see trees rather than a wall helped speed recovery of hospital patients after surgery, reduced the need for painkillers, made people less negative and reduced post-surgical complications.

Trees are also known to improve one's mood, reduce depression and increase self-esteem. From large-scale studies on English and Dutch populations it appears that it is those in the lowest income groups that benefit the most from having green space near their home. In the UK, the Woodland Trust has said that 80% of people live in urban areas and yet only 15% have access to a local woodland within 500 m of their home, so urban trees help to give a country feeling within the city (O'Brien 2006). People show an increasing tendency to walk for recreation and to commute to work on foot the nearer they live to green spaces, which has obvious benefits to the individual in terms of reducing obesity and blood pressure, and increasing life expectancy. Because of the improved health produced by trees, the Campaign for Greener Healthcare in the UK is working on a National Health Service Forest. This 20-year project planted its first tree in 2009 and aims to plant a tree for each of the 1.3 million NHS employees around hospitals and other NHS facilities.

Trees and forests can also have a direct physiological effect on our bodies. Although VOCs (including monoterpenes) produced by trees can be pollutants at high levels, at the low levels normally found in woodlands, they are known to reduce tension and mental stress, and sleeping in a woodland promotes faster recovery from fatigue than sleeping indoors. This is the basis of shinrin-yoku (forest-air bathing), a major form of relaxation in Japan that involves slowly walking and breathing in a forest. It has been shown to reduce aggression and depression, increase feelings of wellbeing, reduce adrenalin and promote the immune system. The psychological effect of lovely scenery, tranquillity and fresh air is an important contributor to its success but the 'natural aromatherapy' is also seen as important through inhaling VOCs. Li (2010) found that a 3-day forest visit can have a positive effect on the immune system up to 30 days later.

Trees also affect us through our diet. These effects vary from the common place to the more obscure. In the former camp there are many claims that nuts

(a)

(b)

Figure 10.2 Trees make a huge difference to the aesthetics of an area and improve our sense of wellbeing and mental health. Contrast (a) the scene around Lichfield Cathedral in central England with many trees to (b) a street in Liverpool which, although architecturally interesting, has just one tree (a false acacia/black locust, *Robinia pseudoacacia*) midway along.

such as walnuts, which have a high content of polyunsaturated fatty acids, may help to reduce the risk of heart disease. Certainly, terpenoids (which include the monoterpenes) are very important in our diet. These are found in the smell of eucalyptus and the distinctive flavour of cinnamon, cloves and ginger (*Zingiber officinale*), and include menthol, camphor and the cannabinoids found in cannabis. Terpenoids are found in many fruit and vegetables and play an important role in herbal medicine; see Wagner & Elmadfa (2003) for more in-depth information.

Urban trees can also make life much more comfortable. Cities tend to be 'heat islands', 5–10°C hotter than surrounding rural areas. These higher temperatures indirectly affect health (for example, by increasing ground level ozone amounts that makes chronic respiratory conditions worse) and directly via heat stress and dehydration particularly among the elderly, very young or chronically ill. The heat wave across Europe in the summer of 2003 was directly attributed to the death of 2000 people in Britain and more than 50 000 across Europe. It would have been much worse without our street trees.

## Monetary value of urban trees

The nature of modern life is that for something to be valued it has to have a monetary value. This has led to any number of studies concerned with just this. For example, it has been concluded that landscaping with mature trees increases USA house prices by 5–20%. Moreover those same trees can reduce air conditioning costs in summer by 30% and heating costs in winter by 20–50%. In hot, dry climates like Greece the air temperature underneath a single spreading tree of paper mulberry (*Broussonetia papyrifera*) can be 5–20°C lower than in the sun.

More systematic studies of the value of urban trees have been possible through the i-Tree programme run by the United States Department of Agriculture (USDA) Forest Service. Using a set of computer programmes, simple data are used to quantify the value of community trees and the ecological services that they provide. So, for example, New York City has an estimated 5.2 million trees and 8.2 million people. The most common trees are tree of heaven (*Ailanthus altissima*), black cherry (*Prunus serotina*) and sweet gum (*Liquidambar styraciflua*) although the London plane (*Platanus* × *hispanica*) makes up some of the biggest trees. An i-Tree study concluded that these trees remove 1998 tonnes of air pollution each year with an annual value of this service put at $10.6 million. They also hold 1.22 million tonnes of carbon worth an estimated $24.9 million, and each year absorb another 38 000 tonnes of carbon worth $779 000.

The much smaller town of Torbay in southern England has around 818 000 trees but even these are valued at over £280 million as a structural asset. They remove over 50 tonnes of pollutants every year, with an annual estimated value of £280 000. They also store 98 000 tonnes of carbon (valued at almost £1.5 million) and soak up another 4300 tonnes each year, worth more than £64 000. While you may not like this sort of valuation of something that is invaluable, it does help when weighing up the economic returns of a new development that would otherwise be displacing 'worthless' trees.

## Problems with trees

Despite the many benefits of trees, they can present a number of problems, especially in urban areas where we live close to the trees, and often mistreat them (Figure 10.3). A major problem is fear of litigation. Roots can disrupt pavements, creating tripping hazards (Figure 10.4), bits of tree can fall and damage people and possessions, and the list could go on almost endlessly. As a consequence there can be a move to get rid of the trees. Between 2002 and 2007 it has been reported that the London Borough of Hackney (just 19 km$^2$) felled 40 000 street trees apparently due to fears of being sued. The insurance industry contributes to the demise of trees through worries about damage to drains from roots, or damage to buildings by roots or trees falling. The sensationalising views

(a)                                    (b)

Figure 10.3  (a) A lime tree (*Tilia* x *europaea*) being used as a living fence post, and (b) a tree growing around a set of railings. The first is deliberate, the second probably inadvertent but resulting in the same problems. Not only does this allow in rot but can also distort future growth, and any metal that remains in the tree is a potential danger to a tree surgeon. Photographs (a) Lichfield Cathedral, England (b) Chirk Castle, Wales.

**Figure 10.4** Brick pavement being upheaved by tree roots in New Orleans, USA. Photograph by Infrogmation, reproduced under Creative Commons Attribution-Share Alike 3.0, courtesy of Wikimedia Commons.

of media add further nails to the coffin of urban trees. It is true that trees can be dangerous things: the US National Institute for Occupational Health and Safety recorded that between 1980 and 1989 at least 207 deaths were associated with the pruning or trimming of trees. But to put this into perspective, in a city of one million people you would expect one tree-trimming death per decade, compared to more than 1000 in the same time attributable to vehicles. As if this wasn't bad enough, large numbers of trees are also being removed due to the huge increase in CCTV use and the desire for clear sight lines.

Some trees are just in the wrong place, usually because they either grow too big for the room available or from such things as throwing up disruptive suckers (particularly bad with the much hated tree of heaven, *Ailanthus altissima*; a quick search of the web will show why!). In the same category in the UK is Leyland cypress (x *Cupressocyparis leylandii*; Figure 10.5). Planted especially as hedging (for which it is very good since the trees don't go brown where they meet unlike with many conifer species) it can grow five times quicker than yew (*Taxus baccata*) and produce a metre of new growth in a summer. If not checked it will rapidly reach 30+ metres and is the cause of many neighbourly problems. In many cases it should probably never have been planted in the first place.

**Figure 10.5** Hedge of Leyland cypress (x *Cupressocyparis leylandii*) renowned in the UK for rapid growth if left unchecked, and the cause of many problems between neighbours. Imagine the effect this hedge would have on light and views if it was close to a house. In this case, the rapid, dense growth is being used to shield the road from the lights of an all-weather sports pitch at Keele University, England.

## Are they worth it?

Although trees can cause problems, these are far outweighed by the benefits. I am obviously biased, but look at almost any normally treed landscape, urban or rural, and it is the trees that add significant value. Or imagine a favourite treed vista with the trees removed and see how much would be lost.

For many of us trees are a dominant part of our living space, and yet because of their size and longevity, and their possession of a woody trunk, they are often misunderstood and can be mistreated out of kindness. If nothing else, I hope this book helps you appreciate how the trees around you are working and why they look and act the way they do.

## Further Reading

Alañón, M.E., Pérez-Coello, M.S., Díaz-Maroto, I.J., *et al.* (2011) Influence of geographical location, site and silvicultural parameters on volatile composition of *Quercus pyrenaica* Willd. wood used in wine aging. *Forest Ecology and Management*, 262,124–130.

Anon (2011) *Torbay's Urban Forest. Assessing Urban Forest Effects and Values. A Report on the Findings from the UK i-Tree Eco Pilot Project.* Treeconomics, Exeter.

Forestry Commission (2001) *National Inventory of Woodland and Trees: England.* Forestry Commission, Edinburgh.

Gilchrist, K. (2012) Promoting wellbeing through environment: the role of urban forestry. *Trees, People and the Built Environment. Research Report* (edited by M. Johnston & G. Percival). Forestry Commission, Edinburgh, pp. 84–93.

Karl, T., Harley, P., Emmons, L., *et al.* (2010) Efficient atmospheric cleansing of oxidized organic trace gases by vegetation. *Science*, 330, 816–819.

Li, Q. (2010) Effect of forest bathing trips on human immune function. *Environmental Health and Preventative Medicine*, 15, 9–17.

Maas, J., Verheij, R.A., Groenewegen, P.P., De Vries, S. & Spreeuwenberg, P. (2006) Green space, urbanity, and health: how strong is the relation? *Journal of Epidemiology and Community Health*, 60, 587–592.

Mitchell, R. & Popham, F. (2007) Greenspace, urbanity and health: relationships in England. *Journal of Epidemiology and Community Health*, 61, 681–683.

Miyazaki, Y. & Yatagai, M. (1989) Effect of components of plant odors on humans and animals. *Man and Environment*, 15, 33–42.

Morita, E., Fukuda, S., Nagano, J., *et al.* (2007) Psychological effects of forest environments on healthy adults: Shinrin-yoku (forest-air bathing, walking) as a possible method of stress reduction. *Public Health*, 121, 54–63.

Nilsson, K., Sangster, M., Gallis, C., *et al.* (2010) *Forests, Trees and Human Health.* Springer, Dordrecht, Germany.

Noe, S.M., Peñuelas, J. & Niinemets, Ü. (2008) Monoterpene emissions from ornamental trees in urban areas: a case study of Barcelona, Spain. *Plant Biology*, 10, 163–169.

Nowak, D.J. & Crane, D.E. (2002) Carbon storage and sequestration by urban trees in the USA. *Environmental Pollution*, 116, 381–389.

Nowak, D.J., Crane, D.E. & Stevens, J.C. (2006). Air pollution removal by urban trees and shrubs in the United States. *Urban Forestry and Urban Greening*, 4, 115–123.

O'Brien, E. (2006) Social housing and green space: a case study in Inner London. *Forestry*, 9, 535–551.

O'Brien, L., Williams, K. & Stewart, A. (2010) *Urban Health and Health Inequalities and the Role of Urban Forestry in Britain: A Review.* Forest Research, Edinburgh.

Pope, C. (1980) The candidates and the issues. *Sierra*, 65, 15–17.

Sharkey, T.D., Wiberley, A.E. & Donohue, A.R. (2008) Isoprene emission from plants: why and how. *Annals of Botany*, 101, 5–18.

Ulrich R.S. (1984) View through a window may influence recovery from surgery. *Science*, 224, 420–421.

Wagner, K.-H. & Elmadfa, I. (2003) Biological relevance of terpenoids: overview focusing on mono-, di- and tetraterpenes. *Annals of Nutrition and Metabolism*, 47, 95–106.

# Further Reading

In addition to the lists of publications at the end of each chapter, the following books will prove useful for finding out more about how trees work.

Arno, S.F. & Hammerly, R.P. (1984) *Timberline: Mountain and Arctic Forest Frontiers*. The Mountaineers, Seattle.

Aronson, J., Pereira, J.S. & Pausas, J.G. (2009) *Cork Oak Woodlands on the Edge*. Island Press, Washington, DC.

Bonnicksen, T.M. (2000) *America's Ancient Forests*. Wiley, New York.

Bowes, B.G. (2010) *Trees and Forests: A Colour Guide*. Manson, London.

Britt, C. & Johnston, M. (2008) *Trees in Towns II*. Department of Communities and Local Government, London.

Burns, R.M. & Honkala, B.H. (1990) *Silvics of North America. Vol 1: Conifers. Vol 2: Hardwoods*. United States Department of Agriculture, Forest Service, Agriculture Handbook 654.

Büsgen, M. & Münch, T. (1929) *The Structure and Life of Forest Trees* (3rd ed). Chapman and Hall, London.

Clobert, J., Baguette, M., Benton, T.G. & Bullock, J.M. (2012) *Dispersal Ecology and Evolution*. Oxford University Press, Oxford.

Coutts, M.P. & Grace, J. (1995) *Wind and Trees*. Cambridge University Press, Cambridge.

Cutler, D.F. & Richardson, I.B.K. (1981) *Tree Roots and Buildings*. Construction Press (Longman), London.

Eaton, R.A. & Hope, M.D.C. (1993) *Wood: Decay, Pests and Protection*. Chapman & Hall, London.

Ennos, R. (2001) *Trees*. The Natural History Museum, London.

Hallé, F., Oldeman, R.A.A. & Tomlinson, P.B. (1978) *Tropical Trees and Forests: An Architectural Analysis*. Springer, Berlin.

Hoadley, R.B. (2000) *Understanding Wood*. Taunton Press, Newtown, CT.

Kozlowski, T.T. (1971) *Growth and Development of Trees. Vol I: Seed Germination, Ontogeny and Shoot Growth. Vol II: Cambial Growth, Root Growth and Reproductive Growth*. Academic Press, New York.

Lanner, R.M. (1996) *Made For Each Other: A Symbiosis of Birds and Pines*. Oxford University Press, New York.

Larcher, W. (2001) *Physiological Plant Ecology* (4th edn). Springer, Berlin.

Longman, K.A. & Jeník, J. (1987) *Tropical Forest and its Environment*. Longman Scientific & Technical, Harlow.

Lowman, M.D. & Nadkarni, N.M. (1995) *Forest Canopies*. Academic Press, San Diego.

Mattheck, C. & Breloer, H. (1994) *The Body Language of Trees: a Handbook for Failure Analysis. Research for Amenity Trees, No 4*. HMSO, London.

Menninger, E.A. (1995) *Fantastic Trees*. Timber Press, Portland, Oregon.

Mosbrugger, V. (1990) *The Tree Habit in Land Plants*. Springer, Berlin.

Pigott, D. (2012) *Lime-trees and Basswoods*. Cambridge University Press, Cambridge.

Richards, P.W. (1996) *The Tropical Rain Forest* (2nd edn). Cambridge University Press, Cambridge.

Richardson, D.M. (1998) *Ecology and Biogeography of Pinus*. Cambridge University Press, Cambridge.

Roberts, J., Jackson, N. & Smith, M. (2006) *Tree Roots in the Built Environment*. Department for Communities and Local Government, London.

Rupp, R. (1990) *Red Oaks and Black Birches: The Science and Lore of Trees*. Storey Communications, Pownal, Vermont.

Sedgley, M. & Griffin, A.R. (1989) *Sexual Reproduction in Tree Crops*. Academic Press, London.

Shigo, A.L. (1989) *A New Tree Biology*. Shigo and Trees, Durham, New Hampshire.

Shigo, A.L. (1991) *Modern Arboriculture*. Shigo and Trees, Durham, New Hampshire.

Thomas, P.A. & Packham, J.R. (2007) *Ecology of Woodlands and Forests*. Cambridge University Press, Cambridge.

Thomas, P.A. & McAlpine, R.S. (2010) *Fire in the Forest*. Cambridge University Press, Cambridge.

Turner, I.M. (2001) *The Ecology of Trees in the Tropical Rain Forest*. Cambridge University Press, Cambridge.

Williams, J. & Woinarski, J. (1997) *Eucalypt Ecology: Individuals to Ecosystems*. Cambridge University Press, Cambridge.

Wilson, B.F. (1984) *The Growing Tree*. University of Massachusetts Press, Amherst.

# Index

Page numbers in *italics* refer to figures and boxes

acacia, 19, *35*, 36, 85, 108, *131*, 161, 286, *327*
  and ants, 326
acid rain, 348
age, 206, 317–22, *318*, 370–2
  parts of the tree, 322
Air-Pots, *115*
alder (*Alnus* spp.), 347
  bark, 89
  flowers, *155*, 168
  fruit, 187
  green, *347*
  grey, *347*
  leaves, 21, 46
  nitrogen fixation, 131
  roots, *131*, *132*, 141, 143
  saplings, 221
  seeds, 190, *241*, 292
  water movement, 68
  wood, 81
alder buckthorn (*Frangula alnus*), *155*, 158
allelopathy, 330
amber, 334
American hop hornbeam (*Ostrya virginiana*), 43
angelica tree (*Aralia* spp.), 13, *14*, 37
angiosperm, defined, 3, *5*
annual rings. *See* growth
aphids, 47, 84, 328, 333
apical control and dominance. *See* shape
  of trees
apical meristem. *See* meristem
apple (*Malus* spp.), 13, 18–19, 126, 140, 142, *155*,
  177, 183, 201, 218, 278, 323, 328, 346–7
ash (*Fraxinus* spp.)
  age, *318*, 372
  American, 84–5

  and ivy, 316
  bark, 84–5
  buds, 258
  environmental tolerance, *346*, 349
  flowers, *155*, 181
  fruits, *285*
  green, 110
  growth, 222, *306*
  leaves, 18, 37, 46
  roots, 110, *112*, 143
  seeds, *241*, 284, *291*
  wood, 357
ash dieback, 366
aspen (*Populus* spp.), 43
  clone, 181, 208
  defences, 330
  flowers, 180
  leaves, 30, 46–7
  nectaries, 326
  seeds, 190, *292*
  wood, 81
avocado (*Persea americana*), 40, 126, 196,
  228, 377

bacteria, 130, 219, 309, 317, 327, 339, 376–7
balsa (*Ochroma pyramidale*), *5*, 161
bamboo, *1*, 1, 205
banana, *1*, 1
baobab (*Adansonia* spp.), 10, 37, 161, *162*, *211*,
  211, 376
barberry (*Berberis* spp.), 162, 323
bark
  and fire, 354
  chlorophyll, 34, 86
  defences, 334

bark (cont.)

    growth, 52, 83–92, *86*, *88*

    lenticels, 15, 83, 86, 90–1, 137

    monocots, 87

barrier zone in decay, 344, 372

bay laurel (*Laurus nobilis*), 376

beech (*Fagus* spp.), 236

    age, *318*

    bark, 88, *338*, 354

    copper, 304

    cut-leaf, *307*, 307

    environmental tolerance, *346*

    flowers, *155*, 177, 240

    growth, 222, *264*

    leaves, 13, 45

    roots, *106*, 107, 110, *112*, 126

    seedlings, *293*

    seeds, *191*, *241*, *291*

    water movement, 70

    wood, 81, 376

biggest trees, 207–12

biomechanics, 212–14, 358–9

    design compromise, 359

    gravity, 253–6

birch (*Betula* spp.)

    age, *318*

    bark, 91

    biomechanics, *213*

    downy, 235

    dwarf, 219, *292*

    environmental tolerance, *346*

    flowers, *155*, *169*

    leaves, 47

    root pressure, 63

    roots, *112*

    seeds, 195, *241*

    silver, 228, *233*, 259, *292*

    water loss, 19

blackberry (*Rubus* spp.), *142*, 323

blackthorn/sloe (*Prunus spinosa*), *155*, 237,

    323, *348*

blue tits, 180

bog wood, 339, *340*

*Boscia albitrunca*, 108

bottlebrush (*Banksia* spp.), 160, 354

bottlebrush (*Callistemon* spp.), 161

bottlebrush (*Melaleuca* spp.), 146

bougainvillea (*Bougainvillea* spp.), 159, 175

box (*Buxus* spp.), 21, *142*, *155*, *347*

branches. *See also* shape of trees, trunk

    angle, 258, 274

    attachment, 73, 94–7, *95*, *358*

    epicormic, 94, *271*, 275, 370, 372

    fasciated, 34

    length, 263–4

    long and short shoots, 263, 273

    orders of branching, 258

    scaffold, 261

brazil nut (*Bertholletia excelsa*), 197, 200, 377

broom (*Cytisus scoparius*), 34, 155, 161, *163*,

    288, 323, *324*

buckthorn (*Rhamnus catharticus*), 87, *264*

buds

    adventitious, 93, 303

    bud scales, 224

    death, 259

    dormancy, 235–6

    epicormic, 92, *93*, 245, 260, *271*, 275, 280,

    303, 365, 370, 372

    flower, 266

    formation, 234

    growth, 226–8

    leaf, 266–7

    mixed, 267

    naked, 225, *226*

    new buds in conifers, 260

    opening, 233–4

    shape of trees, 256–8

    terminal, 16, 227, 267

burr, *92*, 92–4, 136, 370

butchers'-broom (*Ruscus aculeatus*), 34

butterfly bush (*Buddleia davidii*), *155*

California bay (*Umbellularia californica*), 331, *332*

callus growth. *See* defences

cambium
  cork, 85, 88, 90
  vascular, 11, 52, 90

camphor tree (*Cinnamomum camphora*), 40, 342

canopy shyness, *251*

carbon dioxide, 13, 21, 23, 166, 220, 350, 378

carob (*Ceratonia siliqua*), 294, 377

cashew (*Anacardium occidentale*), *196*

catalpa (*Catalpa* spp.), *17*, 37, 82

cauliflory, 178

cedar, 335
  deodar, 262, 272, *273*
  flowers, 169
  incense, 169, 184
  Japanese red, 33, *142*, 180, 260, 352
  of Lebanon, 206
  Tasmanian, 33
  western red, 59, 184, 219, 228, *242*, 342, *346*, 352
  white, 206, 260, 331
  wood, 342

celery-topped pine (*Phyllocladus* spp.), 34

century plants (*Agave* spp.), 56

cherry (*Prunus* spp.), *112*, *142*, 236–7, *267*, *336*, 336, 382
  bark, 91
  black, 326
  defences, 326, 328
  flowers, *155*
  laurel, 328, *329*
  leaves, *17*, 18
  nectaries, 326
  roots, 141

chestnut (*Castanea* spp.), *142*, *157*, 178, *242*, *291*, *301*, *346*, 354, 366

chestnut blight, 366

chimaera, 306

Chinese fir (*Cunninghamia* spp.), 180

chlorophyll. *See* leaves, bark

cinnamon (*Cinnamomum verum*), 87, 331, 377

cladoptosis. *See* shape of trees

Clark's nutcracker, 197

climate change, 220, 352
  phenology, 236–7

clonal trees, 208, 275, 321–2

cloudberry (*Rubus chamaemorus*), 179

cloves (*Eugenia caryophyllus*), 377

clubmoss (*Lepidodendron*), 4

*Clusia* sp., 23

coastal redwood. *See* redwood

cocoa (*Theobroma cacao*), 10, 164, 178, 228, 328, 377

coconut (*Cocos nucifera*), 284, 290–1, 296

CODIT. *See* defences

coffee (*Coffea* spp.), 175, 331, 377

colour. *See* leaves, flowers

conifers
  defined, 4
  evolution, 4
  water loss, 19

cork, 89–91

cosmic radiation, 220

cotoneaster (*Cotoneaster* spp.), *142*

creosote bush (*Larrea tridentata*), 26–7, 34, 261, 321, 329

currant (*Ribes* spp.), *126*, *156*

cuticular transpiration, 21

cuttings, 303–4

cycad, 56, 131, 164, 171, 180, *182*

cypress, *112*, *171*, 183–4, 260–1, 305, *318*, *346*, 350, 352

damage and death. *See also* climate change
  cold, 352–3
  disease, 280, 365–8
  drought, 357
  Dutch elm disease, 366–8
  environmental causes, 345–50

damage and death. (cont.)

    fire and heat, 357

    hollow trees, 359–62

    insects, 369

    starvation in old trees, 369–72

    wind, 359, 362–5, *364*

date (*Phoenix dactylifera*), 288

dawn redwood (*Metasequoia glyptostroboides*), 6–7, *8*, 16, 39, *115*, 261

defences. *See also* wood

    against freezing, 352

    callus, 337–8

    chemical, 328–32

    compartmentalisation of decay in trees (CODIT), *343*, 345

    cost, 330–1

    evolutionary battle, 332–4

    genetic mosaic, 333

    physical, 259, 323–6, *324*

    resins, gums and latex, 334–7, 341

    using animals, 326–7, *327*

dendrochronology. *See* growth rings

desert ironwood (*Olneya tesota*), 36

design compromise, 212–14.

    *See also* biomechanics

dicotyledon, defined, 3

dinosaurs, 190

dioecious flowers. *See* pollination

dipterocarp, 173, 175, 190, *242*

disease. *See* damage and death

dodo, 286

dogwood (*Cornus* spp.), 159

domatia, 327

dormancy, 63, 92, 139, 217, 235, 259, 270, 285, 287–8, 355. *See also* buds, seeds

dove (handkerchief) tree (*Davidia involucrata*), 45

dragon tree (*Dracaena* spp.), 4, 56, 320

durian (*Durio zibethinus*), 161, 178

Dutch elm disease. *See* damage and death

ebony (*Diospyros ebenum*), 82

elder (*Sambucus nigra*), *112*, *156*, 197

elm (*Ulmus* spp.), 13. *See also* Dutch elm disease under damage and death

    American, 134

    bark, 93

    branches, 262

    buds, *266*

    disease, 366

    flowers, *156*, 167, *168*

    fruits, 184

    roots, *112*, 134, 361, *362*

    seeds, *241*, *291*

    wood, *82*, 342

*Ephedra* spp., *2*, 60

epicormic buds and branches. *See* buds, branches

eucalyptus

    bark, 85, 91, *368*

    defences, 328

    flowers, *160*, 160, 177

    genome, 310

    growth, 205

    leaves, *17*, *38*, 38, 369

    roots, 129, 146

    tallest, 207

    water loss, 19

    water use, *124*, 124

extractables in heartwood, 342

false acacia/black locust (*Robinia pseudoacacia*), 36, *142*, 219, 323, *324*, *381*

    buds, 226

    environmental tolerance, *347*

    leaves, 21, 27, 323

    nitrogen fixation, 130

    roots, *112*, 138

    seeds, *241*

    suckers, 300

    wood, 82

fertiliser, 42, 118, 140

fig (*Ficus* spp.)

  and pollution, 349

  banyan, 148

  cloth from bark, 84

  fruits, *179*, 183, 197

  leaves, 32

  pollination, 164–6, *165*, 178

  roots, *113*, 135, *144*, 148

  'rubber tree', 337

  strangling, 134, *149*, 150

  weeping, 148

fir, 79

  balsam, *289*, 289

  bark, 92

  Douglas, 58, *72*, 77, 91, *117*, 130, 135, *171*, 171, 206–7, 210, 228, *242*, 270, 303, *318*, 335, 353, 371

  environmental tolerance, *346*

  roots, 108, 141

  subalpine, *252*, 345

  white, 354

fire. *See* damage and death

firethorn (*Pyracantha* spp.), 323

firewood, 357

'flagged' trees, 252

flooding. *See* roots

floss-silk tree (*Ceiba speciosa*), 325

flowers. *See also* pollination

  age of first, 240–3

  catkins, 168–9, 178, 180

  colour, 154, *158*, 160

  exploding, 162

  fragrance, 160, 163

  mass blooming, 173–4

  nectar, 164

  new and old wood, 266

  opening, 235

  pollinators, *158*

  timing, 175–6

  why so many?, 201–2

forest-air bathing, 380

foxglove tree (*Paulownia tomentosa*), 13, 37, 190, 274

frankentrees, 312

frankincense (*Boswellia carteri*), 336

frost cracks, 79

frost tolerance, 346

fruits

  cauliflory, 297

  development, 184–5

  parthenocarpic, 183

  types, 184–9

fungi, 57, 127, *128*, 129, 286, 327–8, 334, 339, 343, 350, 365–8

General Sherman, 208, *209*

genetic engineering, 308–12, 333

genome, defined, *309*

genotype, 246

germination

  crytogeal, 296

  epigeal, 293

  hypogeal, 293

giant sequoia. *See* sequoia

ginkgo (*Ginkgo biloba*), 7, *8*, 46, 136, *148*, 150, *171*, 171, 180, 184, 228, 345, *347*, 376

giraffe, 161, 315

*Gnetum* spp., *2*, 60

gorse (*Ulex* spp.), *286*, 286, 288, *291*, 323, 356

grafting, 305, 353

grape (*Vitis* spp.), 36, *142*

grass trees (*Xanthorrhoea* spp.), 56

growth

  and cosmic radiation, 220

  and tides, 220

  control, external factors, 219–21

  control, internal factors, 214–18

  lifetime changes, 238–40

  monopodial/sympodial, 261

  primary, 11, 52

393

growth (cont.)

  prostrate, 304

  rings, 60–3, 97, 216, 222–4, *223*, 319

  secondary, 11, 52

  speed, 205–7

guelder rose (*Viburnum opulus*), 159

gums. *See* defences

gymnosperm, defined, *2, 5*

hardwoods

  defined, 4

  evolution, 4–7

hawthorn (*Crataegus* spp.), *55, 55, 112, 142,*
    154, 167, 183, 237, 259, 323, *347*

hazel (*Corylus avellana*), 18, 93, *142, 156,* 168,
    *241, 271, 291,* 294, *346*

healing wounds, 337

health benefits of trees. *See* value of trees

heather, 32, 136, 195, *292*

height. *See* biggest trees

height limitations, 76–7, 212–14

hemlock (*Tsuga* spp.), 97, *296,* 335, *346,*
    371, 377

hiba (*Thujopsis dolabrata*), 260, 278

hickory (*Carya* spp.), *241*

hollow trees. *See* damage and death

holly (*Ilex aquifolium*), 34, 41, *142*

  bark, 85

  environmental tolerance, *348*

  flowers, *156,* 180

  leaves, 21, 39, 323

  seeds, 197, 287, *291*

  wood, 357

honey fungus, *347*

honey-locust (*Gleditsia triacanthos*), 126, 235,
    *241,* 259, 323

hormones, 72, 139, 216–17, 233, 270–3

hornbeam (*Carpinus betulina*), 18, *112, 281, 346*

  flowers, *156*

  leaves, *17*

  seeds, *241, 291*

horse chestnut (*Aesculus hippocastanum*), 18,
    158, 207, *225, 241, 265,* 294

  flowers, *156*

  leaves, 37

  roots, *112,* 114

  seeds, *291*

horsetail (*Calamites*), 4

Huon pine (*Lagarostrobos franklinii*), *318,* 321

hybrids, 176, 183, 305, 308

hydathodes, 22

hydraulic architecture, 70–3

hydraulic lifting, 123–5, *125*

hydraulic network, 73–5

Indian laburnum (*Cassia* spp.), 26, 166

insects. *See* damage and death

ivy (*Hedera helix*), 21, 41, 274, *291, 316,* 316,
    *348*

Japanese umbrella pine (*Sciadopitys verticillata*),
    34

jasmonates, 330

jay, 197, 294, 297, 299

jojoba (*Simmondsia chinensis*), 20

Joshua tree (*Yucca brevifolia*), 56

Judas tree (*Cercis siliquatrum*), 178

juniper (*Juniperus* spp.), 129, *157,* 179, *198,*
    286, *346*

  Utah, 24

kapok tree (*Ceiba pentandra*), *144,* 161, *194,*
    195, 201

katsura (*Cercidiphyllum japonicum*), 119, 263

kauri (*Agathis australis*), 180, *210,* 211, 336

  biggest, 208

Kentucky coffee-tree (*Gymnocladus dioica*), 37,
    126

King's holly (*Lomatia tasmania*), 321

knots, 94, 97

Krakatoa, 317

Krummholz, 251

laburnum (*Laburnum* spp.), *113*

laburnum-Spanish broom (+ *Laburnocytisus adamii*), 306

lammas growth. *See* leaves

*Lantana* shrub, 159

larch (*Larix* spp.), 224, 236, *292*, 303, 305, 335

    bark, 84

    environmental tolerance, *346*

    flowering, 176

    Japanese, 176

    leaves, *17*, 39

    seeds, *242*

latex. *See* defences

layering, 301

leaves

    absorption, 23–6

    age, 39

    antitranspirants, 23

    arrangement, 258, 275–8

    chlorophyll, 16, 18, 33, 45

    colour, 44–8

    compound, 15–16, *16*, 30, 37–8

    cooling effect, 20

    death, 42–4, 235

    evergreen/deciduous, *5*, 39–42

    guttation, 22

    juvenile, *38*, 38

    lammas growth, 29, 228

    leaf exchanging, 40

    light tracking, 26

    movement, 21, 26, *27*, 28

    needles, 33

    number, 13, 370

    orientation, 20

    photosynthesis, 34, 216

    phylloclades, 34, *35*

    phyllodes, 34, *35*

    scale, 33–4

    size and shape, 21, 28–34

    stipules, 15

    stomata, 17–20

    sun/shade, 18–19, 30

    wind, 358

Leyland cypress (x *Cupressocyparis leylandii*), 111, 305, 384, *385*

liana, 55, 138

light, 219

lightning, 89

lignotuber, *135*, 136

lilac (*Syringia vulgaris*), *113*, *142*, *347*

lime/linden (*Tilia* spp.), 207, *383*

    bark, 84–5, *87*, 88, 93

    buds, 225

    environmental tolerance, *346*

    family change, 10

    flowers, *156*, 158

    leaves, *17*

    roots, *112*

    seeds, 201, *241*, *291*

    shape of tree, 245, *247*

lychee (*Litchi chinensis*), 228

macadamia nut, (*Macadamia integrifolia*), 127

magnolia, 6, 36, 40, 110, 154, *159*

mango (*Mangifera indica*), 201, 228

mangroves, 87, 126, 143, *144*, 146

maple (*Acer* spp.)

    age, *318*

    bark, 85

    bigtooth, 182

    buds, 258

    environmental tolerance, *346*

    family change, 10

    field, *119*, 241, *346*

    flowers, 154, *156*, 180, 182

    leaves, 18–19, 48

    Manitoba, 37, *142*, 177, 226

    moosewood, 182

    Norway, 104, 201, *347*, 363

    red, 18, 110, *134*, 139, *318*

    root pressure, 63

    roots, *106*, *112*, *119*, 139

maple (*Acer* spp.) (cont.)

    seeds, *241*

    sugar, 19, 43, 47, 64, 219, 276, 289

    wood, 77

mast seed years. *See* seeds

meristem, 51, 224–5

mespil (*Amelanchier* spp.), 183

mesquite (*Prosopis* spp.), 37, 108, *354*

micro-propagation, 303

migration rate of trees, 351

mistletoe, 315

mock orange (*Philadelphus gordonianus*), *142*

modular organisation, 11

molecular systematics, 10

monecious flowers. *See* pollination

monkey puzzle (*Araucaria araucana*), 39, 58,
    94, 296, *300*, *348*, 376

monocotyledon, defined, 3

mountain laurel (*Kalmia* spp.), 162, *163*

mountain pine beetle (*Dendroctonus
    ponderosae*), 332, 369

moving a tree, 122

mutations, 45, 304–5, 307, 333

mycorrhiza. *See* roots

myrrh (*Commiphora* spp.), 336


name changes, 9–10

New Caledonian pine (*Araucaria columnaris*),
    182

nitrogen pollution, 348

northern rata (*Meterosideros robusta*), 150

nuclear explosion survival, 345

nurse log, 295

nutmeg (*Myristica fragrans*), 331, 377

nutrients, 219–20


oak (*Quercus* spp.)

    acorns, 190, 192, *242*, 284, 288, *291*, 294,
        296–8

    age, *215*, *318*

    biggest, 210

biomechanics, *213*

branches, 258

burr, 92, 201

cork, 21, 39–40, *40*, 85, 89, *90*, 108,
    184, 296

defences, 330

emory, 331

English, 40, 70, 113, *168*, *254*, *291*, *347*, *377*

environmental tolerance, *346*

flowers, *156*, *168*, 240

gambel, 181

genome, 310

holm, *22*, 22, 39, 184, 190, *347*

kermes, 39, 260

leaves, *17*, 18, 21, 29, 39, 368

Lucombe, *40*, 40

Mirbeck's, 41

Oriental white, 297

Pyrenean, 297

red, 69, 92, 107, *116*, *126*, 138, 184, *242*,
    331, 342

roots, 107–8, 111, *112*, 113, 116, 138

scarlet, 18, 184

seedlings, *293*

sessile, *92*, 107–8, *262*, *291*, 377

Turkey, 9, 21, *40*, 184, 206, *371*

water loss, 19

water movement, 69–70

white, 331, 377

wood, 59, 63, *74*, *338*

oldest. *See* age

oleander (*Nerium oleander*), 17

olive (*Olea europaea*), 21–2, 32, 228, 377

Oregon grape (*Mahonia* spp.), 162, *348*

osage orange (*Maclura pomifera*), 82

ozone pollution, 348


Pacific madrone (*Arbutus menziesii*), 91

palm, 4, 7, 11, 56, 317

    Brazilian carnauba wax, 17

    cabbage, 56

flowers, 166, 201

oil, 377

raffia, 13

rattan, 36, 207

roots, 143, *144*, 146

sago, 81

traveller's, 13, *14*, 161

trunk, 56

water loss, 19

palo-verde (*Cercidium* spp.), 37

papaya (*Carica papaya*), 317, 377

paper mulberry (*Broussonetia papyrifera*), 84, 382

parasitic tree, 133

passion flower (*Passiflora* spp.), 326

pear (*Pyrus* spp.), 142

pecan (*Carya illinoensis*), 126

persimmon (*Diospyros* spp.), 181

phenology

autumn, *233*, 235–6

climate change, 236–7

defined, 230

spring, 231–5

phloem, 83–5

defined, 11, 51

photosynthesis, 13

bark, 86

C₄, 23

CAM, 23

roots, 138

sugar movement, 84

phytochrome, 175, 287

pine (*Pinus* spp.)

bark, 89

big-cone, *189*

bristlecone, 39, 206, 223, *317*, *319*, 371

Canary, 94

Caribbean, 228

Chihuahua, 94

cones, 189, 201

Corsican, *347*

dwarf mountain, *189*

epicormic shoots, 94

evolution, 4

grey, *189*

growth, 222

jack, *189*, *355*, 355

Japanese red, *335*

Jeffrey, *118*, 195, 335

Korean, 377

limber, *318*

loblolly, 137, 228

lodgepole, *78*, 201, 356

longleaf, 357

maritime, 107, *347*

Mexican white, 80

Monterey, *189*, 228

needles, 33, 38

pinyon, 33, 192, 197, *198*, 377

pitch, 94

pond, 94

ponderosa, *291*, *318*, 335

red, *251*

roots, 107, 137

Scots, *60*, 98, *98*, *157*, 201, 207, 228, *240*, *291*, *318*, *346*, 365

seedlings, *293*

seeds, 192, 195, *242*

shortleaf, 228

stone, 224, *250*, 250, *291*, *347*, 377

sugar, 80, *189*

Weymouth, *291*, *232*

whitebark, 288, *291*

wood, 63, *78*, 78, 98

plane (*Platanus* spp.), 79, 91, *112*, 207, 226, 262, 345, *346*, 354, 382

plum (*Prunus* spp.), 177, *268*

poinsettia (*Euphorbia pulcherrima*), 175

pollination

animal, 154–66, *155*

apomixis, 183

buzz, 166

pollination (cont.)
   cheating, 166
   costs, 181–3
   dioecious flowers, 178–81, *179*
   monoecious flowers, 178
   parthenocarpy, 184
   pollen size, 170
   pollination drop, 170, *171*
   rewards, 163–6
   self- and cross-pollination, 176–81, 202
   self-incompatability, 176–7
   sexes at different times, 177–8
   traplining, 172
   water, *155*
   wind, *155*, 166–71
pollution, 345–9, 378–80
poplar (*Populus* spp.)
   branches, 262
   environmental tolerance, *346*
   flowers, *156*, 180–1
   genome, 310, 333
   hybrids, 308
   leaves, 16, *17*, 235
   roots, 111, *112*, 114, 129, 138
   seeds, 195, *241*
   wood, 59
privet (*Ligustrum* spp.), 40, 158, 278, 347
proteas (*Protea* spp.), 161, 354
pruning, *96. See also* shape of trees

quetzal bird, 197
quince (*Chanomeles* spp.), 36
quinine bush (*Petalostigma pubescens*), 198
quinine tree (*Cinchona* spp.), 87, 377

rain tree (*Samanea saman*), 26
rays. *See also* wood
   defined, 51
reaction wood. *See* wood
reaction zone in decay, 343–4
red pine (*Dacrydium* spp.), 33

Red Queen theory, 332
red-belt damage, 352
redwood, coastal (*Sequoia sempervirens*), 19, 24,
     58, 76, 91, 94, 169, 184, 206, 214, *228*,
     241, 260, *261*, 292, *303*, 317, 370
   tallest, 207–8
reiteration, 256–7
resins. *See* defences
respiration costs, 230
rhododendron, 22–3, 28, 39, *142, 157, 292,* 328
root pressure. *See* water movement
roots
   adventitious, 57, 143, 302, 361
   age, 140–1
   burial problems, 119
   buttress, *144*, 147–8
   canopy, 126
   control of growth, 139–40
   damage to buildings, 111–16
   depth, 107–10, 117
   dwarfing root stock, 218
   fine, 110, 116–20, 140, 217
   flooding, 141–7, 220
   food storage, 135
   girdling, 114
   grafting, 133–5
   growth, 102, 136–8
   hairs, 126
   heart-root system, 104, *106*, 363
   in nutrition, 130–3
   knee roots, *144*, 146
   loss and death, 120–2
   mycorrhiza, 125, 127–30, 327
   pillar, *144*, 148
   primary growth, 136–7
   root cage, 107
   root plate, 104–7, *106*, 362
   secondary growth, 137
   shoot ratio, 217–18, *218*, 230, 357
   sinker, 107
   soil volume, 122

speed of growth, 138

spread, 102, *109*, 110–11, 118, *134*

stilt and peg, 143–6

suckers, 93, 300, 365, 384

tap root, 104, 107, 125, 301, 363

vessel and tracheid diameter, 71

water table, 123

rose (*Rosa* spp.), *142*, *157*, 323, *324*

rotting of wood, 81, 338–9

rowan (*Sorbus aucuparia*), *142*, 154, 177, *291*, 348

rubber tree (*Hevea brasiliensis*), 139, 175, 200, 228, 294, 337

salt tolerance, *347*

sandalwood (*Santalum album*), 342

screw pine (*Pandanus* spp.), 56, 147

sea-bean (*Entada gigas*), *200*, 200

seeds

    cost of, 201

    defences, 297

    dispersal by animals, 193, 195–9

    dispersal by water, 199–201

    dispersal by wind, 193–5

    distance, 195, 199–200

    dormancy, 284–7, 356

    eliasome, 199

    germination, 285–8, 292–4, 299

    mast years, 190–3, *191*

    oldest, 288

    seed caching, 195, 197, 298

    seedling success, 298–9

    serotiny, 289

    size, 290, 294–8

    storage, 287–90, 354–6

    structure, 284–5

senescense in old trees, 370

sensitive plant (*Mimosa* spp.), 26, *27*

sequoia, giant (*Sequoiadendron giganteum*), 184, 206, *209*, *241*, 260, 290, *318*, 353, 370, 376

    age, 317

bark, 91

biggest, 208

environmental tolerance, *348*

growth, 206

seeds, *292*

size, 1, 205

serotiny, 354–6

Seychelles double coconut (*Lodoicea maldivica*), *290*, 290–1, 296

shape of trees. *See also* biomechanics

    apical control and dominance, 269–70

    branch length, 263–4

    buds, 256–61

    changes with age, 273–5

    characteristic, 245–6, 256, 270

    control, 269–73

    environmental control, 246–53

    flowering, 264–9

    mono- and multilayer, 277

    pruning, 278–80

    shedding branches (cladoptosis), 261–2, *262*

    'Sheffield blight', 282

    topophysis, 275

shedding branches. *See* shape of trees

she-oak (*Casuarina* spp.), 34, 290

shoots

    epicormic, 270

    long and short. *See* branches

size. *See* biggest trees

southern beech (*Nothofagus* spp.), *128*, 206, 299

spalted wood, 339

spindle (*Euonymus* spp.), 34, 154, *157*, 188

sports. *See* mutations

spruce (*Picea* spp.), 335

    Norway, 81, 140, 201, 214, 235, *292*, 321, *346*

    seeds, *242*

    Sitka, *72*, 206, 228, *239*, *346*, 370

    white, 259, 352

    wood, *99*

squirrel, 197, 315, 329, 332, 337

stag-headed tree, 370

stinging trees (*Dendrocnide* spp.), 325

stipules, 225

strawberry tree (*Arbutus unedo*), 136, *157*

*Strychnos* lianas, 331

sudden oak death, 350

sulphur dioxide tolerance, *346*

sumac (*Rhus* spp.), 37, 181

swamp cypress (*Taxodium distichum*), 39, 141, *144*, 146, 261

swamp tupelo (*Nyssa* spp.), 141, 143

sweet gum (*Liquidambar styraciflua*), *29*, 29, 228–9, *241*, 251, 300, 336, 382

Swiss-cheese plant (*Monstera deliciosa*), 64

sycamore (*Acer pseudoplatanus*), 9, 232, 333

   environmental tolerance, *347*

   leaves, 45

   roots, 108, 130

   seeds, 195, *241*, *291*

tallest. *See* biggest trees

tamarind (*Tamarindus indica*), 26

tamarisk (*Tamarix* spp.), 123, *347*

tannin, 329–30

tea (*Camellia sinensis*), 140, 175, 228, *229*, 238, 331, 377

teak (*Tectona grandis*), 201, 219, 240, *242*

telegraph tree (*Codariocalyx motorius*), 28

topophysis, 304

tracheids. *See* wood

transpiration. *See* water loss

tree fern, *2*, 4, 56

tree of heaven (*Ailanthus altissima*), 37, 180, 199, 330, 341, *348*, 382, 384

trunk

   fluting, *55*, 55

   frost cracks, 79–80

   growth, 51–7

tulip tree (*Liriodendron* spp.), 7, 36, 38, 137, 206, *242*, 273

   buds, 225, 227

   growth, 235

roots, 129

Tutankhamun, 21

tyloses, *341*, 341, 343

urban trees, 349–50, 382–4

   water loss, 19

value of trees

   health benefits, 380–2

   pollution control, 378–80

   sacred, 376

   timber, 376

   urban trees, 384

varnish tree (*Rhus verniciflua*), 336

vessels. *See* wood

viburnum (*Viburnum* spp.), *157*, *226*, 327

Virginia creeper (*Parthenocissus* spp.), 36, 330

walnut (*Juglans* spp.), 64, 112, *140*, 142, *177*, 231, 242, *291*, 330, *346*

water loss, 19–23, 33

water movement, 63. *See also* hydraulic architecture, hydraulic lifting

   air problems, 67–70

   cavitation, 67–8, 83

   embolism, 67, 73

   in diffuse-porous wood, 68

   in ring-porous wood, 68–70

   in roots, 137

   Mütch water, 67

   root pressure, 63–4, 67

   speed, 69, *72*, 77, 84

   Tension-Cohesion theory, 64–5

welwitschia (*Welwitschia mirabilis*), *14*, 23, 172

White Pot, *116*

whitebeam (*Sorbus* spp.), *112*, *157*, 167

WiFi and tree health, 349

willow (*Salix* spp.)

   Arctic, 13, *14*

   bay, 143, 235

branches, 262

crack, 181, 328

creeping, 274

defences, 332

environmental tolerance, *346*

flowers, *157*, 160, 180

goat, *179*

leaves, 21

osier, 234

roots, 111, *112, 113*, 113, *114*, 120, 129, 141, 143

seeds, *241*

wood, 59

wind. *See* damage and death

Winter's bark (*Drimys winteri*), 60, 331

wollemi pine (*Wollemi nobilis*), 7

wood

  bordered pits, 59, 76

  conifer, 57–9, *58*

  diffuse-porous, *61*, 61, 229, 233

  fibres, 59

  grain, 77–8

  growth, 233

  hardwood, 57, *58*, 59–60

  heartwood, 81–3, 339–44, 377

  juvenile and adult, 77

  perforation plates, 59

rays, 51, 59, 80–1, *87*, 87

reaction, 97–9, *98*

ring-porous, *61*, 61, 229, 233

rotting of wood, 81, 338–9

sapwood, 81–3, 339, 344

shrinkage, 78

structure, 57–60

tracheids, defined, 57

vessels, defined, 59

xylem, defined, 11

yellow-wood (*Podocarpus* spp.), 180

yew (*Taxus* spp.)

  age, *317*, 318, *320*

  bark, *55*

  buds, 260

  defences, 334

  environmental tolerance, *348*

  flowers, *157*, *171*, 180–1

  growth, 206, *260*, *279*, 384

  leaves, 21

  roots, 361

  taxol, 377

  trunk, 55, *271*

  wood, *5*, 57